WAR LETTERS

The Pennsylvania Hospital
From the frontispiece of *History of the Pennsylvania Hospital Unit in the Great War*

WAR LETTERS, 1917–1918

From Dr. William T. Shoemaker, A.E.F., in France, and His Family in Philadelphia

Dr. Shoemaker outside his operating room
January 1918

By

George J. Hill, M.D., M.A., D.Litt.
Captain, Medical Corps, USNR (ret)

HERITAGE BOOKS
2020

HERITAGE BOOKS
AN IMPRINT OF HERITAGE BOOKS, INC.

Books, CDs, and more—Worldwide

For our listing of thousands of titles see our website
at
www.HeritageBooks.com

Published 2020 by
HERITAGE BOOKS, INC.
Publishing Division
5810 Ruatan Street
Berwyn Heights, Md. 20740

Illustrations

Most of the illustrations are from the family files of William and Mabel Shoemaker, including the front cover portrait of William Toy Shoemaker, by Elias Goldensky of Philadelphia in 1917. Other images are credited as shown in the *History of the Pennsylvania Hospital Unit* and other named sources, Google Images and Google Maps.

Cover designed by Debbie Riley

International Standard Book Numbers
Paperbound: 978-0-7884-0355-2

In memory of those who served in the American forces in the Great War,
and of those who served at home

Captain William Toy Shoemaker outside his operating room in France, January 1918

Contents

Philadelphia - Armistice Day Parade, November 11, 1918
View to North on Broad Street, with Union League Club on Left

A Trailer
for the
Shoemakers' War Letters

In the movie industry, the Trailer introduces a film to a prospective audience, using the end of the last reel of the film. In the early movie industry, the end of the last reel was unwound to produce a "trailer." It was also once called the Preview. This Trailer introduces the War Letters of Dr. and Mrs. William Shoemaker and their children.

The letters written by Dr. Shoemaker are the dominant feature of this book, and this Trailer will focus mainly on his letters. A counterpoint to them will be seen in the letters of Mrs. Shoemaker – Mabel – and letters of their five children. The letters were written during the Great War – a time of great danger with an uncertain outcome – and during the influenza pandemic that developed just as the war ended.

William T. Shoemaker – called Billy by his wife, and Bill or Will by his friends – will reveal himself in his letters to be an interesting man. Indeed, one might say that he was a "character." He was devoted to his immediate family, and he wrote to them with great affection. He was skeptical, and at some times scornful, of some of the rest of his family and Mabel's. He had a mordant sense of humor about them. He called his cousin's wife "Medea," after the Greek temptress. He called one of his brothers "The Pope," who lived in an elegant home that Will called "The Vatican." He called his nephew's wife "bleeding Mary" because she complained so much. He was intensely patriotic, and he scorned the "stay-at-home" doctors who shirked their duty, as he saw it, during the war.

Will's letters show great concern for his family's finances, and he offered detailed suggestions about proper ways to save and spend money. He behaved as he was taught, to be a good Philadelphia City Quaker: always to do the right thing, do it for a fair price, and be paid promptly. Mabel had her own money, but he would not let her touch it.

The letters show Will Shoemaker's exasperation with Army rules and regulations, which he considered to be capricious, and which he believed were administered by men who were intentionally devious. Mabel tried to dampen his comments about the Army, but he continued unceasingly. He had little trust in the government. He was scornful of the Democrats in Washington and the Republicans in Philadelphia.

He had a morbid sense about the possible length of the war, which had consumed the lives of millions of soldiers by the time he arrived with his hospital unit in France in May 1917. He speculated that the war might go on for many more years – two, three, four or more years. He was not allowed to write some details because of censorship regulations, but we catch a glimpse of the danger – he said "Mr. Periscope" (a German U-Boat) missed his ship on the way to Europe; he mentioned that he was "gassed" (and that you only have seconds to get your gas mask on correctly). He watched the aeroplanes battling overhead, and rode in a tank (it was a dirty thing). He recalled a German Zeppelin that flew overhead on a moonlit night, but didn't drop a bomb. He wrote about a soldier who was blinded, but cheerfully said "Good bye, Old Man" to him when he was transferred to England – "Gone to Blighty;" and about a young officer, whose eye he had to remove after it was destroyed by, of all things, a golf ball! He told of the influenza of one of his orderlies, a private, nicknamed "Doc," and how he and two nurses sat silently through the night by his bed, as he died. He was one of many boys who "Went West" in that war. That story brought tears to my eyes as I read it.

There also are many stories in Will Shoemaker's letters that are not as sad as this, and some of his tales are hilarious. How he built a fire and kept it going, how he was in charge of an enormous picnic on the Fourth of July, 1918, and how he was thrilled to visit his son, an Army private, at his camp near the Alps. He wrote about his two visits to Paris, and of his remembering his trip to that Big City many years previously. He told of having dinner alone there, while thinking of dining with Mabel. He returned again, just as the war was ending, and he hoped to be there on Armistice Day. He tells the story of his rotten teeth, which produced an infection of the cheek; of the "discharge" of the offending teeth, and of the timely rupture of the abscess. Will Shoemaker wrote from London when he arrived there in May 1917, and he told about it again when he returned in November 1918. His stories of wartime London are mingled with recollections his visit to Europe in the 1890's, when he spent 18 months in England and Germany – and when he toured the Near East and Egypt.

Dr. Shoemaker tells stories about taking lessons in French from Martha, a young Belgian woman, and how Martha's little girl would love to sit on his lap. And how he went to have a new Army uniform made from a French tailor, with Martha telling the tailor to do what the doctor says, not what the tailor would want. And that Will Shoemaker also took along his nephew, a dentist in his hospital unit, to be sure that the color of the "new suit" was correct – for Will Shoemaker was color blind.

Will Shoemaker's paternal ancestors came from the part of Europe where German was spoken – the Rhineland-Palitanate, on the west bank of the northern Rhine River. In the 1680s, when they came to Pennsylvania, the family name was spelled Schumacher. They soon changed the spelling to Shoemaker, and learned to speak English. Will had spent much of one year in Germany, learning about the instruments and techniques that were used in operating on the eye – of the specialty of ophthalmology. However, he now detested Germany, because of the war. He referred to German as a "dead language" – like Greek and Latin. He spoke scathingly of Germans, calling them Boches, Huns, and Fritz. He wrote angrily about the German people, and he became increasing bitter and depressed about their prolongation of what he saw as a hopeless, useless, and devastating war.

The letters show that Will Shoemaker had a remarkable range of interests. He cited many passages that he remembered from poetry and songs, from plays and books, and on at least three occasions, from the Bible. He remembered the authors' names, and he commended them to his family. He wrote not a word about the "hereafter," in contrast to what he had been taught by his mother – a devout Methodist – and his father, who was a member of the Swedenborgian Church, which was also known as the New Church. Will Shoemaker was especially sarcastic and scornful about what he read in the New Church's bulletin, *The Messenger.* He derided the church's leaders for ignorance and bigotry. This was surely in contrast to what Will had been taught as a child. One of the "ways" of Methodists and Quakers was, "If you can't say anything nice about someone, don't say anything at all." His mother would also have been appalled at his use of what Methodists called "bad words." Not only did Will frequently break the Commandment not to take the name of the Lord in vain (Exodus 20:7), but he also used barnyard anatomical epithets. It is amazing to read this language in letters from a Philadelphia gentleman, but it shows the depth of his anger and grief. His depression deepened in the summer of 1918, and he wrote to Mabel that he thought he might be going crazy. He said that writing to her helped to relieve the anxiety, and he was then able to end his letters with statements of love, of yearning to be with her, and with "English kisses" – XXXXXXX

This collection of letters also shows the Shoemaker family at home in America. It was a great contrast to the awful things which Will was describing in France. Mabel Shoemaker was, by contrast, almost serene. Her letters show that she took care of their home, and their children, and of his patients, without hesitation or concern. She managed all of her business responsibilities with equanimity. She chided her husband for saying negative things about the Army and about many people, but wrote mostly about the good things that she had done, or that had been done by her children and her family. She became slimmer, and was said to be a very beautiful woman – although Will worried about that. Was she hiding something? Sickness? No, she was probably just dieting. The children were all good students, popular, and happy. It seemed almost too good to be true, but it was. They were fine, cheerful children, doing their part in the war effort. Dot already showed her artistic ability – as both a painter and pianist; Warren displayed his skills as an engineer and businessman; Ted was a popular leader and a fine athlete; Barbara was a beautiful young lady, and also a fine writer; Bob loved to tinker, and he was his father's equal as a bad speller with horrible penmanship.

After he returned to Philadelphia in 1919, Will Shoemaker never spoke much about the war. Few veterans do. But for the rest of his life, there would be two constant reminders of that time – the deep scar on his face, caused by an abscessed tooth, which ruptured on the night of July 25, 1918; and the button that he always wore on left lapel of his suit, showing membership in the American Legion.

AMERICAN
U·S
LEGION

Summation

William Toy Shoemaker, M.D., was a highly respected ophthalmologist in Philadelphia in 1917, and he was the senior eye surgeon at the Pennsylvania Hospital when the U.S. entered that Great War. In 1908, he had received the Alvarenga Prize from the College of Physicians of Philadelphia for his study of retinitis pigmentosum and hereditary deafness. His landmark publication on this disease is still cited.

In 1914, the American Red Cross and the War Department began making plans to form so-called Base Hospitals, utilizing personnel from hospitals in major cities, in order to respond to catastrophes, or in case of war. The Pennsylvania Hospital was invited to form one of the first of these 500-bed Base Hospitals, and it was thus ready to go into action when war was declared on April 6, 1917. Designated Base Hospital No. 10, the staff included thirty-four physicians, supported by sixty-four nurses and about 150 other personnel. They were mobilized on May 15, 1917, and they marched to the Pennsylvania Railroad Station the next day to go by train to Jersey City, and then on the steamer, *St. Paul,* to Europe. Evading submarines, Base Hospital No. 10 disembarked in Liverpool on May 27. Over the next month, No. 10 learned in England how to take over the functions of British Expeditionary Force General Hospital No. 16 at Le Treport, France. To evade submarines, they again sailed during the night, and dropped anchor before dawn on June 29 at the port of Le Havre. Within a few days, they had all reached the seacoast village of Le Treport and the 2,500-bed hospital that the British had built on the high cliffs. Shoemaker wrote, "The nearer we get to the theatre of war, the more horrible the whole thing becomes. We are now in it and you can bet we will all be glad to get out of it when it is over."

Base Hospital No. 10 was located behind the front lines, but many of the staff had duties which took them near the battles. Six of the personnel of No. 10 died and were buried in France. The German advance along the Somme River in April, 1918, caused "The Great Evacuation," when they moved thousands of sick and wounded on a dark, rainy night. Their patients suffered from all of the miseries of trench warfare, including rifle and machine gun bullets, shrapnel from artillery and bombs, trench foot, illnesses from poor food and contaminated water, ghastly wounds to skin and lungs from poison gas, and the epidemic of influenza. And then, finally, it was over. At the end of the war, Base Hospital No. 10 would record that it received 47,811 wounded and sick, with 538 deaths.

Dr. Shoemaker wrote to his wife Mabel on November 10, 1918, "Well, we are all waiting anxiously for official announcement of the Armistice. So far, notice that the time set is the 11th day of the 11th month at 11 O'clock. Right? Come 7 – Come 11!! I wish it would come. Of course, I feel that the Boche will sign it and surrender without conditions. I am certain of it, but until it is done, there is a picking up of great suspense. If he does not, it will be a disappointment, but not for long. Marshal Foch is prepared to annihilate – if it may take six more months." In his next letter, on November 16, he wrote, "It is hard to realize that the Great War is over and that the victory for the Allies has been so complete – hard to realize that the frustration of a Great Empire has been so swift and thorough. But it is all true and is as it should be and as it was sure to be." No more men would be blinded.

This book is based on the handwritten letters that Dr. Shoemaker wrote to his wife and children, and some of the letters that they sent to him. His letters were probably all received by his family, and except for a few pages, all were intact, and were preserved. Their letters to him are incomplete, but enough of them were saved to represent a unique collection. No other letters from an American physician in World War I are known to have been saved and published; and few letters are known to exist from that Great War that were exchanged between an American serviceman and his sweetheart or family.

Dr. Shoemaker was ordered into service as a First Lieutenant on May 15, 1917. He was promoted to Captain on September 4, 1917, and to Major on September 3, 1918. In November 1918, he was transferred to Base Section No. 3 in London as Consultant Ophthalmologist to the American Expeditionary Forces. Major Shoemaker was discharged from service at Hoboken, New Jersey, on December 30, 1918, having been in service for nineteen and one-half months.

Prologue

In 1938, one year before war broke out again in Europe, Winston Churchill wrote: "The Great War which we have passed differed from all ancient wars in the immense power of the combatants and their fearful agencies of destruction, and from all modern wars in the utter ruthlessness with which it was fought. … Germany having let Hell loose kept well in the van of terror; but she was followed step by step by the desperate and ultimately avenging nations she had assailed. Every outrage against humanity or international law was repaid by reprisals often on a greater scale and of longer duration … The wounded died between the lines: the dead mouldered into the soil. Merchant ships and neutral ships and hospital ships were sunk on the seas and all on board left to their fate, or killed as they swam. … Cities and monuments were smashed by artillery. Bombs from the air were cast down indiscriminately. Poison gas in many forms stifled or seared the soldiers. Liquid fire was projected upon their bodies. Men fell from the air in flames, or were smothered, often slowly, in the dark recesses of the sea.

"There is no need here to trace the ancient causes of quarrel between the Germans and the French, to catalogue the conflicts with which they have scarred the centuries, nor to appraise the balance of injury or of provocation on one side or the other. When on the 18th of January, 1817, the triumph of the Germans was consolidated by the Proclamation of the German Empire in the Palace of Versailles, a new volume of European history was opened. 'Europe,' it was said, 'has lost a mistress and has gained a master.' A new and mighty State had come into being, sustained by an overflowing population, equipped with science and learning, organized for war and crowned with victory. France, stripped of Alsace and Lorraine, beaten, impoverished, divided and alone, condemned to a decisive and increasing numerical inferiority, fell back to ponder in shade and isolation on her departed glories. … Bismarck would never have taken Lorraine. Forced by military pressure to assume the double burden against his better judgment, exhibited from the outset and in every action of his policy an extreme apprehension."[1]

Churchill went on to describe the shifting alliances of France and Germany in the years from 1871 until the Great War broke out in August 1914. He recalls the history of Britain's longtime conflict with France and its historic cordiality with the German people, and then of the gradual shifting of Britain's relationships with France and Germany; of France's gradual entente with Russia after the failure of Napoleon's invasion of that land; of the growing unease between Germany with Russia; of the duality of purpose seen by both Germany and the Austrian empire; of the conflict in the Balkans that involved proxies of Turkey, Russia, and Austria; of the deep anger in Russia that followed its defeat by Japan; and of the uncertain posture of Italy, as it warily watched the formation of the Double Alliance of France and Russia, and which initially joined Germany and Austria in a Triple Alliance. However, by the time war broke out in 1914, Italy had decided to cast its lot with France and Russia; and Britain had chosen to side with France, having broken with Germany because of the buildup of the great German battle fleet. Japan became a quiet ally of France, Russia, Britain and Italy, intending to be on the winning side when the assets of the German Empire were distributed. The roles of these countries' impetuous rulers and the imperfect visions of their military leaders played important parts in the catastrophe. There was little apprehension by the public in these countries. The United States is not mentioned at all in Churchill's history of the runup to the Great War.

Churchill's point about the crucial role that Alsace and Lorraine played in the conflict left unsaid the reason that the German military insisted on taking this "double burden" from France. I believe it was to provide a buffer for the western border of Germany. It was difficult for Bismarck to defend politically. It was understandable from the military point of view, but the defeated foe regarded it as a land grab. The northwest border of Germany could be fortified, for it extended for some distance to the west from the Rhine River, before the river continued into Holland and flowed into the Channel at Rotterdam. But without Alsace-Lorraine, there was no opportunity for Germany to build a defense from France on the southwest; the Rhine itself was the line of division between the two countries. The largest city in that disputed region, Strasbourg, thus became the center of the conflict. Most of the people of Alsace and

Lorraine are bilingual, but they usually identify either with France or with Germany. France lost the territory to Germany in 1871, but it would never forget it. Alsace-Lorraine became French again in 1918; then Germany controlled it after the French surrendered in 1940; and the French took it back again in 1945. The region, however, is emblematic of both countries: The "Cross of Lorraine" notably defines the French; and the fierce, obedient German Shepherd dog is a breed also known as Alsatian.

The role, or the potential role, of the United States in the prelude to the Great War is rarely discussed. The leading politicians and the U.S. military leaders watched the developments with interest. The U.S. had been quietly developing an empire, in fact, although not in name. Liberia could be said to be its first outpost. And then the idea of Manifest Destiny made it seem inevitable that the U.S. would stretch "from sea to shining sea" with its Conquest of the West and the acquisition of northern Mexico. Over many decades, the U.S. tried many times, but with little success, to acquire territory immediately to the north, in the region now known as Canada. It began in the Revolutionary War and it continued in the War of 1812. Failing in both of those wars, the U.S nevertheless kept pushing (diplomatically) for more territory along its northern border. President Polk declared "54-40, or Fight," meaning a border somewhere north of the 54th parallel, though the U.S. finally settled for 49°. It purchased Alaska, and it subverted local rule to acquire Hawaii. And then, clear vision and sea power (think Theodore Roosevelt and Alfred Thayer Mahan) resulted in a "splendid little war" in 1898 with Spain, and the U.S. suddenly had a world-wide empire. It stretched from Puerto Rico to the Philippines, with Cuba as a protectorate. And after a staged revolution, it then acquired a Canal Zone in Panama. The Monroe Doctrine told European powers to stay away from the Western Hemisphere, and Theodore Roosevelt emphasized this in 1904 with his "Roosevelt Corollary": Stay away, Europe. Europe accepted this, for it saw plenty of territory to conquer for colonies in the Middle East, South Asia, and most of Africa.

Protected by oceans on both coasts from the Great Powers, and with a growing navy of its own to insure "freedom of the seas," the U.S. could watch for the likely outcome in the forthcoming war in Europe. It could join the winning side to divide the spoils. Some American military men probably anticipated a large number of casualties because of the weapons – repeating rifles, Gatling guns, and artillery – that had been employed during the U.S. Civil War in the 1860s. Devastating effects of these weapons were seen at Antietam and Gettysburg. They had been refined and tested in the Boer War and in other smaller conflicts of the Fin de Siècle. But did American military officers realize that the effects of bombs dropped from airplanes, poison gas, machine guns, flame throwers, and trench warfare would far surpass those casualties? I suppose that some did, because many Americans were invited to be observers at Army maneuvers in Europe. However, young officers risk their careers when they appear to question their seniors. Nevertheless, looking back, it is difficult to understand the careful observations and note-taking by military observers and newspaper reporters in the early years of the 20th century, who failed to alert the world to a Great War that might be coming.

When the end came, the Armistice in 1918 was but a pause. The Allies signed the Versailles Treaty with the defeated Central Powers in 1919. President Woodrow Wilson, a Democrat, failed to work successfully with the majority Republican party in the U.S. Senate to ratify Treaty's conditions, which included a League of Nations. Wilson was succeeded by a Republican – Warren Harding – and in 1921 the U.S. ratified a separate Treaty with Germany, without membership in the League. Germany was angry over the territorial losses and the payment of reparations that were imposed by the Treaty. Germany learned that it could defy the requirements of the Treaty, initially with small steps, without fear from the League of Nations or its member states. Its industries were revived and it joined again with Austria by *Anschluss*, and it took western Czechoslovakia, calling it *Sudetenland.* British Prime Minister Neville Chamberlin proclaimed that he had approved this in a meeting with German Chancellor Adolph Hitler in Munich in order to achieve "Peace in our time." And then Hitler and Russian Premier Josef Stalin signed a secret agreement to divide Poland. Hitler invaded Poland on September 1, 1939, and war resumed. It soon became a World War, with some of the combatants having switched sides.

Introduction

The letters in this book, written by Dr. William Shoemaker (hereafter, WTS) and his wife, Mabel (née Warren) Shoemaker (MWS), and their children, in 1917-1918, were saved by Mabel and transferred at some time, probably after William died, to their daughter Barbara. The letters were placed into two small cardboard boxes that originally contained stationery from the firm of J. E. Caldwell & Co., Philadelphia. One box was labelled in Mabel's handwriting "War Letters – W.T.S" and this provides the title for this book. Most of the letters were bound with silk ribbons, and many of them were in chronological order. However, some letters were inserted in and around those that were bound, and also a few other items were inserted in the boxes after the last letter, which was written in December 1918. The letters from WTS are probably complete, except for those lost in transit from Europe. Many ships carrying letters were sunk by submarines, and others were mislaid due to wartime exigencies and never delivered. Censors also deleted some letters completely, without notice to the sender, but this seems unlikely to have been the cause of any of WTS's letters. He and other officers were responsible for censoring letters, and he knew the rules.

Many of the letters received by WTS while he was in service – from home and friends and business associates – are not included in this collection. What we have from them is what he returned with to America after the war, less whatever else was present, but not retained by Mabel. We also know that many items sent home by WTS are not in this collection. The letters show that Mabel was preparing a scrapbook which included such things as menus, photographs, and playbills that WTS sent home; the scrapbook has been lost. Mabel transcribed, or had a secretary type, many of WTS's letters. He commented on this, and warned her not to do it, because it would cause embarrassment to him and it might violate security. The markings to show what was being transcribed appear in a few early letters, but not in later ones. Mabel numbered most of her letters beginning in about 1918, but unfortunately most of the numbered letters are missing. WTS began to number his, but he abandoned the plan because he lost track of the numbers.

After their daughter Barbara, now Barbara (née Shoemaker) Zimmermann, died, the papers that were in her attic, including these letters, were transferred to the attic of her daughter Barbara (née Zimmermann) Johnson. As Barbara Johnson was downsizing her possessions to move into an extended care facility in 2007, she invited her sister Helene – my wife – to look at the items in her attic, in case there were any things she wanted to keep. I went along for the ride, and saw the trove of papers and photographs – including old 16mm home movies – and I recognized that they had potential value to a family historian. This inadvertent collection became the basis for several books that I have written about the Shoemaker family, and about the work of Lieutenant Albert Zimmermann – Barbara Shoemaker's husband, and the father of Barbara Johnson and Helene Hill. Until recently, however, I put off working on the World War I letters of WTS and MWS. A brief glance at Dr. Shoemaker's letters showed me that it would be a very difficult job. He was often so busy, and so tired, that he frequently mis-spelled words, and he lost his train of thought. I also wasn't sure that I could make sense of what he was writing about. After all, I didn't know any of the people that he wrote about, and I didn't imagine that places and names in letters more than 100 years old could be identified. However, when I began to study the letters, I found that I had two wonderful sources to help me. The first item was the wartime history of his medical unit, which was available in print, and also in a down-loadable pdf version that I could search. And the second item is the internet, with its search engines: especially Wikipedia and Britannica, which provided references to confirm the details; and Google – including maps, images, books, and other references.

In this book, I focus on the War Letters themselves, providing as deep a background as I can about items that appear in the letters. I have written a brief background about the family of William and Mabel Shoemaker; and a short summary of the origin of the Great War, its impact on the world, and how it was followed by another great and horrible conflict, called World War II. These views are my own. Except for the simplest items, such as names, and dates, much of history is written as it is seen through the eye of the historian. And even the verity of these simple items can sometimes be challenged.

Part One

Families

William Shoemaker and Mabel Warren

Their Ancestry, Their Early Lives, and Family Life

The Shoemaker and Warren Families[2]

The Shoemaker Ancestry

Near the end of winter in 1686, a recently-widowed mother stepped ashore with seven sons and a daughter on the west bank of the Delaware River. They had arrived in Pennsylvania, at the recently-settled town of Philadelphia. They would soon move north to establish a new settlement, and the town that they created would initially be named for them. They probably said, or thought quietly to themselves, "*Grüss Gott*," for they were German-speaking Quakers, who were emigrants from the Upper Rhineland. They arrived on the ship *Jeffries* on 20, 1mo. 1686 in the Quaker calendar (20 March 1686). On their way north from their landing place, Sarah Schumacher, the "Palatine widow," and her children passed through Germantown, in north Philadelphia. In Germantown, they were greeted by others from their home village of Kriegsheim, in the Rhineland-Palitanate, on the west side of the Rhine. One of those who met them in Germantown was a daughter who had come earlier with one of the other neighboring Schumacher families. They had all followed the leadership of William Penn and the English Quakers who came with him on the *Welcome* in 1682. The small settlement of Philadelphia grew and prospered, and it eventually became the greatest city in the Empire, other than London itself. The Quakers' religious tolerance was like a magnet that would attract others to this colony, and it was soon seen as a good place to do business. The German-speaking immigrants promptly assimilated with those who came to Pennsylvania from England and Wales. They anglicized their names – Schumacher became Shoemaker – and they learned to speak in English. The families from Britain and Germany intermarried, and they all became part of the Quaker business and religious community.

This section traces one branch of the Shoemaker family in its line of descent from the farmer, George Schumacher, who died of smallpox while still at sea on his voyage to the New World in the seventeenth century, to William Toy Shoemaker, M.D., one of the leading physicians of Philadelphia in the early twentieth century. William T. Shoemaker also had many other ancestors in America, of course. In each generation, other European families became part of his genealogy. There were Mennonites and other Quaker families from the northern Rhine valley who were founders of Germantown, in north Philadelphia; Welsh families that settled in what they called Montgomeryshire – now divided between Montgomery County and Chester County; and English Quakers that were dispersed throughout the region, from Bucks County on the north to Chester County on the south. All of those who purchased a tract of 1000 acres from William Penn also were entitled to a house lot in Philadelphia. It was in the City of Brotherly Love where the children met and marriages were arranged. And thus, William Shoemaker's ancestors on his father's side include the German families of Lukens, Tyson, and Op den Graeff; Welsh families of Cadwalader, Jones, and Evans; and English – Leech, Penrose, Wall, Comly and Potts.

Less is known about Will Shoemaker's mother's ancestors, because they appeared in America much later than his father's, and their European ancestry is largely unknown. It appears that the Dippolts and Hesters came from the same part of Hesse, and that they stayed in America after the Revolutionary War. John Dippolt Hester was William Shoemaker's grandfather. He was the son of a successful immigrant Anglican vestryman and tavernkeeper in Trenton, New Jersey. John D. Hester marched the length of New Jersey to defend it in the War of 1812, marched back home, married a Quaker woman, and became a Methodist. It was a familiar pattern; when Quakers married "out of meeting," they often became Methodists. John D.'s daughter – Hannah Hester – married a lapsed Quaker, and their son – Will Shoemaker – took lessons in religion, politics, and behavior from both of them.

The Warren Ancestry

The Warrens were an ancient family in England. William de Warenne was one of the greatest supporters of the William, duke of Normandy, who became King William I of England in 1066. But the line of descent from William de Warenne to Arthur Warren, the emigrant, has not been proved. Arthur Warren's birthplace and his immediate ancestors are unknown, although it can be presumed that he was from England. He was a competent farmer in the Massachusetts Bay Colony, and his children were notably successful in farming, and later in business. They "married well," as the saying goes, and the branch that we follow in this section includes several who were descendants of English gentry.

Many came to New England as Puritans – Anglicans who sought freedom from what they believed were improper Roman Catholic traditions, imposed by the Bishops. In later generations, they became Baptists, Congregationalists, and even Unitarians. But there were others who were simply looking for business opportunities, and they were glad to return to the traditional Anglican form of worship. The descendants of Arthur Warren in the line to Mabel Warren includes men and women whose ancestors were magistrates, military officers, educators, leading members of the clergy, and high government officials, both in England and America. In later generations, the Warrens' farms in New England became large plantations, and the family included successful inventors and businessmen. Mabel Warren's ancestors are all from various parts of Britain. Her father's ancestors came from many counties in England, Scotland, and Cornwall. They include family names such as Moors, Fletcher, Patch, Stratton, Parker, Underwood, Jackson, Pierce, Conant, Munro, and Atherton. Mabel's mother was born in Bath, England, and less is known about her ancestry; her ancestors were probably clustered in the south-west of England, from the English Channel to the Bristol Channel.

Mabel Warren, in fact, was a descendant of William de Warenne, by another line of her ancestry, for at least two of her ancestors are of Royal descent. Elizabeth Stratton, wife of John Thorndike of London, and Thomas Trowbridge, of Taunton, England, are descendants of kings of England and France, and of Charlemagne, Emperor of the West.

William Shoemaker

William Toy Shoemaker, eighth child and seventh son of Julien and Hannah (Hester) Shoemaker, was born in Philadelphia, 22 January 1869; died in Philadelphia, Pa., 26 November 1942. He grew up in Philadelphia, the son of a prosperous businessman who was a director of the J. B. Lippincott Co. – then one of the nation's largest publishing companies. His parents, Julien and Hannah (Hester) Shoemaker, had nine children, seven of whom survived to adulthood, and William was the youngest of these. All of the other children were boys except one, and his childhood must have been strongly influenced by them.

William's Shoemaker grandparents had died before he was born, so he knew nothing of them personally. His grandfather Charles Shoemaker, who died in 1847, had been a bookseller in Philadelphia and Baltimore. His grandmother Shoemaker had brought up their six surviving children, and William had a great many aunts, uncles and cousins on the Shoemaker side of his family. Many other Shoemakers lived in the area around Philadelphia, particularly in the northern suburbs. The Shoemaker family had emigrated from Germany as Quakers in the early seventeenth century and settled north of Germantown, the northernmost part of Philadelphia. Their large community of farms and small villages was then called Shoemakertown, in Philadelphia County. This area is now densely populated and now comprises the cities and towns in Montgomery County of Cheltenham, Jenkintown, Elkins Park, and Abington. Two of William Shoemaker's seventeenth century relatives had been mayors of Philadelphia. Another ancestor, Toby Leech, had given the name "Fox Chase" to his estate, by which that area is still known.

Little is known about William Toy Shoemaker's maternal ancestors, except that his mother's father, John D. Hester, originally from Trenton, N.J., was a veteran of the War of 1812. In his later years, John D. Hester lived in Philadelphia, where he died in 1880 at the age of 86. He was already seventy-five when William was born, and we do not know how close the old soldier was to his daughter's family. However, he must surely have been known to them, and William would have had recollections of his grandfather Hester. John Hester's daughter, Hannah, was a Methodist, and when she and Julien Shoemaker were married, he was "read out of meeting" by the Quakers. Hannah and Julien then joined the Swedenborgian church, also known as the Church of the New Jerusalem, or "New Church," and Julian Shoemaker changed his name to Julien to avoid being identified with Julian the Apostate.

The Shoemakers had many connections with the Lippincott family and other prominent families of the time in Philadelphia. An older brother was named Joshua Lippincott Shoemaker. His brother Harvey, just senior to William, and five years older, became his role model. Harvey Shoemaker studied medicine at the University of Pennsylvania, received the M.D. degree in 1886, and then entered practice in Philadelphia. William followed Harvey to medical school at Penn, and received his M.D. five years later, in 1881. The lives of Harvey and William and the lives of their families were intertwined for many years. Both became physicians to the Southern Home for Destitute Children, on South Broad Street in Philadelphia. William's grandchildren recall their great-uncle Harvey with affection. He was a man who always wore his hat in the house; and, when asked how he could support a scoundrel whose only favorable attribute was that he was a Republican, replied that, "I just hold my nose and vote." As was common for young physicians who wanted specialty training, William studied for nineteen months in Europe after graduating from medical school, to become a specialist in ophthalmology. He traveled to Egypt before he returned to America, and sailed on the Nile before the turn of the 20th century.

It was at the Swedenborgian church in Philadelphia where young William Shoemaker met his bride-to-be, Mabel Warren. Mabel came from a prosperous family in West Newton, Mass., but in 1880 she had suddenly been orphaned at the age of ten. Mabel came with her older sister and two younger brothers to Philadelphia, and they were raised with the large family of her father's younger brother, her uncle Ebenezer Burgess Warren. E. Burgess Warren was a successful real estate developer and philanthropist in Philadelphia, who was also a prominent member of the Swedenborgian "New Church."

William Toy Shoemaker and Mabel Warren were engaged for some seven years before they were married, on 6 June 1895. Their engagement began when he was a medical student, and it continued until he finished his training and was established in practice. Although she was an orphan, Mabel brought a substantial dowry to the marriage, and the Warren legacy has continued to help young people in the family from generation to generation. From the fee paid by his first patient, Dr. Shoemaker bought a piece of jewelry for Mabel in the form of an elephant. This was the first piece in what eventually became a fine collection of elephants, carved in ivory and other semiprecious materials.

Within only a few years after entering practice, Dr. Shoemaker had achieved considerable success. In 1903, when he was 34 years old, he was described as an "ophthalmologist, graduated M.D. at the University of Pennsylvania, 1891; is a fellow of the College of Physicians, member of the Philadelphia County Medical Society, Academy of Natural Sciences, American Ophthalmological Society, Goettingen Society, etc.; resident physician, German Hospital, 1891-92; assistant ophthalmologist, German Hospital; dispensary ophthalmic surgeon, Presbyterian Hospital; oculist to Pennsylvania Institution for the Deaf and Dumb; ophthalmologist to Home for training in Speech of Deaf Children before they become of School Age; ophthalmologist to Southern Home for Destitute Children ... Residence, Philadelphia."

Five years after this summary of his life to date was published, Dr. Shoemaker won the Alvarenga Prize of the College of Physicians of Philadelphia in July 1908 for his work on retinitis pigmentosa (RP). His essay on this disease was published as *Retinitis Pigmentosa with an analysis of Seventeen Cases Occurring in Deaf-Mutes* (Phila.: Lippincott, 106 pp., 1909). As recently as 1993, RP continued to be a puzzle to scientists.[3]

Young Mabel Warren

Mabel Warren, eighth child and youngest daughter of Herbert Marshall and Eliza Caroline (Copp) Warren, was born at West Newton, Mass., 5 January 1870; died at Philadelphia, Pa., 8 July 1932. Mabel Warren was born five years after the end of the Civil War in the house where her parents had lived for eight years, a country place known as "Blithedale." This house had been built by Horace Mann, and Hawthorne was said to have lived there when he wrote "The Blithedale Romance." Her family was large and prosperous, and there were many aunts, uncles and cousins living nearby.

Mabel's Warren grandparents had lived for many years in Fitchburg, Mass. Her grandfather, Jesse Warren, had died sixteen years before she was born, but Mabel was six when his widow, Betsey Jackson Warren, died. Mabel must have known her grandmother Warren, although they were probably not close, since she was very young, and Fitchburg was then not within easy traveling distance from West Newton. Mabel's maternal grandparents, the Copps, were English and Welsh, and her mother was born in Bath, England. We do not know if Mabel knew her Copp grandparents – James Copp of Exmouth, Devonshire, and Caroline (Bigwood) Copp, of Swansea, Wales, who were married at Bath, England in 1824. In any event, Mabel certainly would have known that her mother was English, and she probably heard the "Queen's English" spoken at home.

Mabel's mother died at the age of 50 in West Newton, presumably at home, when Mabel was only nine. The following year, on the night of 11 June 1880, Mabel's father was lost at sea when the *Narragansett*, on which he was a passenger, collided with the *Stonington*, on Long Island Sound, and burned and sank. There is a legend that, "the family silver was lost on the 'ill-fated' *Narragansett,* and indeed there is apparently nothing from Mabel's parents that has passed down to her grandchildren. The loss of both of parents in close succession when she was so young must have been traumatic, as well as being potentially a financial catastrophe for the children. Mabel's only sister was then 25 and unmarried; her future was surely in jeopardy. Her six older brothers ranged in age from 24 to 38. They appear to have been more secure. All were successfully established in business and in society, although one, Jesse, died at age of 40 at Saranac Lake, N.Y., where a well-known tuberculosis sanitarium was located.

Mabel and her two younger brothers, then but five and six years old, were taken under the care of her Philadelphia relatives. The head of the Warren family in Philadelphia was her father's youngest brother, Ebenezer Burgess Warren – known as Burgess – who was then 47, a wealthy, charitable man with three daughters ranging in age from 12 to 21. Two sons had also been born to E. Burgess Warren and his wife, but they had died when very young – one at one month of age, and the other eight years earlier at one year of age. Burgess had originally been in the oil and coal-tar business with his older brothers, and he had made a fortune from the refining of asphalt from Trinidad Lake and by laying asphalt paving in Washington, D.C. His real estate activities in Philadelphia were also a great success. By the time he was 32 years old, Mabel's Uncle Burgess had built more than 40 homes worth over two million dollars near Walnut and Spruce Streets, in the vicinity of Nineteenth to Twenty-first Streets. He was a pillar of the Swedenborgian Church – the Church of the New Jerusalem, or the "New Church" – in Philadelphia. His brothers were also active in the New Church, and his brother Samuel was an ordained minister of the New Church in Massachusetts.

Burgess Warren was also a leader in the management of the Hahnemann Hospital, which was then at the center of the homeopathic medicine movement in the United States, and he was a prominent collector of fine art. Burgess Warren played a major role in the care of the three youngest orphaned children of his brother Herbert. His principal priorities in that regard would have been to find a suitable husband for Mabel, and to ensure that her young brothers, Walter and Ralph, got a good college education and a start on their careers. It also appears that Mabel's older sister, Ella, was in Philadelphia at this time, too. Ella had been born in Philadelphia, and she apparently returned there at about the time of her father's death on the *Narragansett.* She was nearly thirty-three when she married, in Philadelphia, Dr. William Franklin Arnold of Moodus, Conn., a chiropractor and a homeopathic physician.

Mabel's two younger brothers attended college together in Philadelphia. They were classmates at the University of Pennsylvania, graduating B.S. in 1895 and M.E. in 1896. Both were members of the same social and engineering fraternities, Phi Gamma Delta and Mu Phi Alpha, and both returned to West Newton, Mass., where they were living in 1903. They were by that time junior partners in the Warren Brothers Company, which had been founded by their older brothers and sisters in 1900. As a young woman, Mabel supported herself by teaching young children in a kindergarten that she established.

Bill and Mabel Shoemaker and their Children

William Toy Shoemaker married, 6 June 1895, in Philadelphia, Mabel Warren, eighth child and youngest daughter of Herbert Marshall and Eliza Caroline (Copp) Warren. He was 24 years of age and she was 25. William and Mabel (Warren) Shoemaker had six children. All of the children except William Jr., reached adulthood, married, and had children.

Mabel and William Shoemaker made their home in the city of Philadelphia, near Rittenhouse Square. Their first child, Dorothy, was born the next year, on 27 March 1896. Dorothy was followed by Jesse Warren in 1898, by Theodore in 1899, and by Barbara in 1902. Then two more sons, Robert, in 1903, and finally William, Jr., in 1904. Tragedy befell the family in 1916 when "Little Willie," as he was called, was struck and killed by a train while William and Robert were walking across the tracks. They had become confused by the passage of two trains from opposite directions.

Mabel, then 29 years old, was present in 1899 when her sister and seven brothers met for several days to discuss the possibility of organizing a Warren Brothers paving company. The meeting was held at the summer camp of her uncle, George Warren (1863-1940), on Lake Ontario, near Oswego, New York. One brother, Jesse (1856-1897), had died, but the others were already successful in their own businesses. They came together to discuss establishing a family business based on the recent discovery by Mabel's uncle, Frederick J. Warren (1866-1905), of an important advance in asphalt paving, which he patented as the "Bitulithic" pavement.[4]

Mabel Warren Shoemaker was very interested in the origins of her ancestors and of their lives in England and in America. The definitive genealogy of the Warren and Jackson families was written by her cousin, Betsey Warren Davis, and published in 1903. Betsey was Uncle Burgess' daughter, two years older than Mabel, and they must have been close friends for many years. It is said that "the presses for Betsey's book were held to include Mabel's baby," Barbara, who was born on 13 February 1902 (not 3 February, the date in Betsey's book), and who does seem to be the youngest entry in the book. Mabel was a member of the Society of Colonial Dames and the Daughters of the American Revolution, and many of her siblings and cousins were also members of the D.A.R. or the Sons of the Revolution.

Part Two

War Letters from Dr. William Shoemaker and His Family

Prelude to War

"June 28, 1914, is one of the most infamous days in world history. On that day, gunshot rang out on a street in Sarajevo and reverberated around the world. The assassination of Archduke Franz Ferdinand and his wife, Sophie, Duchess of Hohenberg, set off a chain reaction that triggered the First World War." The Archduke was heir to both the Austro-Hungarian Empire and the Hapsburg throne. "When Austria annexed Bosnia-Herzegovina in 1908, many Serbs were dissatisfied, and this led to continuous tension and clashes between the Serbs and the Austro-Hungarian Empire. It was this ongoing tension that ultimately led to the assassination … by a student named Gavrilo Princip." Austria immediately blamed Serbia, and with the support of Germany, they declared war on Serbia on July 28. Serbia called on Russia for support, based on a treaty with the Russian Empire, which in turn invoked its treaties with France and Britain. The major countries of Europe then mobilized for war, four Allies versus the Central Powers. Other nations began choosing sides, and Germany declared war on Russia on August 1, 1914.[5]

Germany declared war on France on August 3 and sent troops into neutral Belgium. It refused Britain's demand to withdraw, leading Britain to declare war on Germany. Based on a treaty with Britain, Japan declared war on Germany, and on October 29, the Ottoman Empire joined the war on the side of the Central Powers.

For the next three years, Germany fought on two fronts. In the west, a line of trenches extended west of the Rhine from the North Sea to Switzerland, and the Allies struggled to defend Paris. In the east, the battle line extended from the Baltic Sea to the Black Sea. Millions of soldiers were engaged, and the casualties were stupendous. American public opinion was divided as to which side to support, but U.S. ships used the principle of "freedom of the seas" to send food and other material to Europe. Germany was self-sufficient, France could feed itself, but Britain was desperate for assistance. The shipments benefited Britain more than Germany, which used its submarines and began attacking neutral ships.

America Prepares to Enter the War

The following letter was written by Coningsby Dawson to his mother from the Somme battlefield in France. Dawson grew up in England and graduated with honors from Oxford in 1905. He emigrated to the United States with his family, and began a study of theology at the Union Seminary. He decided instead to be a writer, and he had published his first novel in 1913. He volunteered for the Canadian Army in 1916 and was commissioned as an artillery officer. The letter shows the deep sense of duty that defined him and others, including Dr. William Shoemaker, whose letters continue in much the same vein as Dawson's, albeit often with some bitterness, surprising wit, and sarcasm. Neither writer – Dawson or Shoemaker – intended for their letters to be published.

February 6th, 1917[6]

My Very Dear M[other]

I read in today's paper that U.S.A threatens to come over and help us. I wish she would. The very thought of the possibility fills me with joy. I've been light-headed all day. It would be so ripping to live among people, when the war is ended, of whom you need not be ashamed. … Somewhere deep down in my heart I've felt a sadness ever since I've been out here, at America's lack of gallantry – It's so easy to find excuses for not climbing to Calvary; sacrifice was always too noble to be sensible. I would like to see the country of our adoption become splendidly irrational even at this eleventh hour in the game; it would redeem her in the world's eyes. … I hate the thought of Fifth Avenue, with its pretty faces, its fashions, its smiling frivolity. America as a great nation will die, as all coward civilizations have died, unless she accepts the stigmata of sacrifice, which a divine opportunity again offers her.

April 1917

America Declares War

The Cunard liner, *Lusitania*, was sunk by a German U-boat on May 7, 1915, killing 1,195 people, including 123 Americans. This event was a public relations disaster for the Germans, and Germany therefore halted attacks on ships flying flags of neutral countries. Until that time, public opinion in the United States had been divided. Many Americans of German ancestry were sympathetic to the German cause, and Irish-Americans had little sympathy for the English. Others in America were not interested in events in Europe. The sinking of *Lusitania* tipped the balance of public opinion, and it allowed the U.S. government to begin preparations for America to enter the war on the side of the Allies. American ships carried goods that were needed by England and France, but Germany was largely self-sufficient, and it regarded the American trade as benefiting the Allies. Woodrow Wilson had been elected to a second term as president in November 1916 by a narrow margin. He promised to keep America out of the war in Europe, but nevertheless he continued preparations for war. For various reasons, he had ordered the U.S. Army into Mexico and several other Latin American countries. In the early months of 1917, Germany renewed unrestricted submarine warfare, and the U.S. declared war on Germany on April 6, 1917.

May 1917

Departure from Philadelphia and Arrival in England

The Pennsylvania Hospital Unit, Base Hospital No. 10, U. S. A., leaving Philadelphia, May 18, 1917.

In the Port of New York – On the S.S. *Saint Paul*

"I think of you always."

U.S.M.S. "ST PAUL"
10 20 P.M.
My Dear Mabel,
We went straight without

[Saturday, May 19] 10:20 P.M.*
[New York harbor]
U.S.M.S. "St. Paul"
My dear Mabel,

We went straight without stop to Jersey City – then out to a special ferry boat to the St. Paul. Nice boat, with good accommodations. My state room fellow is Norris Vaux of Germantown – one of our Germantown Hospital men. About 5 o'clock Charlie Jack, Ed [Shoemaker], Dr. [Frank] Knowles & myself went ashore; went up town to the Bellevue, where we had dinner, listened to some fine music, and then returned to the boat. They say we sail tomorrow at 12 o'clock but I doubt it. A St. Louis Unit is going with us but has not yet arrived.

Everybody is happy.

Our boat carries 5 formidable guns – 2, 6 in. fellows forward, 2 stern, and a Gatling gun. The hatch is guarded by a soldier with a rifle to discourage any one who might like to throw a bomb into our hold. Nice little and big shells are coming aboard for the guns to feed upon at meal time.

All together – Everything is very business like

I hope you have had a successful day. Now that you don't have to "look after" me, you can settle down and see some of the results of your marvelous efficiency. Will write – again – tomorrow, if I get a chance.

By the way, I think we are going to Liverpool.

Much love to you and the children, and remember that I think of you always.

B.

---

* WTS to MWS, ML [Manuscript Letter], 4pp. on one page of folded note paper with the ship's letterhead; postmarked NYC Times Sq. Station, May 1917, 10:30 A.M. In this letter, WTS shows the affection that he holds for his wife and children, and he introduces the often unexpected and amusing language that will characterize his letters for the next 18 months. Although he planned to write again on the next day, the only other letter that has been preserved from his voyage to England was written six days later. He planned to number all of his letters, but he failed to keep track of the numbers and he soon abandoned that plan.

At Sea – On the S.S. *Saint Paul*

"Today we entered the U-boat zone and are completely prepared to abandon ship."

May 25th, 1917* [Thursday]
At Sea[7]
My dear Mabel,

The letter I think will come to you uncensored as it must be given here tomorrow morning and will not go ashore. We have had a good trip but a tiresome one. A great deal of rough weather with many people sick for a few days. I was not sick, but I do not like the sea. Gun practice several times and "setting up'" drills have been held – also talks by the adjutant on how to be a soldier, etc. We went very far north, and it was cold. The run has been published, but not the course, so no one but the Captain knows where we are – he won't tell. I have wished many times that I could tell you that we are all right for I know you must be having an anxious time thinking about us. It is a curious sensation, travelling under these circumstances. Great care and watchfulness are exercised – At night, there are no lights on deck, and all port holes are covered – no regular ship lights even. We can smoke in the dark, but must not strike a match or light. The complete darkness is appalling. Watches of course are everywhere and we go ahead as fast as we can.

Today we entered the U-boat zone and are completely prepared to abandon ship. Everyone is assigned to a life boat. My boat is No. 7 – It is to carry about 40. We had a drill this afternoon with our life preservers on. The nurses line up against the rail facing the water. Back of them are lined up the Enlisted men, and back of them the Officers. Three officers to each boat. Major [Matthew A.] DeLaney, C.O., Captain me, and Lt. [J. Howard] Cloud are in my boat. I will be the last one on, as I have to call the roll and put everyone in before I am through work. The boats were lowered even with the deck this afternoon, ready for immediate use. There they will stay until we are safe in our haven. Our guns are now loaded and manned all the time by 22 gunners. All 22 must stay right by their guns until it's over. Most of us will remain on deck all night, ready to decamp. Little chance I would have in my cabin which is 3 stories below deck and below the water line.

Contrary to expectations, no convoys have arrived – perhaps we will get there tomorrow. We would have a feeling of great safety had we a couple of convoys to help us should we get into trouble. Our wireless can only receive messages, but cannot send messages except in emergency, and then, it must be in code. The Germans have been sending many S.O.S. signals and when answered, have lay in wait for the good Samaritans. Fine work, I don't think. We have a bully Captain – Capt. Mills. I crossed on the *Westernland* with him many years ago, but remember only the name. He seems to know his business, has inspired great confidence, and tells us nothing.

We are going to Liverpool and to London – then to France. Where in France no one knows. It will be some stunt for me to walk the streets of London in a U.S.A. uniform. The food has been excellent. At our table are [Norris W.] Vaux ("Bursey"), Charlie Jack, Ed [Shoemaker], Dr. [Wm. J.] Taylor & Dr. [Frank C.] Knowles. First day or so out – typhoid and para-typhoid was given – I claimed exemption on account of my age – it is not compelled after 45 – I hate it. It made nearly all sick – vaccination – arms sore. I will take that, but it will probably not take me.

Dr. [Richard H.] Harte just told us if all goes well, we will reach port in 48 hours. It is now 11:25 P.M. (Thinking room time to see from my left wrist.) Charlie Jack is a special travelling companion – he is fine – cheerful, interesting, interested and alive. Ed is Ed, and a little more Ed. I do hope you will be all right and will be happy in the thought that in helping me by being with me, you are doing a great part in a great work. Tell the children that they must take good care of you, and do everything to help you. My feeling is that we will be home in about 6 mos. Of course, I know nothing about it. If the war stops you may look for me sooner. I hope it may stop.

I found your books in my bunk and read some of the *Path of Life* – I will read more later. I also found chocolates. In fact, you were all so good to me that I am very grateful. Good night. With lots of love to you and the children.

Written but not read B.

---

* WTS to MWS, ML in pencil, 7pp., double-sided, on 4pp. of plain note paper. Return address: Dr. Wm. T. Shoemaker / Base Hospital #10 / U.S. Army; 1d English stamp, cancelled "POSTED AT SEA 29 MY 17 LIVERPOOL; received May 25, 1917.

Safe and Sound in Liverpool

"Mr. Periscope had a sinking spree before he could give us one"

Dr. William T. Shoemaker's wartime letters continue after he reached England with the Pennsylvania Hospital Unit, Base Hospital No. 10. In this letter, addressed to his wife Mabel, he mentions his children, Dorothy "Dot" (age 21), Warren (19), Ted (17), Barbara (15), and Bob (13). Dr. Shoemaker was then 48 years of age; his wife was 47. His writing is often difficult to decipher, perhaps because – like many physicians – he wrote rapidly and cared little for perfection in penmanship; and it also made it difficult for the censor to read. He shows a sense of humor in his letters, as he mentions a German submarine, "Mr. Periscope," who had a "sinking spree."

S.S.* Saint Paul, *outfitted for wartime service in WWI

May 28, 1917* [Monday]
Liverpool / North Western Hotel[8]
My dear Mabel,

Here we are Safe and Sound in Liverpool. Entered the Mersey last night and docked this morning at 7 o'clock. The time that goes here. Just what it really is I do not know . . . The officers of #10 are here, the Nurses are at the Adelphi Hotel a block above, and our Enlisted men with some of our junior officers and the Adjutant were sent at once to Blackpool some 30-40 miles from here. Tomorrow morning, we all go to London, where will stay about a month, then to France and to work. The care and treatment by the Pennsylvania Hospital Friends interested in our Unit has been wonderful. We have a very considerable sum of money – a large sum to be used without accounting, to get anything that the Unit wants or needs. Dr. Harte arranged to have Cable messages sent, transmitted to the family of every member of the Unit. That is why I did not Cable you personally. His message would have probably reached you first. We have had a very fortunate trip – without incident of importance and with mishap. Only one periscope was seen, but we were so much on the job, and a nearby destroyer was also so much on the job, that Mr. Periscope had a sinking spree before she could give us one. You may remember that others who came out of the New York harbor with us have not been so fortunate.[9]

---

* ML on hotel letterhead, 3pp., double sided. To: "Mrs. Wm. T. Shoemaker / 109 So. 29th St. / Philadelphia, Pa. / U.S.A." From: "Capt. Wm. T. Shoemaker / Base Hospital #10 U.S.A." 1d stamp.

There is much that I could tell you if the Censorship were not so strict and efficient. As it is, we have to hand out the ordinary tourist stuff. We had with us on the *St. Paul* a B.H. Unit from St. Louis with personnel that most of us thought do not compare very favorably with that of Base 10.[10] Also, a small Orthopedic Unit to be stationed in England. It consists of 20 men without any Enlisted personnel. There were in addition, a few – 3 or 4 – independents. The people here are glad to see us and it is a reminder that while in England we are to be the guests of the British Government. And so, these exchanges will be beneficial.[*]

We understand that the Officer must pay his own hotel expenses, etc,, therefore to work in London might clean Capt. S. out of funds. Will write from London next, and as the mails are few, your letters will probably come in together. Will also start in on the Children very soon. I am glad Dorothy got her scholarship and that Barbara is going to like her Camp, and am especially glad that Warren does not hate us. To be in the first cabin with all the comforts of home (except Marblee) and have him in the steerage with all the comforts that pertain, through very rough weather, would have made me very unhappy. Ted of course is at Becket and Bob is getting ready for this. Am anxious to hear about the Convention. It must have been a great success, as no mail instructions were given. I presume I will not hear from home for several weeks yet. That is a long time. However, I will think not about it.[†]

With much love,

Wm. T. Shoemaker

P.S. I am told that the censor wants the full name signed to letters.

TELEPHONE No 6449 MAYFAIR, 6 LINES.
TELEGRAMS, HOTEL CURZON, LONDON.
S. HARWATH PROPR

HOTEL CURZON,
CURZON STREET,
MAYFAIR,
LONDON, W.

May 30th, 1917[‡] [Wednesday]
London
Hotel Curzon / Curzon Street / Mayfair
My dear Dorothy,

A funny feeling indeed to return to London office after 23 yrs. and in an Army uniform. The surrounding country and city are full, all the people have got a good spirit, and except for certain machines and the presence everywhere of soldiers, one could not believe that there was a war in progress. I would think that at least 40% men seen are in uniforms. Most of them have been to the front and many of them show unmistakable signs of their experiences. Today we went to the Embassy to get passports, which the French insist upon. Our uniform was sufficient for the English. (We must also be supplied with several photographs. The camera business is breeding girls mature.) We met the Ambassador, Mr. Page. He told us in his diplomatic way how glad they are to see us.

A service at St. Paul's Cathedral was most interesting. The American Legion of Canada delivered to the Cathedral the American & Canadian flags for safe keeping. To hear the great organ & choir of St. Paul's rendering the Battle Hymn of the Republic, Onward Christian Soldiers, The Star Spangled Banner, and God Save the Queen (King) was very inspiring.[11]

---

[*] B.H. No. 10 sailed to England on the *S.S. Saint Paul*, an American Line ship that was on contract for the U.S. Navy.
[†] "Marblee" was probably their maid. Ted was at Camp Becket in the Berkshires. The "Convention" is a mystery.
[‡] ML, 3pp., double-sided, on hotel letterhead, to "Miss Dorothy Shoemaker / 109 So. 20th St. / Philadelphia, Pa., U.S.A. Envelope: "Capt. Wm. T. Shoemaker / Base Hospital #10 U.S.A." 1d stamp.

I walked up to my old boarding house on Trafalgar's Square to see Miss Maughn. Miss M. was not there, but the place was, known as the Maughn family hotel. It is kept by a relative. Miss M. is still living but I guess I will be unable to see her. We will be here about a month before going to France. In France, the work begins.

My address, remember, will be: -

Captain "Me" / Base Hospital #10 – U.S.A. / France. C/o B.E.F. via London

B.E.F. means British Expeditionary Forces.

Or, in an emergency, I can be reached at any time – C/o American Embassy – London – with of course the letter stating that you give the particulars as to the question.

Now, about yourself. I hope you have a good summer. If you go away, see that your mother does not need or want you. She is the most important one in the world to us. When letters from home get started, I hope you will all keep her informed and frequently supplied. It will probably be 3 or 4 weeks before I hear a word from home. Now that is not nice. I like it not, but you know what War is.

Good bye, love to all,

Your father,

Wm. T. Shoemaker

The censor wants to know who is writing. Must use full names.

May 31st, 1917* [Thursday]
London[12]
Hotel Curzon / Curzon Street / Mayfair
My dear Warren,

I have been thinking much whether I am glad or sorry that you are not with me. On the boat you would have had rather a marvelous time. You would have been in the steerage and probably sea sick, as many were, for we had a number of rough days. That would not have been nice. If not though, you would have found it quite interesting, and extremely novel. Passing through a life boat drill with a lot of life boats trying their win and race. This is a new experience for an American. But we were fast and very well armed and had a splendid Captain and efficient officers. Three attempts were made to get us. I personally did not see a periscope, but a number of our people did. From Liverpool, our Enlisted men with 7 officers, including Lt. Edwin Shoemaker, were sent to Blackpool where they still are. Blackpool is a Coney Island. They will later come to London. We came straight here and will leave for France next Tuesday. Then the real work begins, and I hope to earn my pay. This far, I have been cared for by the American & British Governments. After we get to France, both will probably their bait back. I have been through a great time with Dr. Rowley – he is a D.K.E & of the first in the Chapter – and of cousin Jack – Dad.†

We are so glad that you are also a D.K.E. On the boat, saw Hudson Chaps in one and another fellow you answer whose name I cannot remember. This afternoon went to tea at the home of Ambassador and Mrs. [Walter H.] Page. Great swells all right. Previously enough, I attended a function at the Dance House with your grandfather and uncle Charlie before you were born. Thomas Bayard, also a D.K.E. from Delaware was there – Ambassador.

I cannot tell you any more stuff because I would be censored. There is however much that might be told. You I presume are or soon will be at W.B. Co. If you have decided to go in for military business, and your mother approves, I am extremely satisfied. Everyone here thinks you all will have to get in before the end of it. If so, the sooner the better.‡

With love,

William T. Shoemaker

---

* ML, 2pp., double-sided, on hotel letterhead, to "Mr. J. Warren Shoemaker / 109 So. 20th St. / Philadelphia, Pa., U.S.A. Also, another ML, 2pp, double-sided, to Ted. Envelope: "Capt. Wm. T. Shoemaker / Base Hospital #10 U.S.A." 1d stamp.

† 1st Lt. Edwin Shoemaker, D.D.S., and 1st Lt. Charles Shoemaker Jack, D.D.S., were the dentists with B.H. No. 10; they were cousins of William T. Shoemaker's children. Charles was promoted to Lt. Col. at the end of the war. I cannot identify Dr. Rowley or the other named men who were in D.K.E. (Delta Kappa Epsilon). Some of WTS's sentences were incomplete.

‡ W.B. Co. = Warren Brothers Company, founded by Mabel's brothers, in which she was an equal partner.

May 31st, 1917* [Thursday]
London[13]
Hotel Curzon / Curzon Street / Mayfair
My dear Ted,

I think of you often and picture you at [Camp] Becket having a fine time and being of great use. I hope, before you are old enough for military duty, the war will be over, but if not, your time will come.

We had a good but rough trip across the Atlantic. The last 3 days were anxious ones, as the pesky U. boats were trying their best to sink us. It was a Great Game, and one which you have to learn how to play, unless you want to pass on at the bottom of the Ocean. We were very fortunate and our crew members knew how to play the Game. Three U. boats tried to get us. The finest thing we saw on the whole trip was on the second day from Liverpool. A destroyer picked us up and helped us along to another destroyer. She looked mighty good with her American flag, and we felt much relieved. She had a speed of 35-40 miles [per hour], and played around us like a bee. It would have been a bad time for a U. boat to poke his nose up. However, one did, but it ducked before we or the destroyer could get a shot.

We [are] going to London for a week and will leave for France next Tuesday. While in England we are the guests of the British Government. Which is very nice and thoughtful of John Bull. Officers, you know, have to pay their own expenses. Which would amount to something of a small fortune.

It is interesting to renew my old association here after 23 years. I lived here for 6 months in 1894 and became quite a Londoner. Strange that today, I can see the same things that became favorites with me. It is unquestionably the greatest city in the world, in spite of New York. Which of course is some city.

Write to me soon and often, and let me hear how you are coming.

Address – Base Hospital #10 U.S. Army France B.E.F. via London

With much love –

Wm. T. Shoemaker

May 31st, 1917†
London[14]
Hotel Curzon / Curzon Street / Mayfair
My dear Barbara,

It is very difficult to write so many letters to the same house without saying in them the same things. However, I can't miss you just on that one now.

If I had you here, I would show you lots of interesting little things, interesting things, all different. Now, for instance, sitting near me in the writing room is a lady – very refined and quite elderly, above 65 yrs. and she looks so much as Grandma looked at that age. She is calmly smoking a cigarette. Ladies here do it gracefully and it looks funny but all right. The bus drivers on the busses and cars are girls and like the bar Rupees are also girls. Real nice ones, most of them. In fact, the women in England are just as busy as the men. The streets here are full of soldiers – Wounded soldiers or those who have been wounded wear a ribbon on their arms. I would have to tell your mother the color. There are ambulances – New Zealand, Scotch, Irish – Belgian and American soldiers in evidence. And thousands of Canadian fellows. They are all tired of the war and wounds. The Eats are very amusing. Sugar especially is very scarce, and is given out in little bits, just enough to not make your tea or coffee sweet. In Liverpool, I had to eat my oat meal without sugar. In France I believe we will get no sugar, or sugar on a on a basis of 1½ lbs. per week. Yet everybody gets enough to eat. It is very different in some ways from the London of peace times.

---

* Another ML, 2pp, double-sided, to Ted, enclosed with the letter of May 31 to Warren. However, WTS usually sent individual letters to each child, so the envelope to Ted may have been discarded.

† ML, 2pp., double-sided, on hotel letterhead, to "Miss Barbara Shoemaker / 109 So. 20th St. / Philadelphia, Pa., U.S.A. Envelope: "Capt. Wm. T. Shoemaker / Base Hospital #10 U.S.A." 1d stamp.

Now the next fellow is Bob, but not tonight. Tell him that he has not been forgotten, but in order to get through systematically, I have to go straight down the line. Please write – so that I may imagine what you are doing at home. Well must – Love, Dad

----

Thursday, May 24, 1917* [Thursday]
Philadelphia, Pennsylvania[15]
Dear Billy,

According to my calculations, you should reach Liverpool either today or tomorrow, so I am looking for a cable message to tell of your well-being. Oh! I hope it will come soon.†

Our Convocation guests have all departed. Mrs. Dickson was the last to go, and she certainly did have a beautiful time. Ella is returning with George's family by automobile. They were to spend last night with the Miles family in New York, and they made an early start this morning. If they are on the Mohawk Trail today, they must be enjoying it, for it is gorgeous weather. I wonder what kind of weather you are having, going through the submarine zone, if you have not already landed. The paper yesterday told of the first United States Hospital Unit to arrive (I suppose it is the one from Cleveland) being received by the King and Queen. I wonder if they will honor you so too, or if it is only the first one.

We are just settling up the Convocation bills. The banquet bill was $2500, about $1000 of which was paid in for tickets. The Hospitality committee bills are all in but one. They will amount to about 360, while the Luncheon Committee will be out of pocket about $600. The lunches are to be paid by the Ladies' Aid. Everything went off wonderfully well, but I am glad we won't have to do it again for a few years.

Luler came in this morning with Lucretia. She is a darling, and looks quite well and plump. I forgot to tell you that Em Reece gave me such beautiful roses before he left. Wasn't it thoughtful of him! This evening, Barbara and I are going to the commencement of the School of Industrial Art. Dorothy told you that she won a prize scholarship, but I don't know whether she told you that she is the only one whose scholarship is reserved. It is extremely gratifying to have her do so well. A card from Ted yesterday said that he reached Becket safely and was about to begin his four-mile climb. A tiresome climb it must have been with a heavily loaded suitcase. There doesn't seem to be any more news, so I will say good-bye and take a ten minute nap before getting supper, as Bessie is out.

The enclosed letter you may be interested in. I sent them to all those dead-beats whose bills I found on your desk. Don't you think they will feel pretty mean if they don't pay up? You see I used blank stationery, so that the letters would be opened. Lots of love,

From
Mabel

May 23, 1917‡ [Wednesday]
Mr. John E. Pierce
5337 Wyalusing Ave. / Philadelphia
My dear Mr. Pierce,

Dr. Shoemaker has gone to France to serve his country. It will be of material assistance if the money that he earned when here is promptly paid for the use of his family while he is away. If you will pay the enclosed bill as soon as possible it will be greatly appreciated.

Thanking you for the prompt response which I am sure you will be glad to make, I am,

Very truly yours,
[s] Mabel W. Shoemaker

---

* MWS to WTS, ML, 3pp., on both sides of letterhead note paper of "Dr. William T. Shoemaker / 109 South Twentieth Street / Philadelphia; enclosing a copy of TL to Mr. John Pierce.

† MWS's calculations were optimistic. The S.S. *St. Paul* did not arrive in Liverpool until the night of 27-28 May 1917.

‡ TL, 1p on plain note paper, marked "Sample Copy," enclosed with MWS's letter to WTS, May 24, 1917.

5-27-17* [Sunday]
Dear Father –

I arrived in camp safely after eleven and a half hours traveling counting one hour stop in New York, two and a half hours in Springfield, and about an hour walk up from Becket.

Of the six days I have been here it has rained some part of five of the days. And has been about as cold as Greenland all the time.

The main part of our work has been piling lumber for the new recreation building which is to be built this summer. Also we have done a lot of painting of the buildings and floors of the building around camp

There are five other fellows up here all of which are old campers except one. Also "Pop" Gibson and the baker.

Things look pretty good for a good season every thing is alright except the lake has gone down so far that we will not be able to use the pier for diving as the water is only about four feet deep off the end of it.

Seventeen of the new house tents are completed and the floors and uprights of the remaining ten are up.

Mr. Gibson was up on Wed. and left again Thurs. afternoon after giving us enough work to keep us busy for about two months. Of course we will get it done by the time he comes back next Thursday.

Yesterday we went into the lake for a twenty second swim the water was like water in an ice cream freezer.

For the last three nights we walked down to Becket to get the mail and maybe it wasn't some tiresome walk coming back over the hill.

Things are so dead up here I can't think of anything more to say. So I guess I'll quit.

Lots of love
Ted

May 27, 1917†
Philadelphia
Dear Billy –

We are still waiting for news of Base Hospital No. 10, but as Charlie says, the *St. Paul* is rather slow, so we are not unduly anxious yet. Owen and Greta and Martha were here to dinner today, and now Dorothy is driving them out to Wallingford [Penna.]. Later we will take the Harveys for a little ride in the park, and then Dorothy, Edwina and I are going to their house for supper. Martha is a trump, just as brave and cheerful as she can be. I wish I could say the same of the wife of your cousin. She is behaving like a selfish chump. Everyone is disgusted with her.‡

Some of the wives of the doctors of Base 10 are meeting together on Fridays at St. James Parish House, to do Red Cross work for the Unit, and I am going to join. I shall be glad to know better the women whose greatest present interest is the same as mine. Edwina went over to New York in the machine with the family on Wednesday, and returned yesterday, to visit with us for a week or so before Bryn Mawr Commencement. She was down in Washington yesterday at Laura Delano's wedding. She married a young diplomat who has been to Petrograd, and is going soon again, taking her with him. Helen Shaw is engaged to Will Crosby, whom we met at Greenbush last summer. Edwina says she is now the only one of the six girls who were so intimate at college to be neither married nor engaged. Last evening, I spent three hours at your desk over the bills. I am happy to say that I paid practically all of them, all except two or three very small ones and Rowan's bill. That I thought I would let stand until we took the car away at the end of the month and settle all at once. Warren asked about the advisability of taking the car away several days ago, but Rowan said it would cost the same as to leave it until the end of the month, so we left it.

We are getting along very well at home, but of course we miss you. Returns have been good from your bills. Mr. Auspack paid his bill by return mail. We have sent patients every day to Dr. Creighton, some days as many as

---

* ML, 2pp., from Ted to WTS, on notepaper of "In the Berkshires / Becket, Mass."

† ML, 4pp. on plain note paper, from MWS to WTS; marked on envelope, rec'd 6/15/17.

‡ Owen Shoemaker was WTS's older brother; his wife was Margareta "Greta" (née Jack); Martha was their daughter-in-law. Their son Edwin Shoemaker (WTS's nephew) and his cousin Charles "Charlie" Jack, were dentists in B.H. No. 10. The "selfish chump" is probably Mrs. Charles Jack. Wallingford, Penna., is about 15 miles SW of their home in Philadelphia.

four. I sent your pay voucher for May to Washington yesterday. I held it as long as I dared, hoping to receive the Captain's commission. I hope it will come in time for the June voucher.

Dorothy's commencement was very nice, and the exhibition of work most interesting. They certainly do good work there. Mr. Copeland is having an outdoor sketching class during June, four days a week, at a charge of Ten Dollars. I told Dorothy to join, for I feel that it will be of great benefit to her. She has never had any instruction in that line. Well, that seems to conclude the news. Oh! I forgot to tell you that Dr. Frank's ____ failed and died last week. Warren wrote to Frank and then he and Allen called this afternoon, but Frank was out.

All send love, but "no-one loves you quite so much as me." Write as often as you can, and take good care of yourself. Love to Ed & Charlie. Yours,

Mabel

May 28, 1917* [Monday]
Philadelphia, Pennsylvania
Dear Father,

We were all so glad to learn this morning that you had a successful trip, without any mishaps. Rather we heard last night, because someone from the Pennsylvania Hospital phoned. It was in the paper this morning, also, and Dr. Furbush sent a telegram that arrived about lunch time.

Gertrude Russell, Connie Westcott, and I all spent Friday night with Louise Harding. Mr. and Mrs. Harding took us to the movies that night, and on Saturday we went up to the Hill-Haverford game, at Haverford, in Helen ____'s car. The cheering was great on both sides but the game wasn't very exciting. Hill won 8-1. We were all going to Carol Smith's dance Saturday night, but Carol got sick and the dance was postponed till next Friday night. ___ Coffin was coming here but we postponed her, too. She and Louise are coming here next week after the dance tho', and then on Saturday we're all planning to go to the Sunday School picnic.

Wasn't it great that Dot got her scholarship renewed! She was the only one, too. Mother and Olive and I went to the Commencement – you never saw such drawing, and paintings, and modeling, and stencilings, as were in the exhibition in your life! I won't dare go to that place after Dot makes her rep! I wouldn't mind living in the rooms that they make at that school, but I guess I'll never have an interior decorator if they make such elaborate, and expensive rooms! We met Mr. Copeland, and he certainly is nice.

The afternoon you left Philadelphia, Flora ___ and I went to Franklin Field to see Ted in the Interscholastic meet (track). He and George Andrew (C.H.S.) out-vaulted all their schools, and then went up, tying all the time – So the score keeper got tired of waiting and they tossed up for first. Of course, Andrew won the toss, but Ted vaulted just as high, and got all the honor just the same. He got a silver medal, too – they both jumped 9 ft.10in. Pretty good!

It's getting late so I'll have to stop – Mother and Dot and Edwina have gone to the movies. Bob sends love.
With lots of love,

Barbara

P.S. Helen Shaw is engaged to Bill Crosby. She'll breakfast here tomorrow. Only 7 more days of school! Everyone here feeling fine – Write soon and tell us what war is really like! Were you seasick?

---

* Barbara to WTS, ML, on both sides of 1p. of plain note paper.

June 1917

From England to Le Porte, France

"Base 16 – B.E.F. (British Expeditionary Forces) and has about 2000 beds"

Sunday, June 3rd, 1917*
London / Hotel Curzon
My dear Mabel,

This is London but hope to leave shortly for France. It has been a very pleasant and enjoyable time, but tiresome, because we are not earning our keep. The facilities here are fine and London parks are in the height of beauty. I have renewed very old residents and have been surprised at the few changes in the general topography and appearance of the town. The same buses, motor driven, instead of horse driven. Practically no hansom cabs. We had nothing to do but visit and come and see.

At least 40% of the men are in the streets are in uniform, and many are from all of the colonies. Canada is much in evidence. We are looked at and cheerfully greeted. An advisory instruction results from our uniform. All English officers wear a beret. It is not regulation in our service, so a great many officers here cannot tell whether or not we are officers. They also wear their insignia on the sleeve – cuff; and they cannot understand our insignia on the shoulder. The units that arrived ahead of us were furnished with purchased belts *ala* English. This was becoming, but we were warned not to do it. We individually purchased, by units, anything for our uniforms without order. But Harvard and the orthopedic bunch could not get belts and arms enough yesterday.[16]

I visited Dr. Dunston's Hospital for the blind. It was most interesting. Founded by Sir Somebody, whose name I have forgotten and who is himself blind, it is doing a great work, almost 300 inmates are there. All corridors have the same floors: in the center, a slip of linoleum, on each side of which is carpet. Those that can see are expected to use the carpet so that the blind fellows can keep the center and go straight ahead without fear. Approaching the steps, they came to a piece of metal molding. So they can travel all over alone and so fast as they care to go. They are taught Braille, shorthand and typewriting. And complaining brought work of another kind. It requires about 9 mos. to learn the work they are fitted for, after which they have only to work with materials that are nearby, and are found for them. They have a learning capacity of above 2/3. They are happy and interested which seems to me surprising. These people are remarkable; not a complaint or anything but a quick respite – the fact that then they are blinded means maimed for life. Limbs gone, eyes gone, hands smashed – etc.[17]

It is dreadful to see the results of the war. Of course, they [the British] all wish it would end, but they are going to stop it. They can no more be conquered than you could drink all of the water in the Atlantic Ocean. Empire Day was yesterday – a very impressive service was held at St. Paul's – we were given seats in the choir. The American flag & the flags of the Empire, 45 in number, were presented at the altar and left in the Cathedral. The music was fine. I enclose the program.

The Chicago Unit that came out of N.Y. on the ship with us, arrived at this hotel tonight. I do not know when they left ship. They had minor troubles but are now safely here. I have received no word from home yet and am getting anxious – (I hope you have entirely recovered from your "indisposition" which was bothering you when I left. I have thought much about you but find it most pressing to feel now that you are all right. It is the only thing I can do that does not worry me. Please write often. When you want a breather, get your car and a drive and take Owen & Greta to Radnor or any place else you can do it. Dorothy can take you. I would feel <u>so much</u> better if I knew that you and the children were having a normal time.)†

I have a new address for you. Addresses seem to be a little uncertain, but here is the latest.

Capt ___, U.S.A. / Base Hospital No. 10 / American Expeditionary Force / on Active Service.

---

* WTS to MWS, ML, 12pp. on both sides of 3pp., double sided, of 9x7 in. Curzon letterhead.

† The parenthetical marks in pencil were added later, probably by BWS, perhaps to mark out part of the letter that would not be transcribed for others to read. The word "Omit" was written between the lines above the first mark.

All of our mail would go first to Washington, be put into a separate pouch and sent to us where ever we are – (Or, it would probably reach us by any other address I have thus far given. They seem to change the address every few minutes. If you want to cable me, do it thusly: Capt ____ Amex force {Western Union} That's enough. You see it is made from American Expeditionary Force. Aux is our cable code name. If my commission arrives, note the date in it and see that you get proper pay from that date. Also notify me of the date. I have had some trouble with my shoes and putties. The combination has given me a lovely sore on my *tendo Achilles*. Bought a new pair of putties and think I am getting better. Marching helps an open sore on your foot very little).*

Food here is a little *jeane* in spots. There is enough to eat, but sugar is handled like gold. One very small very small piece for coffee – the waiter sprinkles just a little on your oatmeal, and there is no *crème de lait* with the sugar bowl. The bread is miserable, made of whole wheat. Also delivered in small quantities. Butter is fairly plentiful. Meats and eggs are good – fish is good. There is no sale of alcoholic drinks – wholesale or retail – except between 11:30 -2:30 P.M. & 6:30 and 9 P.M. This applies to Clubs also.

Now, I have just enough room to say good bye, and send you lots of love. Must write to Bob, so tell him it's on the way. B

Wm. T. Shoemaker

June 3, 1917†
109 So. 20th St., Philadelphia
Dear Billy,

Your picture appeared this morning in the Pictorial Supplement of the *Ledger*, a wretched copy of one of Drover's pictures, and adorned by the caption "First Lieut. Edwin Shoemaker, D.D.S., Dental Department. The whole page was devoted to Base No. 10, including Drs. Wilmer, Austin & Swatley in civilian clothes. I will save the sheet for you to see. Wednesday, in *Bulletin* told of an impressive ceremony in St. Paul's at which the recently arrived hospital units from Phila. and St. Louis were present. Were you there, and also were you at the presentation of medals to soldiers by the King yesterday?

A few days ago, we received announcements from Dr. Frederick Dillon Owsley, of Rydal, that his daughter Leonora had been married on May twelfth to Dr. Leon Hermann. They are to be at home at The Bleuheim. The announcements came two weeks after the wedding, so it must have been hastily planned. All his friends were much surprised, never having heard of the engagement. We had quite a breakfast party yesterday. Barbara and Warren had been to a postponed dance of Carol Smith's Friday evening and brought Louise Halding and Kathryn Coffin home with them. Edwina was already here, and Helen Shaw arrived from Boston on the *Federal*, and came to breakfast. It was Sunday School Picnic Day, but we planned to go on the eleven o'clock train, all except Bob and guess who went at eight. We had breakfast at nine o'clock with Serena to wait on the table and do the upstairs work, so that I could go with the children to the picnic. The day was fine, and it was a good picnic, but I was so tired after it that I have done nothing but rest all day, not even going to church.

Edwina and I called on the Burnhams one day last week – but I think I told you that in my last letter. On Friday, Martha and I went to St. James Parish House to work for Base Hospital 10. Mary Jack was there too, and she seems brighter now that she knows that the *St. Paul* reached out. She had a presentiment that it never would, which shows how much good presentiments are.

To-morrow Dorothy begins her sketching lessons with Mr. Copeland. I think they should be very useful. They are to take their lunch with them and stay all day. Tomorrow they go up to the Wissahickon. The idea is to go every clear day until they have been four days in a week, this to continue during the month of June. Did I tell you that Caroline is to be married June 30th and Rachel probably in July?

There doesn't seem to be any more news. Oh! I forgot to tell you that a letter came from the Pennsylvania Hospital, signed by John J. Lewis, Jr., President – saying that the "managers want not only to give their support to those who, in the name of the Hospital, have answered the call to duty in foreign lands, but they desire also to be of

---

* The second set of parenthetical marks were annotated similarly in pencil as follows: "Omit" after the opening; and "Begin here" after the closing. The "putties" that WTS referred to are usually spelled "puttees" – a cloth wrap of the leg from knee to ankle.
† MWS to WTS, ML, 4pp. on 2pp of plain note paper, double sided. Envelope to Dr. Wm. T. Shoemaker / London / England/ Base Hospital No. 10 / U.S. Army, marked 7/6/17. She apparently wrote a previous letter; not preserved, probably lost in transit.

any possible use to their families remaining at home." They hope that we will all keep well during your absence, but in the event of any of us being ill, they extend to us the use of the private rooms and any other service the hospital may be able to render. It is surely very kind of them, and I shall write and thank them. We are all quite well and hope to remain so, but in the possible event of being sick, don't you think we would feel more at home on Girard Ave. in the care of Harvey or Henry Page? Probably they would extent us the same courtesy if the occasion arose.

Dr. & Mrs. [Lawrence] Turnbull phoned today, and offered their service in any capacity. So did Dr. Hollaway.

The children all send love. We will begin soon to look for a letter from you. If it were not for the delays incident to War, I suppose one should come to-morrow, but – as it is, I will try not to be disappointed if no letter comes for a week.

Lots and lots of love from

Mabel

June 6th, 1917 [Wednesday]*
Hotel Curzon
London
My dear Mabel,

Look at that date and try to remember what it means. I started the day right by sending you a message of love and congratulations which I hope you received shortly after your breakfast. But that is not all the thinking of you which I have done. I wish I could drop in on you for a call and fly right back to my job, which is getting to France. We go shortly; if I told you when, you would not get the letter. That would be sad. No news from home has reached any one as of yet. When the line of communication is established, I hope your letters will come often. Took lunch today at the Old Cheshire Cheese, sitting on the bench where Laine Johnson used to sit. I want to go there quite a little. It does not look a day older. Enclosed is the menu which is interesting. Do not destroy it. Note the Public Meals Order. It is not a joke, but a strict regulation enforced throughout the country – Hotels, Clubs and all. I could not order half a chop, because it exceeded the regulations. In fact, there was no grill. The supposed scarcity of sugar is curious in that the candy shops are going full blast, while you can't get enough to sweeten the coffee. One would think that they would stop the making of candy.†

Visiting my old hospital, the Royal Westminster Ophthalmic – the certificate from which hangs in my office – not a single member of the old staff remains. But a few men have at advanced age returned to help out in the Emergency, as most of the regular men have gone to the front. Mr. Jules & Mr. Frost are working there. The man at Moorfields that I worked for and liked best was Mr. Stanford Morton. Called on him yesterday at his office. He was very nice and was glad to see me. Same lively English questions, and in his same back office. We are all getting tired of doing nothing in London. It is expensive too, because we have to go around a great deal, and pay out money in comparatively small but very frequent repeated doses of crunch. We are not prepared for that. Perhaps I should say, I instead of we, for many of our Units are used to money troubles.[18]

Charlie is happy but misses his children very much. He apparently misses his wife also, but not so seems Ed. Since we landed, because he is at Blackpool. One of the boys came down to London (or Enlisted) and reported that Ed had bought a new suit – was combing his hair very carefully and was quite the real thing. I am glad he got a suit for he looked like the devil. Most of our men are buying clothes, and are having great trouble getting the English tailors to make regulation U.S. uniforms. After about 6 fittings, I think Charlie has a pretty good looking suit - $40. - however. Quite an expensive uniform, but it will probably wear well. It a sort of whip cord; color not so good as mine. Be sure and get my pay as soon as it comes – however away at the Quartermaster General, Washington, D.C. Our Quartermaster (What I would L- to have been!!) wants $30,000. for June, so you can see that Base Hospitals are expensive – at least 1000. – per day. Now I must kiss you good night, and always wish you many happy returns of the day, the day must in the future be spent with me. Yes? Lots of love, B

Wm. T. Shoemaker

---

* WTS to MWS, ML, 8pp., on 4pp., double-sided, on Hotel Curzon letterhead, with hotel envelope, 1d stamp, cancelled London. They were married on June 6, 1895; this was the 22nd anniversary of the marriage.

† The menu was nevertheless not preserved.

June 6, 1917[*]
Philadelphia
Dear Billy –

The first direct message from you came this morning, carrying your love and congratulations on our anniversary, across the Atlantic. I was just a little blue this morning because you were away, and I was hoping that maybe a letter would come, and it hadn't. Mrs. Failkrod had called for me in her car to take me to the Shakespeare Club "Formal Closing," which took the form of a luncheon at Valley Green, and we were in the car in front of the house when the telegraph boy came with your cablegram. It did me good through and through, and so I had a lovely day. I was sure that you would <u>think</u> of the day, but it never occurred to me that you would cable. Thank you so much for your thoughtfulness. Now I know that you are alright and in London to-day. I wonder how long you will be there, and how soon I should send letters elsewhere, and where else.

The photographs from Mrs. Sagen came Monday, but the one from Goldensky's has not come yet. I sent Miss Miles down to see about it, and they said it would be here in two or three days. I gave Miss Miles one of your pictures, and she was certainly delighted. Business continues to be good, both as to patients sent to Dr. [William J.] Creighton and as to receipts. Your salary from the government has not come, however, and neither has your captain's commission. Don't you think you should write to Washington about it?[†]

By the time I get an answer to this letter, probably I shall be at Wallingford, but I think you had better still address letters here, for on the days that I do not come in, I will instruct Miss Miles to take them down to Owen and he can take them out to me. The car is up at the hospital, and Warren has been working with it all the afternoon. We want to get it on Sunday to take Mr. & Mrs. Harvey and the baby out to Wallingford to dinner. Bobby is through school now, and Barbara's last day is the commencement to-morrow. I will let Bob go out to Wallingford on Sunday to stay. Then there will be only four at home – Barbara, Dorothy, Warren and myself. Kind of lonesome, yes? I will keep Bo here until we really vacate the house, it seems so much safer here with a dog, even if he is little.

Warren will be through at college about the fifteenth, and then he will go away in a day or two, probably to Bridgeport. I am telling you this ahead so that you will still know where to think of us. Bob goes to camp the twenty-eighth, and Barbara the twenty-ninth, Caroline is to be married on the thirtieth (I think) and Dorothy goes to Main with Cousin Emma and Cousin Will about July first. I forgot to tell you that invitations are out for Mildred Jack's wedding on Saturday, June twenty-third at 12:30. The invitation came to Dr. & Mrs. & Family. I suppose it will be alright to take Dorothy, Barbara & Bob, as Warren will in all probability be away.[‡]

A letter from Ted yesterday says that farming isn't so easy as it sounds, and that his hands are blistered from holding the plough handles practically all day Saturday and part of Friday. There doesn't seem to be any more news, so I will say good-bye. I hope there are many letters from you on the way. The cable was fine, but a letter can tell so much more, even though it is not such fresh news. All the children send love, but I send even more than they do.

Lovingly your wife,

Mabel Give my love to Ed & Charlie

June 10th, 1917 [Sunday][§]
Havre, France
My dear Mabel,

At last we are in France and are gradually approaching our destination. We all became very tired of London, because we had nothing to do but pose around. Our trip across the Cannel was the best that could be. The weather was perfect and the water perfectly smooth.

---

[*] MWS to WTS, ML, 4pp., on 2pp. of plain note paper, double sided; unless otherwise noted, her letters are all from her home. Envelope marked 7/6/17, surely meaning received on that date by WTS.

[†] Elias Goldensky (1867-1943) "The Wizard of Photography" was at that time the most prominent photographer in Philadelphia.

[‡] The company office of Warren Brothers Co. was in Bridgeport, Conn.

[§] WTS to MWS, ML, 8pp., both sides of 2pp. of folded graph paper, postmarked Le Havre, 11 June, French Republic, 25c stamp.

We left South Hampton in the afternoon on a small boat, and a few miles down were put on a hospital ship, and brought to Havre. The hospital ship was fitted to carry 500 wounded. We were quartered in the hospital beds below. Two British destroyers convoyed us all the way. Many aeroplanes were playing about above. And all together it was a most interesting trip. A submarine was sunk by a destroyer just ahead of us, and a large vessel coming behind us was torpedoed. She reached port however under her own steam, but had a large hole in her side that gave her a list just short of turning her completely over. We saw the destroyer that sank the submarine. She rammed her and spoiled her for doing it. We passed through a great quantity of oil on the water afterward, which came from Mr. Lot. While waiting to get off and away from the dock, our boat was loaded with 496 wounded coming from the front in a hospital train, or batches of several hospital trains. It was a very depressing sight – pathetic.[19]

They started back late this afternoon. The nearer we get to the theatre of war, the more horrible the whole thing becomes. We are now in it and you can bet we will all be glad to get out of it when it is over. I cannot tell you where we are going, but our base hospital has 2090 beds. Some many for us to care for. Of course, we will only pull up after a drive – behind drivers, with so many beds, and have so much to do. A crew does drive and is under us as often, of course. Readers from our papers so far will notice that the Allies are tearing the Boche (the Germans) to pieces. In fact, on the Western front, the Germans are beaten every time. General Pershing and his staff created quite a sensation in London. They were the real articles and made a great impression. The papers here say that 10,000,000, some claim 2.8 have registered. From that number, I fancy a very large Army can be made. I also hear that 3 Divisions are coming this way. My but they will be welcome. The French to my mind are not very good looking soldiers. The English and Colonial troops are magnificent. The American troops will look all to the good, and when they get here, Crazy Bill might just as well stop.[20]

I am trying to figure out where you all are, but can't remember your various dates. Ted is at Becket. Have him there. I am anxious about Warren. If he goes in, he will almost surely get over here, and I am afraid that would be a little rough on you. Howard and Orlando will probably be coming over; in fact, there will be a general exodus of the Shoemaker family[*]

["Omit to end" added here in pencil] I am most anxious that you should have everything you want (except me yet), so if you want to help me to be mighty good, [take] care of yourself. Try to have my practice kept in place in so far as possible. Much can be done by seeing that my patients are properly referred and that they get the needed service. Answer all letters properly, just as if we were still in business, etc. For a time, much can be done – if I am away too long it will be more difficult.

Good bye for the present. Love to all.

<u>B</u>

Wm. T. Shoemaker

June 11, 1917[†]
Philadelphia
My dear Billy,

According to the papers, you are now in France, so I suppose your real work has begun. It was fine to have such an ovation as you all had in London (papers again) but – I guess you are glad it is over. The July number of the Red Cross Magazine gives a plan showing the arrangement of the hospitals, and Miss Miles, and the children, as well as myself, are all delighted to see that the base hospital is situated some miles from the firing line, so we will consider that you are now working hard, but in a safe place. Yesterday, Dorothy and I took the Harveys out to Wallingford to dinner. We had a very nice time, but while we were gone, Dr. A. B. McKee called up on the phone. He was passing through town, but if we had been there, I suppose he would have stopped in to see us. He wanted to be remembered to you, Warren said. They are getting up another ambulance unit at the University [of Pennsylvania], and Warren wanted so much to offer to do his bit – that I thought I ought to agree. He has given in his name, but I don't believe they will take him, as he is sixteen pounds under weight for his height. Bob went out to Wallingford to-day to stay until it is time to go to camp. That leaves only Dorothy, Warren, Barbara, and myself at home, and Warren will

---

[*] Howard (b. 1885) and Orlando (b. 1891) are his nephews, sons of his older brother Charles Shoemaker.

[†] MWS to WTS, ML, 4pp on both sides of 2pp. of plain note paper; envelope marked 7/5/17 [received].

probably go to Bridgeport soon, that is in about a week, unless he qualifies for the Ambulance Corps. The Ambulance Corps are being trained at Allentown [Pennsylvania]. Walter Rodman is up there now.

Your picture has come from Goldensky's. It is perfectly fine, indeed it couldn't be better. I am absolutely satisfied with it, you look so soldierly and manly, and withal so kind and pleasant, that it is a real pleasure to look at it. You will be sorry to hear that Cousin Will Burnham was operated on this afternoon for gall stones. He is in the German Hospital, and Dr. Deaver performed the operation. I just telephoned the hospital, and Dr. Stokes says that Cousin Will is doing very well. He has been suffering a good deal the last few weeks, but now he will be better. There were only two little stones which were removed with no complications. Dr. Furbush came from Washington for the operation, but returned after supper. The operation was at four o'clock and it is now about eight.[21]

Aunt Caroline Jack came in to see us this evening, and brought some of the loveliest roses I have ever seen. We have them all over the house. I gave Aunt Louise one of your pictures, which pleased her very much. I have promised one to "Molly C.," too. She says (over the phone) that she has never had a picture of you since you were a boy, and so she cut-out the wretched one from the paper last week. She was immensely pleased when I promised her a good one. She sends her love to you. Miss Jane brought me some lovely flowers a few days ago, and so did Mildred Boericke. We couldn't have more flowers in the house if we had our own garden.*

We are all well and busy, and the office business continues good. Of course, there are not nearly so many patients as when you are home, but still there are always some. We are looking anxiously for letters. It is two weeks and one day since you landed, and Mrs. Harvey says their letters from England take three weeks, so I must be patient.

Dorothy sends love. She is the only one on hand just now. Take good care of yourself for me.

Lovingly your wife,
Mabel

[redacted], France†
June 13th, 1917 [Wednesday]
My dear Mabel,

I have learned that nobody cares if you do know that I am at Treport so that is where I am. Look it up; you will find it right on the [redacted], about 12 miles from [redacted] and about 60 miles from [redacted] It is a beautiful location. The hospital is about 350 ft. above the town, on a plateau overlooking the Channel, on top of a chalk cliff. There are about 5 base hospitals here, amounting to in all some 6 to 7000 beds. Ours is Base 16 – B.E.F. (British Expeditionary Forces) and has about 2000 beds. Of course, it is only full when there is fighting – When there is no fighting there will be comparatively little to do. We arrived yesterday and have not gone to work yet.

The unit here leaves as soon as possible, some have already gone. They hate to go for they are not working as a unit, and will be split up, going every which way. Some go to Salonica – others perhaps to Egypt, etc. To get here from [Le] Havre was the most trying thing I have ever done. It is only about 70 miles, but in the train we more or less crawled from about 11 o'clock Monday night to 3 o'clock Tuesday afternoon. Sitting up with our clothes on, going about 10-11 miles per hour, and stopping any old time for half an hour or more was very hard. Then, when we arrived, I learned that we would have to put up in the tent. My worst fears were realized. Me in a tent. No conveniences, all of the inconveniences; cold and a headache. My first night in a tent was not a pleasant one.

Two in a tent did not make it any more comfortable. Tonight will not be much better, because I again have my headache. When the other fellows leave, I hope to have a room in what is called a hut. I can at least then put up your picture and unpack my things, which have remained pretty much as you found them. In London, I had your picture and the children's up, and slept a little, but one can't live in a tent. Everything around the hospital is English, including the patients. In fact, the part of France where we have been looks as if France and Great Britain had been at war and G.B. [Great Britain] had won, and was in possession – so many British soldiers. Our nurses are already at

---

* In 1920, Dorothy Shoemaker married Edmund R. Boericke; they were later divorced. Mildred was married to his uncle Gideon.
† WTS to MWS, ML, 4pp on both sides of 2pp. of plain note paper; in envelope with 2, 2¢ U.S. stamps, cancelled at ARMY POST OFFICE 14 July 17; marked OPENED BY CENSOR 3331, and PASSED BY CENSOR, signed by Shoemaker. The censor did not spot the place name, Treport. The name in French is Trépont. WTS spelled it both as Treport and Treporte.

work and seem to enjoy it very much. They have been having the time of their lives. Most of the nurses here, "Sisters," they are called, are not trained nurses, but they have just done the best for some time, don't you know?

Our staff when it gets under way will do much better work than has been done by our predecessors. They apparently have been rather lax. Englishmen never seem to have anything more to do than have their tea when the time comes. However, we will have to wait and see about that. Not a word from home yet; I am getting quite anxious. I wonder what is going on and all about you, but nothing comes. Am sending postals to the children.

Good night – Much love, B Wm. T. Shoemaker

June 13, 1917 [Wednesday][*]
Philadelphia
My dear Billy,

I was certainly delighted to receive your letter from Liverpool yesterday morning. You might with impunity have said it in anything you wanted to, for it was not opened, but I know that you are so loyal that you would not say anything prohibited, even if you knew the letter would not be opened. A cable was received at the Pennsylvania Hospital Monday saying that Base Hospital 10 had arrived safely in France, with all well. Mr. Tost sent me a card telling the news, and also giving the address that is on this envelope. You are quite a joker, saying that you are "glad Dorothy won her scholarship" before you had received a letter at all. I told you in my last letter (which as it was sent to London, perhaps you will not receive until after this) that Will Burnham was sick. He was operated on by Dr. Deaver on Monday afternoon. Two gall stones were removed, and he is now quite comfortable, and doing well.

Yesterday Dorothy and I chose wedding presents for Mildred and for Caroline. We chose a dozen afternoon-tea spoons for Mildred, and a lovely silver sandwich plate for Caroline. Dorothy and Warren want to buy Liberty Bonds with the money in the Saving Fund. I felt sure that you would approve, and I want to arrange it for them, but the Power of Attorney is locked in the box at the Aldine, and you have the key on your key-ring, unless perhaps you have mailed it to me. I went up to the Aldine to-day and they say that for $2.50 I can have a locksmith open the box, so unless the key comes in the morning mail, I will have to do it, as there are only two days left in which to buy bonds. To-morrow, the Liberty Bell is to be rolled out of Independence Hall, and Mayor Smith is to hit the rim with a hammer to celebrate the Liberty Loan.[22]

Dorothy has a caller, Lloyd Weaver, whom she met at Margaret Boericke's dance. He seems to be a nice boy, but it is ten o'clock and I wish he would go home, so I can go to bed. We usually all go to bed at ten o'clock now, but I read until I get sleepy. I have an electric light beside the bed now, and it is very nice. Last night I read about half of Walter Rodman's poems. They are beautiful, and full of pathos. Some of them written about his wife, since she died, really made me cry, he misses her so.[23]

Monday night, Gimbel's stable and garage burned down. It was a very spectacular fire, though not so large as the warehouse fire that we saw last February. All the horses were saved. Dorothy and Warren got up at 2:30 A.M. and went to the fire, while Bessie and I stood on the window box in Bobby's room and looked at it. No one else was at home, Barbara being at Gertrude Russell's and Bob at Wallingford.

I guess that is about all the news, so good-night. Write as often as you can, and take good care of my Captain.

Lovingly your wife,
Mabel

----

[*] MWS to WTS, ML, 4pp on both sides of 2pp. of plain note paper, to WTS at American Expeditionary Force / Base Hospital No. 10 / U.S. Army; envelope marked 7/6/17 [received]. Her letters will all be addressed in this way until otherwise noted.

6/15/1917 [Friday]*
Le Treport, France
My dear Mabel,

Great joy, I received three letters today. The first word from home since I left, and what is more they were the only ones received, so I have it all over the others. Dr. Gibbon did receive one & Dr. ___ one the day we arrived, but I think they came over on the boat with these letters. Two from you and one from Barbara are however apparently not your first ones. They were postmarked Phila. May 28th – 29th & 30th – and do not talk like your first ones. They told me nothing of the Caravan time – or of our departure. It is being thought by some now that a lot of our mail has been torpedoed. At all events, it has been a bad mix up.[24]

(Fountain pens never work. Now for your questions. The key to the lock box is somewhere on or in your bureau. I remember that distinctly. It is not with me. We stood by the bureau and arranged that. So look again, and if you cannot find it – they will open the box and give you a new key for about 2.50. Don't forget also that Charlie has as security a large share of your stock. Question 2: I ordered the book you mentioned, but think that I sent a check for 1.50 with my order. If not, send the 1.50.)

Begin: Barbara tells me that Dorothy got her scholarship. I am delighted and very proud. She is making good. (Omit: Sorry Frank MacFarland died – he bloomed and flourished early – and perished lastly. He had been of little use for a number of years. The several weddings mentioned are fine. What you say of my cousin's wife is true, but it would not apply to my cousin. He is fine, and we are of great use to each other. He is an exceedingly fine man and a great travelling companion.)

Begin: Getting down to work – our location is delightful – on a vast plateau on up of the chalk cliffs. Four units are here with a maximum capacity of 12000 beds. Our section has 2000 beds. We will fill only after the drive. One is now and patients are supposed to be landed at the rate of 3 or 400 per day at any time. Then there are times – in the intervals we are not so busy, but we have many hundreds of patients. Our food is prepared by our French chef – whom we brought with us from the St. James Hotel, Phila. It is very good. Absence of running water – drinking water in abundance, etc., is my greatest annoyance. I hate metal basins, tin cups – and especially when they are in tents. However, I will soon be in the shack. Baths are impossible here so when I got too dirty, I went down to a hotel in the town and had a fine one. A bath with two large pitchers of hot water and of course all of the cold water I wanted. I got clean all right, and I needed it. Everyone else is dirty too.

Of course, we don't look dirty because we keep our faces clean, and our clothes always buttoned up according to regulations. But I discovered the way to get clean, so much for that.

Official directions have again been issued regarding letters. I feel they are

1. Name and title (Capt. W.T.S. ~~U.S. Army~~)
2. Base Hospital No. 10 U.S. Army
3. American Expeditionary Force

That is all; do not say where. Do not say France, England, or any other place. My letters are to come to you without postage, and will be censored by myself. They should not be opened, except by chance – 1-1000. When a letter is opened anyhow, just to keep it as it were. I think I told you that our trip across the Channel was much more dangerous than that across the Ocean. I was surprised at that, but the subs were both just ahead and just behind us. We slipped through in the middle without being touched. It seems to be a very dangerous trip, and we counted ourselves very lucky. Good Bye for a short time. With lots of love from here.

B

----

* WTS to MWS, ML, started in ink, then pencil. Envelope markings similar to 6/13/17; without postage; passed Censor 3963. "Passed by censor," signed Shoemaker. Two parts of this letter, marked with parentheses, were marked "Omit."

6/18/1917 [Monday][*]
Le Treport, France
My dear Barbara,

Your nice letter came through ahead of others, and told me things that I did not know. Had for instance not heard of Dorothy's scholarship. You seem to be having a great time, and I trust it will continue. I am not so much in love with Army life yet that I care to continue beyond the necessity. Today I am Officer of the Day, and have many duties aside from the treating of my patients. At 12 o'clock, I inspected the food in the kitchen, that was to be served to the patients; then passed through the mess rooms of patients (walking patients) and also orderlies, etc., and gave the once over asking in each place if there were any complaints. At 4 o'clock, the tea is inspected and passed on.

After supper, I must again go to the kitchen and see if it is clean and in order. At 10 or 11 tonight, I must visit the night supervisors – medicine & surgical – and get complaints if any. If anything happens during the night, I have to be routed out. At 6 o'clock tomorrow morning, must get to the parade ground and see that the men going back to the front at that time have their equipment & rations. About 8 o'clock, inspect the meat coming by transport. Must visit the dead and give certificates for them, making sure at the time that they are dead. In between times, I take care of my patients just as usual. We take turns at this; the majors are exempt. It will come around about once in two weeks. The day preceding the one just referred to I was "supernumerary in waiting." I was that yesterday. My only duty yesterday was to attend a funeral at 2:30 P.M. That was very sad. The Padre (Minister) walks ahead of the "lowery" (motor truck or ambulance) containing the coffin, draped in the British flag. Behind in double file are about 15 men, and at the rear, two officers – an English officer and myself. I walked over to the mortuary and there found the procession. Then we marched at a measured tread to the cemetery about 1½ miles. There the English burial service was read – the bugler sounded "Final Taps" ["Last Post"] and we saluted and came home in the lowery. I nearly collapsed on account of the heat – boiling sun about 100° – and of course I could not open my coat. When on duty every button must be buttoned. Two heartbroken women were at the grave – probably a mother & wife or daughter.

Another duty that I have for a week starting last Saturday is censoring letters. This I do immediately after breakfast. Glance over them, stamp them as this one is stamped, and sign my name. It is more or less perfunctory. This morning I read one from a chap in one of our wards to his mother. He asked for money to buy cigarettes. I found his name and location, and when I found time, took him some cigarettes. He was surprised, but I did not tell him how I knew he wanted them.

This is the end of the paper. Tell your mother not to be jealous. I will write to her next. Much love,

Dad Wm. T. Shoemaker

6/18/1917 [Monday][†]
Le Treporte, France
My dear Dorothy,

Your letter arrived today after being Censored, seemed insufficient address and other things. I am very much pleased that you got your scholarship and I congratulate you. Of course, I am not surprised. Now, I know just what I expected from you, from the work you did during the winter and the good reports that followed through. I am very proud of you, and also of your mother and myself for having you. A pretty safe address would seem to be – British Expeditionary Force (B.E.F.) Base 16. Base 10 – A.E.F. is also good. France might be added, but is not necessary – but Le Treporte must not be in evidence.

We are kept very busy, and have not enough Doctors, nurses or Enlisted men for 2000 patients, which we have, when we are far short of sufficient staff. At present time we have about 1000 – but a drive can fill us up without notice in a very few hrs. And such injuries & wounds; many of the wounded are literally shot to pieces – with

---

[*] WTS to his daughter Barbara, ML, 4pp. on 2pp. of plain note paper, double sided; with two U.S. 2 stamps. Stamped "PASSED BY CENSOR" signed [s] Shoemaker; OPENED BY CENSOR 3378.

[†] WTS to his daughter Dorothy, ML, 2pp on 1p of plain note paper, double sided; envelope as above, Censor 3744. This letter and the preceding letter to Barbara were typed and sent in an envelope marked "Mr. J. Warren Shoemaker / Bridgeport, Conn./ Box 203, c/o Warren Bros. Co. The envelope was also marked in handwriting "For Greta Shoemaker."

shrapnel. Arms, legs, bodies, heads, eyes, and every portion of the anatomy destroyed by the Devilish stuff. But the most remarkable thing is the wonderful spirit of the men. There are no complaints, and all are game and regard their misfortunes as simply ordinary casualties. They never talk about them, and are interested in what's going on.

It is rumored here that our good friend S.S. [redacted] that brought us here so safely has been sunk. If so, you would know it before we would. With us it is only a rumor without foundation. Take good care of your mother.*

With lots of love,

Dad Wm. T. Shoemaker

June 18th, 1917†
Philadelphia
Dear Pa,

You can bet that we were mighty glad to hear from you a few days ago. Your letter dated June 4th just arrived a short time ago, so I guess that it takes about two weeks for the mail to cross. I am finished with college now for the term and think that I passed everything alright, I haven't as yet gotten the marks but feel that I passed all of the exams which I had to take. I was exempt in Chemistry. Last Thursday I received a letter from Dean Frazer (he has charge of the National Intelligence Bureau at Penn) saying that they were forming two more ambulance corps from the University, I immediately enrolled but was rejected because of my weight or rather lack of weight. I therefore am going to work with Warren Brothers again. I got a letter from Uncle Ralph the other day saying that I was to be in Bridgeport again and that my pay was to be raised to eighteen dollars per week. I expect to leave next Thursday morning to start work in Bridgeport.

The day after the Convention was over I took the automobile out of the Aldine and drove it up to the German Hospital, it was making a great deal of noise in the rear end and as I drove in Mr. Williams heard it and said that we had a broken main driving gear. Therefore the next day I went up there and took the whole rear end apart, axles, gears, differential and everything. Upon examination every part seemed all right and Mr. Williams decided that it must have been a bearing out of adjustment. He helped me the next day and we got it fixed up fine as silk. That was Saturday and the next day Dot took the Harveys out to Wallingford with Mother. The car was running finely when all of a sudden a great engine knock developed, she drove down to the next garage at Lansdowne and the man said that she had a broken engine bearing and that the car could not be driven home. She accordingly put it up there and they took the trolley home. The next day Mike took me out in one of the ambulances and towed me into the German. Mr. Williams looked the car over and decided that we had a burnt out crank bearing (that is a bearing which connects the piston and the crankshaft and costs five dollars), I could, with Mr. Williams' help, fix it in a day but Mother and I both think that as I am going away so soon we had better put the car up for the summer and either have the Studebaker people overhawl it, or do it myself, in the fall or maybe wait until we hear of your starting home and then fix it up in good shape. Mother told Uncle Josh how things stood and he decided that he would just as soon not have the car as have us go to the trouble and expense of repairing it at present or to do so himself. I think it is a darn good idea because the car seems to be badly in need of a general thorough overhawling and if he took it the way it is it would be a constant course of trouble and expense to him and he would be sorry that he ever took the car. The arrangements are, I think, best for all concerned and the car will certainly be more serviceable for you when you come back and need it.[25]

The boarding conditions in Bridgeport, I hear are very good, there is a house which takes only Warren Bros. men and has arranged for meals at the necessary hours. That sounds pretty good to be able to have table board instead of restaurants. I will try and write to you every week during the summer to let you know how things are getting along. Good bye,

[s] J. Warren Shoemaker

---

* They came on *S.S. St. Paul*; St. Paul was redacted. Hereafter, I will normalize the spelling of Le Tréport as Le Treport, regardless of how WTS spelled it.

† From Warren, age 19, to his father, WTS, TL, 3pp., on 8 ½ x 7in. plain note paper; envelope marked 7/16/17 [received]. The punctuation is unchanged; three typos corrected. Sentences were longer and had fewer commas that are now in use, and the word "overhauled" was then spelled with a "w."

June 18th, 1917 [Monday]*
Philadelphia
Dear Pa,

Thanks ever and ever so much for your letter – It surely was interesting & the whole family enjoyed it, and four others which we jointly received. Mother got two – Dot one, Warren one, me one – and as Bob is at Wallingford we don't know about him. England must be a peculiar place – with lady's smoking & driving trolleys.

The family gets smaller and smaller. After Warren leaves on Thursday, just Mother & Dot & I will be left – It seems peculiar now with only four – but with three ___!

Mother, Dot & (~~my~~) I went to Overbrook (Aunt Florie!) for Sunday dinner, yesterday – Florence was away. Louis took us out in his car – and we made Overbrook in about ten minutes, nearly loosing our hats on the way. They gave us good eats & time, tho' – Aunt Florie doesn't seem so well, but the dog (two German dogs – liked Bo) were cute as the dickens. I took them out for a little while & they were sweet – stuck right to my heels all the way – but Bo – Oh dear! I took him, or at least tried to take him, to Butler's – at 19th & Market the other night – but incidentally & not by my choosing, we went to 18th & Chestnut, 19th & Market, twice – 20th & Samson - & after a little run – home! It wasn't too much fun!†

I'm becoming very domesticated – you wouldn't believe it – but I've made, by myself – two "articles of clothing" – the ___ on a smock, and am still working on a sweater which is the old scarf that I started two years ago – and which is now about ¼ done. Clever?

Dot is doing beautiful work with Mr. Copeland – some of her sketches you wouldn't appreciate, because you'd "have to ask what are the colors" in them! Mr. Copeland is about as heartbroken as Dot, that the Burnhams aren't going to Suttons – because of the lovely chance for sketching that she is losing. It is hard luck – It seems that all the auto trips for the Shoemaker family are off – That friend of Mrs. Harding's is going up earlier so that Louise & I will have to go to Pocono in the train!

Do you suppose it would be possible for us to supply you with sugar? Mother says it will be good for your health do without it – but even so!

Well good-by now – I'll write again soon – The whole family sends love –

With loads & loads of love from

Barbara

P.S. – Please don't mind saying anything about the war because of the censors – they haven't even opened any of your letters! B.S.

June 18, 1917‡
Camp Becket / Becket, Mass.[26]
My dear Father –

Believe me I was glad to get your letter today and I think it was great of you to answer to quickly when you are so busy, I suppose. We have been working quite hard up here and have gotten quite a bunch of work done. We planted about a half an acre each of corn, peas and potatoes, a bunch of lettuce and also cucumbers. In addition, we have painted nearly all the buildings in camp. Our bunch has been getting larger slowly. Since I wrote you last our gang has gone from six to thirteen. Last week Mr. Gibson came up here with his wife, she seems very nice although I think she must be a little off or she wouldn't marry an old salt like Gib. Six carpenters have been working on our new $10,000 recreation building for the last two weeks. They have the frame work up as far as the second story, it's really beginning to look like a building and believe me it's going to be some building when it's finished.

---

* To WTS from his daughter Barbara, age 15, ML, 3pp., on 1p. of 9x6 in plain note paper, folded. Envelope marked 7/16/18.
† "Aunt Florie" was Florence (née Hey) Shoemaker, wife of WTS's brother Joshua Lippincott Shoemaker.
‡ To WTS from his son Theodore "Ted," age 17, ML, 5pp. on 2pp. of folded plain stationery.

Did I tell you that I got a tie for first in the pole vault in the Junior Middle States Athletic meet the Friday before I came up. The other fellow and I tried to jump it off for about fifteen minutes but neither could beat the other, so we finally tossed up and I lost the gold medal however I got a silver one but better still I won my school letters.

It certainly was nice of that old Johnnie Bull to entertain you and I'll bet they did it right. I saw in one of the Springfield papers that the base hospital from Phila. and Chicago were at some cathedral in London, I forget the name of it now, with the American flag. That was the first I knew that you got on the other side safely, it sure was a relief. Sat. night we went down to see "Uncle Tom's Cabin" played by a little traveling company of about five people. The thing was about the crudest production I've ever seen barring none. A character would make his exit and pull the curtain with him.

I think Mr. Gibson is going to give Bob a job in the printing shop next season, that is he spoke something about it. Becket is simply full of soldiers in a little one-horse town of about 1,000 people there are 80 soldiers just bumming around. I wonder why they can't give them something to do. It's getting late and I am nearly falling asleep, so I think I'll stop.

Lots of love and good luck

Ted

6-22-17 [Friday]*

Le Treport, France[27]

My dear Mabel,

This time I must write to you. Am now quite comfortably fixed sitting in a good room, nicely furnished *a la* woodsman. Charlie is with me. Two cots, a wash bowl, large table, 2 swell seats, basins, sets for nails, hands, a stove, 2 chairs, and a camp stove, etc. It is a great relief to be able to move around, and set our things out – your picture, and the kids hang at the head of my cot, and believe me they are the best looking things in the room.

Work now is very slack; we have been evacuating back to the front, convalescent camp, and to England, as the case might be, and are about ready now for a convoy of newly injured. It will be along in a day or two and then everybody will be busy all the time. War 12 – my eye ward – looks quite the part. Have operated some, but not much as yet. Every patient here wants to get back to "Blighty." Blighty – means England; it is a corruption of an Indian name for England and was brought here by the Indian troops. It has never become quite the usual and proper name for England. The nurse in my ward told one poor fellow that he would be all right, that we had here one of our best eye doctors from Phila. He said he was afraid he was too good, and would get him well before he could get to Blighty. They are so tired of the game, that a wound serious enough to get them to Blighty is almost a blessing. The British government is wonderful in the way it is arguing over its war operations, pertaining to the sick & wounded. When a case comes in of seriously wounded, it is marked S.I., if <u>dangerously injured</u> D.I., the relatives are notified, and if D.I. they are brought here at once and kept here, if necessary, for a week at the expense of the Government.

Once a week every patient and every doctor gets four packages of Ruby Green Cigarettes and a box of matches. They are distributed regularly – as a "gift from the Queen." Some were put in my tent. Smoking tobacco also for those who smoke a pipe. Everybody smokes cigarettes. The patients smoke them all day long in bed. Each has a little can for ashes & butts. Socially, a Tea, two or three times a week would say it all. I went to the first and dodged the second. We are expected to go whenever invited, so I guess I can't duck any more, or I might draw a court marshall [martial]. That would be sad. For once in his life, yours truly cannot do as he pleases. That's another sadness, but I don't mind it much getting used to it. Salute – Aye, Aye, Sir! And where you are – and me a doing it.

Charlie and Ed are just getting settled and will be ready for work tomorrow. They have a very extensive and complicated outfit – which required 14 packing cases to get it here. They expect lots of work. Poor Ed has not received word from home yet, nor has Hodge or Newlin. I am glad you are working for or with the Pa. Hosp. people. You will enjoy it. I am sure, and I believe they will enjoy having you with them. You ought to know them. I have only met Mrs. Packard at tea, and Mrs. Sweet at our dancing class. The others I have never even seen – except

---

* WTS to MWS, ML, 4pp. on 2pp. of plain note paper, double sided; with two cancelled U.S. 2 stamps. Stamped "PASSED BY CENSOR" signed [s] Wm. T. Shoemaker, with redacted mark across Wm. T.; stamped ARMY POST OFFICE 23 AP 17; OPENED BY CENSOR 3481. No sender is shown on the envelope; it is simply from A.E.F.

Mrs. Hart, just before leaving. The dandy fellows with us are Newlin, Hodge, Cadwallader, Vaux, Sweet, and a few others. They are all human. All of us are of course nice, but the ones I mentioned are especially nice. Once again, take awfully good care of Mabel and give her anything she wants.

Much love, B.

June 25, 1917 [Wednesday]*
Bridgeport, Conn.
Dear Pa:

I am now in Bridgeport, Conn., with Warren Bros. This year I am a time-keeper on the street and receive eighteen dollars per week. I have to keep time on three different gangs all in different parts of town so have considerable running around to do. I am staying at a boarding-house at which there are three other W.B's men (two foremen and Lewis). We all take meals at the house and we certainly do get good food and lots of it. It is a great improvement over last year's mode of living. I haven't yet heard from College about my term and examination marks but I'll send them to you as soon as I do. This is a very short letter but I have little to say and am just writing to let you know what I am doing. I suppose you are pretty busy but if you get time please write to me. My address is P.O. Box 203 / Bridgeport, Conn. / c/o Warren Bros.

6/26/17 [Tuesday]†
Philadelphia
Dear Father:

The letter you promised hasn't come yet and I have hopes it will be here soon, the other letters haven't been censored but probably mine has. My trunk left for camp and I will leave day after tomorrow. I got a very good camp outfit which will probably last next year too. We take the five thirty A.M. train to New York where we will wait an hour then to Springfield for a two hour wait. We arrive in camp about five o'clock. I just came from Wallingford where I had a bicycle and had a great time with it. Barbara is out at a surprise party Allen Anderson's mother is giving him. Dot is at Aldines, Ted is at Camp, and Warren at work for Warren Brothers so nobody is home but Mother and me so she is the only one but myself to give you her love.

Bob

P.S. Speaking about censors, Mother just censored this as you can see.
P.S. Not very much, as you can see. M.W.S.

June 27th, 1917 [Wednesday]‡
Le Treport, France
My dear Mabel,

I am getting tired of being without you, and for that I read your many, many, letters during the day. At present, the days are long and there is no work. I am up before 7 o'clock – breakfast at 8 – lunch at 1. Tea at 4 and dine at 8 P.M. Light goes out at 10 – and I am in bed by 11:30. So you are, when there is little that works, my day is lousy, and the opportunities for the seeing of Mabel are many. Again, the mail does not come. I review, as I love you the most there is in the Company, a couple of letters, then nothing for the past week or more.[28]

---

* WTS from his son Warren, ML, 3pp. on 1p. of folded letterhead stationery with monogram of JWS; postmarked Bridgeport, Conn; received 25/7/1917.
† To WTS from his son Robert "Bob," age 13, ML, 2pp. on 1p. of folded plain stationery. Nothing was omitted, but MWS made a few small corrections in spelling and punctuation. Envelope addressed by MWS, marked 7/27/17 [received].
‡ WTS to MWS, ML, 6pp., on 3pp. of plain note paper, double-sided; envelope markings similar to 6/22; Censor 3600, cancelled 6/28. The letter refers to Charles Jack's wife Mary, and Edwin Shoemaker's wife Martha.

Newlin, Hodge and Ed as well as practically all of the nurses have not yet received a word from home – almost 6 weeks. Many of the nurses are in grief themselves. Poor Ed says little, but of course want to hear from Martha. Charlie's letters are coming right through. The wife writes him every day and he writes her every day. That of course is unnecessary. The best address would seem to be Base Hospital No. 10. U.S. Army – France.

Try it out. It is ridiculous the way different addresses have been given to us. Nobody seems to know definitely what is what. In a letter to Charlie, Mary says that Martha told her she would have to go [to] work, as she found she would not have enough money to get through. Is this true? If so, I am sorry, because Ed of course has no such word. He lives in the clouds, and thinks he has a million dollars. Optimism is great if you like it. I do hope you are to bring her on Fridays with the "men" of Base 10. I think you will enjoy it, when the ice is broken, and I am sure you will find some nice women among them. Of course, I mean the way you and I were part of it. I am also sure that they will find one nice "lady," looked at from any standpoint.[29]

Before I forget it, I must apologize for sending uninteresting letters. I am not a letter writer, as you know. So you must be satisfied with disjointed junk. And have worried a good deal about your nights. Please do not stay in the house at night without a man or a servant, even though you have one of the girls. That would not do, so you must not try it.

It is about time to plan to send me some money – I will not need any for a month or six weeks, but mails are moving slowly and with great uncertainty. Look it over and if possible send me American Express Notes. Get ___ to help you when ready – or he will even do the whole thing for you. My expenses are not very heavy. Have not received the first mess bill yet, but fancy I will be able to do everything while here for 1.50 – 1.70 per day. Can tell you more definitely after a try out. I also heard that the British Government was going to give us the regular 35 cts. a day allowance. Not much to be sure, but every little helps.

Did you get $500 for _____? That was to be so I wouldn't charge him. Have you heard from the Lukens – whose boy I saw many times at the Pa. Hosp.? Unfortunately the name is to be mentioned, otherwise there would have been a real fine bill forth coming. Of course, if he insists, or comes across at all, whatever you charge him was enough. What is happening to Ross – and Page & Whiting and Forst, Halburg and many others? I would like them all to be in uniform, but of course some will "escape" it, I suppose. Some are still saying "if needed" with the customary reservation, of "needing" for themselves, when they are needed. If Ed Shumway or Max Landon ever come across with it – send me a cable. Don't worry however, as you will not be called again for that expense. I hope you are preparing and using the cento, just as pence. Just the load that my office was to work on is entirely up to you. So do not hesitate to organize things to just your own way as you work.

Only one thing was left out in packing – the little lamp that Bob got for me. They have one in the Clubroom, and guess they are still there, as I can't find them.

Bought a sweater just today for $6.75. My suits are not warm enough for this weather. To be sure, one is wool, but is summer serge in weight. Then, I must keep that to come home in, if I ever come. The gloomies out here would make one think that home is 2 or 3 yrs. off. But I can't see it.

Well, I feel better, after writing all of this. Good night, my love. With much love,

B.

Postals sent from France by Dr. Shoemaker, mentioned in his letter of June 13, 1917

A British Ambulance shattered at Verdun

Le Quai at Le Treport

July 1917

Le Treport – Gas Attack

"Was gassed yesterday. Very interesting."

July 2nd, 1917 [Monday]*
Le Treport, France
My dear Mabel,

Not a word from home for about 2 weeks. Now, don't you think that's tough. Have learned a little information from Ed & Charlie. For example, that Warren has joined the University of Pa unit and that you received a cable from me and a few little things of that nature – but not a word from you. I am getting quite peaved. Also, heard by way of Philadelphia's gossip that Henry & Anne had no girls and that their own girls are spending the summer cleaning house. But that does not tell me anything about you and yours.[30]

Today was Dominion Day. Celebrated throughout Canada and wherever Canadians are. There are a great many here so we had a big Field Day with all kinds of spirits. There is a Club House and tennis grounds – golf links – surrounding it, right on our area. The tents were decorated with beautiful flowers and flags and everybody was out who could get out. It was a very pretty sight. Our men and officers participated in some of the sports and did very well. There were 100 yd, 220, & 440 yard dashes, mile runs – pole vaulting – jumps of all kinds. Pillow fights on a horizontal greased pole, and to think of something fantastic – some very funny – both unusually fell, but sometimes only one. Then there were egg & spoon races & needle & thread wars, also obstacle courses – baseball throwing – tugs of war – there were rules – lady & gentlemen rules – men blindfolded and dances by nurses with lines attached to arms – "horsey like" – through obstacles, etc., etc. What's the use – I would just enclose the program. It was magnificent, after two days of cold rain.

Now, a little business – I am losing money on my Express notes – and to be sure really losing, because the Company will pay me the difference in exchange when I get home, but now is when I want it – not when I get home. So, get me a Letter of Credit from Drexel or Brown Brothers. You had better do it soon some, because I am running low, and it takes a very long time for the mail to reach me. It has to be that I must get a Letter of Credit from Drexel through J. L. L. & Co. [J. L. Lippincott Co.] without pulling up the money all at once. That is when I would draw and the draft reached home, Drexel would send around to J. L. & Co. and get the amount of the draft. If that can still be done, it would be much more convenient for you. Tell [brother] Charlie about it and let him fix it up for you. A few days ago I wrote for Express Notes – but had not at that time tried them here, so did not realize that I could not get current exchange. I hope you have not already bought Express Notes. Now another thing – It may be that you can't get a letter of Credit without me to sign it. Remember that you have a Power of Attorney, and also you have many samples of my signature. If necessary, but a necessary one, have them paste it in the Line of Credit. You may have some trouble, so please start at once. The decision, I will leave with you, because you know how the finances are, I don't. By the way – you may not get foreign pay yet. It will come in time, but do not be disappointed if your first pay seems flat for a 1st Lt. Have not yet received my Commission, so I may really be a 1st Lt. If so all the rest are in the same boat. However, they say – (they the base ones) that it will come some in time.

---

* WTS to MWS, ML, 6pp., on 3pp. of plain note paper, double-sided; envelope markings similar to 6/22; Censor opened 3678, cancelled 3 JY 17. "On Active Service" written on upper left of envelope. Two U.S. 2¢ postage due stamps cancelled on it.

Letters that seem to come through best are addressed Base Hospital #10 U.S. Army France. I may have told you this before, but as it seems to change every day, I forget what I did tell you. So do not mind if I repeat.

Not much business yet – but it's coming –

I am very anxious to learn of your summer plans. Perhaps letter time some tomorrow. I hope so.

Much love –

B

July 6th, 1917 [Friday]*
Camp Owaissa, Pocono Pines, Penna.[31]
Dearest Father,

Came up to camp by train last Saturday, June 31st. Louise Harding was already here and she met us in someone's car – a little (?) maroon roadster. It was great!

Camp is very nice this year – There are 35 girls, 3 counselors & 9 tents. We have lots of fun swimming, boating, canoeing, "tennising" & basket-balling. To-morrow the basket ball teams are supposed to be made up – Dot Young (counselor) told me that she was going to put me on the team as side-center – I'm awfully poor, but she couldn't get anyone else. The camp is divided into two divisions – reds & blues, so each side has a team. When they're ready we'll play real games – So far, we've only been trying out.

Before I left home, I saw the enlisted men of Base Hospital 10 in the movies. They were shifting freight at the Armory, and there were no officers there, much to my disappointment. I also saw a whole bunch of nurses. I imagine they were yours because it said they were the first ones to arrive in France. Everything but Captains, etc.

There is a Mrs. Foote up here – the mother of one of the girls, who says she knows Mother – Or, at least did know her before she was married. She says she used to go with her a lot – She also knows Miss Anna Gilchrist very intimately. Cousin Ethyl was her best friend, too. I didn't dare ask her maiden name, but she's lovely.

Well, we're going to tennis now, so I'll have to stop – and finish later. – – –

We had two peachy sets of tennis – And oh! I don't believe I told you what I got with my five dollars. I bought a great tennis racket and a water proof case & one ball – I had .20 left from that and I put it in my dime saving bank. So you see I'm all prepared to play tennis. At first I couldn't decide whether to get a camera or racket, but I fished out a Brownie No. 2 camera that Ted had out West. Ted gave it to Billy & he gave it to me when he got a new one. It was considered passé but I took it to Mr. Ferguson and he fixed it up. So now I have a camera & tennis racket.

I brought your picture up to camp with me, and all the girls think you're peachy looking – and a hero & everything else! Oh, I tell you it's worth while wearing a uniform!

Oh dear! Another interruption – But this time a rather pleasant one – Supper! I'll write some more when I have something to say – I think we're going to have a party to-night – It's one of the girls' birthday – We should have ice-cream & cake & a marshmallow roast. Poor you – having to sneak Philadelphia sugar in your French coffee!

Well, we didn't have ice cream or cake – but we are going to have the marshmallow roast. There is a girl here named Dorothy Haines. Her father is the head of Lineburner Co., Opticians. When Mr. Lineburner died this last winter, Mrs. Haines was the next in line for "boss." – I hope you have been nice in sending patients to Lineburner's because Mr. & Mrs. Haines are coming up!

Mr. & Mrs. Harding are sleeping in Pocono about 7 miles away. At least Mrs. Harding is visiting there, & Mr. Harding comes up week-ends. They both came over to camp yesterday - & you never saw so much food as they brought! Cake, candy – everything. And every speck is gone!! That just shows what appetites we have, or else how little (?) we have for meals! But you needn't think we ate it all – Proof – we're still alive & kicking & not the least bit sick! I can't think of another thing to say, so I guess I'll have to stop – But I'll write again soon. Please write to me while I'm here. I believe a letter from France would thrill the whole camp.

Please come home soon – We all miss you terribly! While you are away, don't work too hard. Lots & lots of love from Barbara.

---

* WTS from his daughter Barbara, ML in pencil, 6pp. on 3pp. of watermarked stationery without letterhead. To Capt. Wm T. Shoemaker, M.D. / American Expeditionary Force / Base Hospital No. 10 / France / United States Army. Received 7/23/17.

7/7/17 [Saturday]*
Le Treport, France[32]
My dear Mabel,

This is important.

I have written you about money and suggested that you send me a Letter of Credit on Demand. All well & good – do it at your convenience. But I am getting low – have $50. Just enough until Aug. 1st when Mess bill comes due. So there! My nurse Miss Amima Fuhrmann, is paid here and has money which she wants to send home. It is rather difficult to do, and the only safe way would be to cable it, which costs about $2 or $3. Therefore, to help both Miss Furhmann and myself, I arranged or am arranging to have you mail your check for $75. at once upon receipt of this to Mr. J. W. Rutan, 1305 Arch St., Phila. Write or phone and explain the transaction.

I thus take Miss Furhmann's $75. In this way, there is no risk and no expense. Then you take the $75. into account in making out the Letter of Credit. Your check of course make payable to Mr. Rutan. Miss Furhmann wrote to Mr. Rutan today explain the transaction and when he receives your check, he is to send her a week end cable stating "money received." I hope I have made myself clear. Now about my pay – you will not receive it, or any notice of it, from Washington. It will be placed to my credit at the Girard Trust Co., and the Girard should notify you that such has been done, but, inasmuch as I neglected to ask them to notify you, they may not do it, as it is not at all necessary. So stop there and ask them to notify you whenever a Government check is deposited to my credit. Then they will do it. I told you this but I also told you so many other things that it would be practically impossible to remember them all. So the money is probably in the Girard on your balance. Stop in and see about it.

Your nice letters came yesterday, – last one June 12th. The Liberty Bond for Dorothy & Warren is fine. It is a good proposition and allows them to participate in one of the most remarkable money raising schemes I have ever heard of. The *Ledger* supplement you spent of adores our Miss Hale. By the way – I wish you could mail me the *Literary Digest*, and the *Red Cross Magazines* when you are through with them; also any clipping of interest, or anything else that looks good. That kind of mail gets through better than letters. That best way for you to do it is to give them to Owen and he will send them from his office. They can do that better in a business office than you can.

I am very sorry to hear of Mr. Burnham's illness. However, from what you say, he will be alright shortly. Deaver was right; he is of more use at home if he will stick to work of that kind. I appreciate the efforts and interest of the Pa Hospital management and trust you have acknowledged Mr. Levin's note. Personally, however, I agree with you, that in our case, sufficient "Courtesy" will be rendered by our institutional friend, the Girard Bank, to meet our requirements. Pennsylvania is certainly coming across in great style on this Base 10 perspective.

Have you met Mrs. Gerhard? Dr. Gerhard says his wife repeatedly mentions Mrs. Shoemaker but has not said whether she means Mrs. Edwin or Mrs. you. I must now get to the base ball game. Was gassed yesterday. Very interesting, so will tell you about it in the next letter.

With much love,

B.

Wm. T. Shoemaker

July 8, 1917 [Sunday]†
Camp Becket / Becket, Mass.[33]
My dear Father –

Received your postal yesterday and in the same mail a copy of the letter that you sent mother telling you had just received the three letters, two from mother and one from Barbara. It seems you didn't receive either of mine, here's hoping you get this one. You surely must have had some time crossing the Channel. Gee, I'll bet it feels good to be safe over there. And you can bet we were glad to hear it also.

---

* WTS to MWS, ML, 6pp., on 3pp. of plain note paper, double-sided; envelope markings similar to 6/22; Censor opened 3300, cancelled 8 JY 17. "On Active Service" written on upper left of envelope. Two U.S. 2¢ postage due stamps cancelled on it.
† To WTS from his son Theodore "Ted," age 17, ML, 4pp. on 2pp. of letter head of "Camp Becket / In the Beautiful Berkshires."

Camp started in full swing a week from yesterday and things are going quite smoothly. There is a new system of activity, in the mornings, the morning is divided into two periods of an hour each for instruction clubs. You have quite a choice of clubs and you are supposed to take at least 6 hours a week. Some of the clubs are Life Saving, First Aid, Manual Training, Gardening, Printing, Chemistry, Photography, Nature Study, and Athletics. The printing is run by a leader who doesn't know any thing about printing at all. Bob dopes out the stuff and tell it to Sturgis Hunt, the leader, who is in charge, and he conducts the class. Did you know that Bob is getting half board for taking charge of the printing? He has quite a bunch of kids helping him and they are getting along pretty well. The $10,000 recreation building, better known as "Gibbie's Dream," is nearly completed and believe me, it's some building. I showed Mr. Gibson the copy of the letters that mother sent me, he was extremely interested and quoted you in a few places about the scarcity of food, while giving one of his talks.

I am assisting in the Life Saving class. We have about thirty fellows in our club. We're teaching them the breaks and carries – how to throw a life buoy, and a bunch of other stuff. It's time to fix up for inspection. I'll write again in the middle of the week.

Remember me to Ed, and Charlie Jack.

Love,

Ted.

7/8/17 [Sunday]*
Philadelphia
Dear Father:

I am in camp safe and sound, they changed the swim from eleven o'clock to five. It is much better because the water is much warmer. I am on a life saving crew again this year.

When we came to camp we got out of the train in a pouring rain and had to take a biffy and jitney to camp. My trunk had arrived and strange to say the lid was still on. We were put in tent thirteen that is Earl and me last night. Earl opened his trunk and found a can of sugar had upset and was all over his stuff. I am not much on long letters so I must close.

Bob.

P.S. I hope you can desipher my writing.

10/7/17 [Tuesday]†
Le Treport, France
My dear Dorothy,

Was glad to get your letter telling me of your summer plans. They are all good. The sketching class you will enjoy, and it will help you a lot in next year's scholarship. I thoroughly approve of the Liberty Bond also. You do your bit and also have something just as good as you had in the Saving Fund. Was very sad to hear of Mr. Burnham's illness, but from what your mother said, he will be all right, and better than ever.

Last year, you were speaking of driving an automobile on a War basis. Out here, all of the Ambulances are driven by English girls from the best families. They are certainly good sports. They do everything themselves, and are wonderfully good drivers. Of course, there are no private ambulances, either in France or England. Ambulances or "Lounges" only also are for a very few large limousines for high officers – generals, etc. Gasoline costs almost a dollar a gallon, and is all Government controlled. I could not drive any car here 10 ft. without getting punished. No one is allowed to have a private car. Fancy Philadelphia – without its automobiles – with nothing but big trucks and ambulances, etc. It seems to me that when this war is over, it will take at least 5 yrs. for things to get natural like

---

* To WTS from his son Robert "Bob," age 13, ML, in pencil, 2pp. on 1p. of folded plain stationery. Envelope marked 7/27/17 [received?]. Postmarked Becket, Mass., July 9, 1917; to Capt. Wm. T. Shoemaker / B.H. 10 U.S. Army / A.E.F. / France.

† WTS to Dorothy Shoemaker, ML, 6pp., on 3pp. of plain note paper, double-sided; envelope markings similar to 6/22; Censor opened 3150, cancelled 11 JY 17. "On Active Service" written on upper left of envelope. Two U.S. 2¢ postage due stamps.

again. It will of course seem much longer for the complete restoration, but I mean, not just ordinary living, so that you can come & go when you please, buy anything you like and be again comfortable.

I have started French lessons, 3 a week from a little Belgian girl. She is a nice girl, and was driven out of Belgium with her family 3 yrs ago. She is using the Berlitz method which she thinks is the best for any purpose. I am afraid that any method would not work for me though. However, there is no reason why I should not learn something about it. Tomorrow, we are expecting a visit from the King & Queen and the Prince of Wales. I will be glad to see old friends, whom I have not seen for at least 48 years. I might recognize them, but I doubt if they will remember me. Queen Mary sends me 4 packages of cigarettes every Tuesday. They are pretty bum cigarettes, but Her Gracious Majesty is doing the best she can, so I do not complain. They say she is some nice lady. Most Queens that I have known are nice. Not of course near as your mother, but for Queens, pretty good.[34]

You will be interested in knowing that the Y.M.C.A. [Young Men's Christian Association] has dug outs in the very first line trenches, built by themselves. When not actually fighting the men can go there & read & write – and buy things and do all of the things that Y.M.C.A.'s usually do. I think it is wonderful that the way they established themselves everywhere here on the "top" where we are, there are two or three of them working all the time – moving patients, and other entertainments are given every day – to which the patients are altogether without charge. In fact, they give this show several times a day – once or twice for patients only and then for officers and others. It is a great thing they do. Then they sell tobacco, chocolate, candy, and many other things at a very low price.

Well, I have done pretty well this time so I will now go to bed, and on the moment first find myself just one day nearer the end of this pestilence. W. is H. – W. is H. – W. is H. – and then some.[35]

Much love,

Yours,

Dad

Wm. T. Shoemaker

July 10, 1917 [Tuesday][*]
879 Stockton Ave., Cape May, N.J.[36]
Dearest Dad: –

It's quite a while since I've written to you and I'm afraid you haven't gotten any of my letters. We were certainly glad that if only one member of the Unit could get mail, that <u>you</u> were the one. However, I guess three letters didn't give <u>very</u> much news.

I came down here on Saturday to stay for about a month. Mother went out to Wallingford the same day ahead of me. I guess you must be kept terribly busy all the time with 2000 beds instead of 500. I should think you'd need an assistant eye man. It must be horrible to see all the awful sights you must be seeing, but I guess it's awfully interesting. Your pictures still hold forth in Goldensky's and Dodner's windows. We're might proud to have you one of the first to go and to have you <u>really</u> in France instead of "expecting to go to France" soon. At least Walter does.

Cape May is just full of sailors. There are several thousand at Sewall's Point. They've completely taken over the place and you can't go up there. It's quite the thing down here to toot around with a sailor but it's really getting too common. <u>Everyone</u> has one but Florence and Ethyl and me, and we don't want one!!!

I'm knitting Mother a purple sweater for Christmas. Just started it yesterday. Believe me, I have some job. I guess Mother told you that Dr. Le B. is at Fort Oglethorpe, Ga., "expecting to go soon." Did you hear that the name of the German Hospital had been changed to the Lankenau? I meant to send you the clipping from the paper but neglected it in the mail rush before I left. Mother and I served like <u>mad</u> for a week after the sketch class stopped.

Well, now that I'm not so frightfully busy, I'll write oftener. Hope you are enjoying the contents of the package Mother sent!

Remember me to Ed and Cousin Charlie. Also to Lawrence Jones if you get a chance –

Lots and <u>lots</u> of love –

Dorothy S–

---

[*] WTS from his daughter Dorothy, ML, 5pp. on 2pp. of folded watermarked stationery without letterhead.

7/12/17 [Thursday]*
Le Treport, France
My dear Mabel,

Four letters this morning – May 22nd, May 29th & two June 19th. So you see, June letters are longer coming – May 22nd, enclosing Howard's article of early ancient history. I sympathize with you when you try to read my records. They are some records – and I often have difficulty myself in deciphering my signs and symbols. I think it is time to write to Mrs. Johnston. I am so sorry that Mr. Johnston's eye is doing blurry. I was afraid of it but hated to think about it. I hoped for the best, and still think the eye care must be to see if the condition is due to cataract, or if it is Lazy eyes business.

May 27th. Ted's first letter, about which I had heard, but which did not arrive until today.

June 19th. Ted's second letter, will answer both as one tomorrow. Then the best of all, a nice letter from you dated June 19th.

Now my dear, if there is one thing that I can't do, it is to write an interesting letter, so please do not circularize my <u>simple</u> effort to keep you posted. Of course, I hope they interest you, but I am quite sure they would be of no interest to others. Tell your friend Mrs. Melhuish that I could not think of calling on her so long as she speaks my name with C's & U's. "Shumacher" is very distasteful to me, and would not go far over here just now. Kind of her though, and I appreciate it. Good that "little Eva" is pleased with her eye. It is better than a blind one or a glass one, although some of the glass ones are quite pretty. Caroline must be really married by this time. Her letter is always terribly nice. I hope she will be happy and think she will be. It is too bad that after such extensive "social operations," costing much money and many hours sleep, that M. Mc., Jr. only landed in Cedar Ave. or Willow Ave. Charlie says he met Mr. B. once but he did not seem like "our kind."

It's nice you know to have a "friend" all of your own. Very satisfactory. Dorothy is evidently doing very good work., which is gratifying. Sorry she can't go to Maine, but have her take you and Owen & Greta some Boston way in the machine. It is a perfectly easy trip and without me to cuss, it might be very enjoyable. You could take your time and travel 125 miles per day. Stop and see Warren. Go over the Mohawk Trail again & have a fine time. Owen & Greta would think it "grand."

I have been worrying for fear you forget to enjoy such summer courtesies but we have Louie, and it seems that he has a key, and I am sure he will work out the rest to his satisfaction. You told me about the Liberty Bonds, and of course I approve, but do you realize that Dorothy needs no help from us any now. Her extensive belongings should be in her own name and right as Alas! She is 20 yrs. of age and more.

Do not worry about <u>sugar.</u> We have plenty of it – no restriction. In England though, it was personally handed out in minute pieces about the size of split peas. One or two such pieces, for tea or coffee, would be the limit – and for cereal, the waiter would sprinkle about a teaspoonful where he thought it would do the most good. His one hand would securely hold the sugar bowl, and his other would amputate the afore said teaspoonful. Here we have the largest grains – dirty sugar in profusion. It is very sweet sugar and the dirt does not taste bad. Do not worry about the Commission or the pay. The pay dates from the date of the Commission and if dates back, the Quartermaster will send along the different one without question.

The Commission will come along, and all money will as you know now, be placed in the Girard Trust Co. – without bothering you. They should notify you. Major DeLaney has passed up Washington in re Commissions, and I am assured that they will come all right in time. Have sent to Quarter Master General today – my pay vouchers for August, Sept. & October. They have to be paid as they become due. Of course, if after they get started, they do not come right for accident, etc., we can risk, but I assume the system to be a most perfect one, and the chances are they will be right. Good of Casey Ward to send me his book. I will write and thank him, and am glad you did. Now I have really <u>answered</u> your letters of Jan 19th. Something quite unusual for me.

---

* WTS to MWS, ML, 12pp., on 6pp. of plain note paper, double-sided; envelope markings similar to 6/22; Censor opened 3814, cancelled 13 JY 17. "On Active Service" written on upper left of envelope. One U.S. 2¢ postage due stamp cancelled on it.

Was going to tell you about my gas experience, but am a little afraid that might come under "Military Information" and that is prohibited. I will say that I have my gas mask and am prepared to go to the Front if sent for. In Emergency, we are liable for duty at the Front, immediately back of the shell fire – in what is called the Casualty Clearing Stations. Gas masks are absolutely necessary, for when the gas comes over, it is fatal within a few seconds, unless you are protected. I don't think any of us will get up, and I hope not, but some of the younger ones may draw the order.

Our Enlisted men are a curious bunch in many ways. Many of them are very rich and are at home members of our Clubs, etc. Among our waiters at Mess and Striker (take care of our rooms) are Will Downs, Dr. Martin Downs' son, Tom Barratt – son of Judge Norris Barratt, several millionaire Main Line kids, etc. Nick Dawes is the ugliest kid I have ever seen; he looks just like his father looked, and just like his uncle Bob Downs look – all the Downs are ugly, but they are very nice not with-standing. Our Striker is young Barratt. He is absolutely useless as a Striker, but a very nice boy – 2nd year U. of P. Wharton. It is unhandy to have your Social equals as your Servants, but nobody minds it much.

Rumor has it that we are to go to Paris in Sept. The place is full of rumors but I rather fancy Base 10 will eventually get to Paris. The work here is not just what we are best fitted for. It is not gruesome etc. work but reconstructive and repair work that anybody can do. It is not really a Base Hospital, but an Evacuation Hospital. We want [to be] a Base Hospital. You see, there are Casualty Clearing Stations, where only necessary surgery is done – i.e., necessary to enable transportation. Evacuation Hospitals, where arms & legs are cut off, eyes taken out, bullets and shrapnel removed, etc. And Base Hospitals, where new arms & legs are made, new eyes and eyelids are made – plastic work, and where the poor fellows are made so they can return to life that is more or less worth living.

Now for another line of prattle. The flowers here are beautiful. Last Sunday, Charlie & I walked along the cliff to Mesneral [Mesnil Val], the next town below – about 1 ½ miles. Charlie picked 17 or 18 different varieties of flowers. All colors and in perfect condition. For a while I began to think that Normandy was a garden, and I wandered about that apple blossom time that we sing about. It was cold and wet and very disagreeable. Now it has gotten hot & sunshiny just like July ought to be. The country is beautiful. I started out on my way home the other day to pick a bouquet for our room. It was pretty, of course, but I picked mostly heads. Nice heads, you realize, but the pretty flowers I seemed to miss despite their profusion. I bought a dandy big Orangia in a pot. It is standing on our table and looks fine. This is a whale of a letter, and if I do not stop you will be spoiled and will not approve.

Circa – the Shorties you are apt to get later. So ly. ly. – With much love –

B

Wm. T. Shoemaker[37]

July 14, 1917 [Saturday]*
1912 South 15th St., Philadelphia, Pa.[38]
My dear Dr. Shoemaker,

I received the postal card you so kindly sent a few days ago and I assure you it was indeed appreciated. Your interesting letters to Mrs. Shoemaker (I have had the privilege of making carbon copies of them for the children) make me wish I could be with you, with my eye dropper and homatropine.

When we heard of your arrival in England, I wrote you, but as you did not receive any mail until a short time ago, I presume the letter went [down on] "the voyage." Everything I believe is going satisfactorily at the office. Of course, not many patients are coming in now, still those who do, are not getting away from me. I think I'll be qualified for a salesman soon and if Dr. Creighton knew the nice things I say, he would probably be surprised to know I have got him on about the highest strata. We have also had some new patients, so I think things are going quite well. But it is very lonely at the office and especially now that Mrs. Shoemaker is away, too. I will be glad when you return and I have some real work to do; when you will take a look at the thermometer and say "My goodness, do you want to die! It's too hot in here" (said heat caused by many patients).

---

* WTS from Ray P. Miles, ML, 5pp. on 2pp. of folded watermarked note paper, without letterhead; marked 8/8/17 [received].

In case you did receive my other letter, I wish to make a correction regarding a statement concerning the Stetson Hospital. The Board of Directors had a meeting and decided that the hospital be closed, simply go out of business entirely. Well, for about four days there were some flustered people on that staff, but another meeting was called, so I wanted to now tell you, the Board reconsidered matters, and things are running smoothly once again.

Mrs. Shoemaker very kindly had the "Press" continued just for me, so I am sending a clipping which will probably interest you regarding Dr. McGinn. I want to thank you for your kindness to me. I hope you are well and that we may soon hope for your safe return.

Very respectfully,

[Miss] Ray P. Miles

15/7/1917 [Sunday]*
Le Treport, France
My dear Barbara,

You are in Camp on top of the Poconos according to my schedule as issued by your mother. – Here I was disturbed and sent it in the next day. A nice letter from you arrived this morning. You are all getting along so well without me that I have decided to say here some time. Just how long I cannot say, but everybody but myself thinks a very long time. The country here is beautiful – flowers are everywhere and blooming so well that everything seems to be perfect.

Every afternoon & night almost we hear the guns from the front – not loud you understand, because we are quite a distance away, but often very distinctly. Aeroplanes are passing over head every day and convoys of ships are going up the Channel frequently. The other day an aeroplane passed over very low, not over 150 or 200 ft. up. I had never seen one so close – the man in it waved at us. It was very interesting. They are large affairs and make a great noise. It seems to me that you can hear them no matter how high they are.

I am taking 3 French lessons a week from a little Belgian girl – a refugee – She has been here three years. I do not think I will ever catch on to the "blooming" language, but I am going to learn something about it so look out when I get home & I will put over the French. Do not worry about the sugar. It is very thoughtful of you to consider sending me some, but in France, it is plentiful. Only in England was there a shortage. Here we have plenty of very sweet sugar. Not clean and white – to be sure, but dirt doesn't seem to hurt it at all. Our eats are pretty good but tiresome. I would like once to get what I want and not something else, but here we get what come along and after a short time it was not difficult to know what was coming along. I am enclosing two postals. One shows a part of the town and the cliffs on top of which our reservation is located. The cliffs are about 350 ft. high. Occasionally someone goes overboard. It is of course good night for him. Then great pieces break off at times and fall on the houses in the town. It is then good night for the houses. The other card shows some of the ambulances with the English girl drivers. – Each ambulance holds 4 patients on stretchers. These girls are great drivers – they care for their cars entirely alone, change tires when necessary or fix their engines. They are good Scouts all right.

Well, this is enough until next time. Write often and tell me how you are getting along with your summer.

With much love,

Dad

Wm. T. Shoemaker

----

* WTS to daughter Barbara, ML, 4pp. written on both sides of 2pp. of 9x6in. plain note paper; opened by Censor; APO 17 JL 17.

July 17, 1917 [Tuesday][*]
879 Stockton Ave., Cape May, N.J.[39]
Dear Dad: –

Your interesting letter came the other day and I was mighty glad to hear from you again. It certainly is too bad that your staff is so much too small – You say you now have 1000 patients – I should think half of those would have something the matter with their eyes – Isn't there another man in the whole Unit who can assist with eye work?

One day on the beach, Florence and I "picked up" a little boy – After we'd talked quite a while, we asked him his name, and he said Arthur Gerhard – I recognized the name of course – Then ~~yesterday~~ several days ago I met him and he asked where I was staying –

Yesterday I came up from the beach with Lucretia looking very sloppy and sandy and Mrs. Gerhard was here to see me. She wants me to go to her house to tea this afternoon – There are going to be some more people there whose families are in the Unit – I'll tell you about it later.

Aunt Florie's been having a tooth ache and went up to town this morning to see Cousin Low –

I'm going to Freehold [N.J.] to visit Ellen Campbell for a week on the 21st – then I'm coming back here and stay a couple of weeks, and then I guess I'll go to Wallingford for a while.

The bathing is simply great – The other day there was great excitement down here and all along the coast – Way out in the ocean a big, long, black object floated by – Of course everyone thought of a submarine, but it turned out to be a dead whale 70 feet long. Some men landed it several miles down the coast – There have been no trolleys running in Cape May this summer so far – Now the government has taken it over and the sailors stationed at Sewall's Point started to fix the tracks yesterday –

Just had a letter from Ted – He says Bob is getting half board at camp for working in the printing shop – He is chief and has about a "thousand assistants!" Not so bad for young Bob, eh what? Well, I think I'll stop now and finish this when I get back from Mrs. Gerhard's tea, so that I can tell you about that.

(Next day, July 18) –

Florence and I had a very nice time at Mrs. Gerhard's. Mrs. Cruse was there, Dr. Cruse's mother, and Miss Cruse, his sister. Mrs. Cruse read some very interesting letters. There was a girl about my age or maybe younger, the wife of one of the orderlies – Mr. William Hoffman – She was very nice too and horrified at the thought of the orderlies' letters being censored by the officers first! I told her I'd tell you not to read his letters if you came across them – and she felt deeply indebted!! She gave me the address on this letter – She says she's used it for all her daily letters and her husband had received 18 last time she heard – She had a dandy picture of Le Treport, in sepia – like the postal you sent, but only larger – Have you taken any pictures with your camera? Wish you would, and send some – Well, I'd better say Good-bye for this time – We miss you a lot and enjoy your letters just loads so write real often.

With a great deal of love –
Dorothy

Friday, July 20th, 1917[†]
Camp Owassia, Pocono Pines, Penna.[40]
Dearest Father,

Maybe I wasn't tickled to get a letter from you! It was a peachy one, too – but very spooky. I sent it on to Mother because I know she'd love to read it. I'm still at Camp as you can see by the heading. I've found several people who either know you or their Fathers do. Marie Rumpp's Father is one of the Trustees at the German Hospital, & Marie knows loads of the doctors and sisters there. Do you know Mr. Rumpp? Marie's a peach & I used to know her at F.S.S.

We have the funniest pair of lemons in Camp. They're more fun to tease, too. One night after we had very mysteriously upset their washstand & tinkling a bell under their beds, we hit upon a new & evil idea. It was Friday

---

[*] WTS from his daughter Dorothy, ML, 6pp. on 2pp. of folded watermarked stationery without letterhead.
Envelope shows postmark Cape May, N.J., 18 July 1917, to Capt. Wm. T. Shoemaker / U.S. Base Hospital No. 10 / London – England / c/o Surgeon General. 8/8/17 [presumably the date of receipt]

[†] WTS from his daughter Barbara, ML, 6pp. on 3pp. of watermarked stationery without letterhead; postmarked Pocono Pines, Pa.

the thirteenth & they were very excited & superstitious. About five girls & two counselors got up at midnight & sneaked into their tent, at least V. Bartlett & I did – (the other were stationed around to help us) – We got them outside & tied towels around their faces – intending to dip their toes in the lake. But Mrs. Paxon heard us & oh! but she was sore. She pulled a few of us into the house to sleep for the rest of the night – She was ripping mad at the counselors especially, but we didn't care because she has no sense of humor anyway.

Mr. & Mrs. Evans & Edward Evans (little brother) came up to Pocono day before yesterday (Wednesday) & took us over to the Inn to eat a real dinner. We got all dressed up - & felt so queer. But maybe we didn't have one glorious time & maybe we didn't eat! Five vegetables – consommé, two large orders of meat, salad, crackers & cheese, & ice cream & pie! Believe me, it was some dinner!

Some of the girls walked yesterday to Tobyhanna & back – about sixteen miles. To-morrow more girls & a counselor (Mary Williams, Louise Harding, Margaret Bartlett & Amina Joyce & I) are going to walk to Pocono Knob – It's about 14 miles there and back but it's up & down hill all the time and is a very hard walk. Some others are going to Locust Knob to spend the night.

Eleven of us went up to an island way up the lake to sleep. We thought it would be fine, & we got some deliciously thin bacon & cheese crackers & bread & butter & oranges and peaches to eat for breakfast. There was a dandy place to put our beds, or rather blankets. Before we went to bed we took a dip – it was glorious! But when we got in bed the mosquitos came in positive clouds! It was fierce & no one slept more than an hour! Aside from the mosquitos it rained all night & we got soaked. We got up at 4:30 the next morning & cooked our breakfast in the rain. We walked home about six & took a dip before the rest of the camp was up. We pretended we had a wonderful time, but oh dear!

I was writing out under the trees, but a terrible storm came up so I had to come up on the porch. It's leaking now, so you'll have to excuse the blots & smears. Here's hoping you'll be able to read this epistle!

Good-by – I'll write soon again.

Loads & loads of love,

Barbara

21/7/1917 [Saturday][*]
Le Treport, France
My dear Mabel,

The automobile. I think you should have it repaired at once and take it to Wallingford where it will cost only gasoline and will help a lot in your summer pleasure, especially as Dorothy is with you to run it. It has only been run about 13 or 14 000 miles, and I know needs overhauling. A bearing is probably gone or some little thing. Have Mr. Williams look it over and advise you. Perhaps he and Frank will fix it as they did several times for me last winter. – pay them of course, but do it quietly. If it is too much for them, have our people send up and get it, and get Williams to keep tab on him so that they will fix it right and not sting you. But I do not like a broken or idle machine, and would like you to get use and pleasure out of it. It will not cost you much I know.

Received this morning announcement of Caroline's wedding. I's all right but I know that Harvey has not reduced Ed's rent. He should do it, of course, but probably "can't afford to." Taxes & upkeep for 3 houses are very heavy you know. But then it's worth it to have a summer place at 20th and Chestnut, and a winter place on 30th St. Oh well, such elaborate domestic arrangements as H.P. needs must have a far disadvantage. I hope you will be able to endure the insurance deluge. One little suggestion, when you pay a fat one, do not sigh and think that the trouble is over, because the next mail is apt to bring another fat one. They pile in every few days until you have about got cleaned up, and then they will stop for some few months. Dividends I use toward payment of premiums, and in some cases you will find that an almost impossible premium is not quite respectable looking by the dividend. Mr. Frank Campion is my broker and will tell you anything you want to know about insurance. Watch him, however, on the automobile insurance. A car depreciates in value every year, and must not be issued up so high for this time - $500 - $600 – will sufficiently cover our car for fire & theft. Malpractice insurance – costing $15 might if you think well of it be enough this year, as I am in no danger over here. The agent for that is in Harry Lewis' office and is a friend of Henry Page. I can't recall his name at present – Howard Field is the name – but if you do nothing, he will

[*] WTS to MWS, ML, 6pp. written on both sides of 3pp. of 9x6in. plain note paper.

simply renew, and send you a bill for $15. It comes along some time in summer or fall. Look into it and save the $15. Policies relating to auto are in my desk – right hand pigeon hole. Sent you by mistake a bill from University Club, about July 1st, $30.00. Write or call up and ask if I am not on Army & Navy list, without dues. That is my status, but I said nothing to them about it, and it is possible that a mistake might be made. You of course know that you are free to use the League Ladies restaurant as usual. Just pay the monthly bill as it comes along. If the amount is too small for a check, have Owen or Josh arrange to pay it in cash.[41]

Taxes must be paid before Aug. 31st. The "meanies" raised it $45. Do not forget that I owe J.L. & Co. [J. B. Lippincott Co.] $150 borrowed just before I left home. Also $1000 – renew the note. When the time comes, you will find the date that this loan is due. Remember that this is business. Will write you a love letter next time. You probably know that I have told you, but my system although perfect is sometimes a little difficult for another to understand. Nothing is ever lost but time trying to find what you want now. But it's all there.

Good bye.

Much love, B. –

7/21/17 [Saturday]*
Le Treport, France
My dear Warren,

Your letter from Bridgeport [Conn.] arrived yesterday and I was glad to notice that you were so comfortably located and had prospects for a good summer.

I am satisfied to have you in Bridgeport than in the Army but of course glad that you offered yourself and were ready to respond if you were wanted. I am seeing a good deal of soldier life and while it is necessary and we would all accept it, it is not nice. A great many of our young men who came over here will never get back. Some will get back ruined for life. When one sees these things at close range, one is not so anxious to hurry with it, but more than ever anxious to get with it when the time comes. When the time comes you will be in it, and that will be the right time, so do not worry because you were turned down by the U. of P. Ambulance Corps. I am satisfied that your examinations will be all right, and I want you to return to college in the fall, unless you are needed in defense by the U.S.A.

So far as I can dope it out, Germany is licked, and the war will stop before the end of this year. I am almost alone in this opinion but I am sticking to it. 90% of the people say 2 yrs., but my guess is just as good as their guesses. Nothing during this war has been accurately guessed out. You will remember how it was. Remember the impossibility for the U.S.A. to invade Europe. "We had no ships; we had no trained men," etc. And yet, we already have a considerably Army right here in France, and will now have even more of an Army. Also, it was to be a naval war; the Army would not figure. The Army figured first and will continue to figure. And many other guesses were shattered in the same way. Nobody knew a thing and the unexpected happens. Therefore, I'm for an early ending.

I would like to tell you what is going on here, but the censorship is getting tighter. I can't give military information, and that is what would interest you. You know from previous letters, about our hospital. It is a big one all right, and a good one. We expect to be very busy from now on. Aeroplanes are very interesting. They pass over us every day. Some are scouts and some are armed. Occasionally they pass over low so that we can see and wave to the drivers. No matter how high they are we can always hear them. They make a great noise which is quite characteristic. One had engine trouble the other day and planed down on our reservation. I saw him close. The first time I ever saw one close. Great.

Write often, and let me know the news. Much love,

Your Dad

----

* WTS to Warren, ML, 4pp., on 2pp. of plain note paper, double-sided; envelope markings similar to 6/22, but not opened by censor; ARMY POST OFFICE 22 JY 17. "On Active Service" written on upper left of envelope. One U.S. 2¢ postage due stamp cancelled on it. Addressed to J. Warren Shoemaker / Bridgeport, Conn. / P.O. Box [illegible] / try Warren Bros.

July 22, 1917 [Sunday]*
Camp Becket, Becket, Mass.[42]
My dear Father –

The last letter I sent you was returned with a rubber stamp on it saying "Returned for better address." So I put it in a new envelope and added Wash. D.C. to it. I thought that while the postmaster at Becket didn't know what to do with it the postmaster of Wash. would. Have you received that letter yet?

I received your two postal cards and thanks ever so much for them. That ambulance sure was shot some. The enclosed "Seen and Heard" was printed under Bob's direction and most of the work was done by him. Not so bad for the first attempt do you think?

Have you seen anything of Hudson Chapman or Frank Chalk?

Of the leaders in camp three have been drafted. Did you know that the draft was made yesterday? And from the numbers that were drawn there are enough to make an army of over a million men.

For the third time the baseball team that I play on was beaten by the score of 10 to 0. I'm beginning to think we have a jinx following us as we are supposed to have the best team.

I never saw so many visitors in camp in my time up here. Last week end and this one also we are filled to the limit. There are about ten auto loads here now, and I suppose more will come.

At last the country has awoken that the soldiers at Becket weren't doing any good, so they took them away. Before this time they had eighty soldiers guarding three bridges.

Yesterday about 30 girls from Camp "Yokum" played Becket a visit. Their camp is about five or six miles from here. They came in hay wagons and brought their lunch and stayed until about four o'clock. It was quite a new experience for Becket.

As I haven't much to say I think I'll quit.

Lots of love,
Ted

7/27/17 [Thursday]†
Le Treport, France
My dear Mabel,

The last time I wrote to you, I wrote a business letter. I promised you a love letter next time. Well, I love you enough to send a dozen letters telling you about it but business of a war nature has started and I have lost. The Censor will not allow me to tell you what we have been doing but we have done plenty of it, and most of us are deep in it. We have no empty beds just now, so you can put A to Z together, if the fair Censor will allow it.

Have received 4 letters today – one from you, one from Bob, one from Catherine Warren, and one from Dorothy. Have also nice letters from Ralph and Walter, Barbara, Ted & Warren. Barbara writes the most interesting letters; so does Dorothy. All the children seem to be in fine shape for the Summer.[43]

It certainly was good and kind of my brothers to help you with the rent. I will write and thank them. I appreciate it very much. Was someone to read my letter to you in the "Alice Reader"[?]. Please be careful because I am really not looking for publication. George is so "progressive" that one must be careful. Glad you enjoyed the Harte's dinner. You are not much older than most of them. My seniors here are Harte, Taylor, Jeffreys. Then there are a dozen or more men in the 40's, some in the 30's, and 2 in the 20's. I am still between 3 40-50 yr. and those are most of us. Everybody admits that you are the best looking of the wives, will not worry about ages. Before I forget Ed & Charlie are now well, although Charlie was quite sick for a short while. Say nothing about it, because nobody knows it, and he is all right now. If Mary knew it, she would raise H—. I do not see a great deal of Ed because he just strolls around silently as it were. Both he and Charlie are working very hard – the mandibles of every one are in bad condition. Charlie is a splendid worker and is very popular. He is in service not a glamourous one. He is doing bully good work.[44]

---

* WTS from his son Ted, ML, 4pp., on both sides of 2pp. of Camp Becket letterhead stationery. To Capt. William T. Shoemaker / Base Hospital #10 U.S. Army / American Expeditionary Force / Washington D.C. / (Please Forward). Received 8/21/17.

† WTS to MWS, ML, 10pp., on 5pp. of plain note paper, double-sided; Censor opened 3378, cancelled ARMY POST OFFICE 27 JY 17. Two U.S. 2¢ postage due stamp cancelled on it. Stamped on back: J. L. SHOEMAKER & Co. / AUG. 26 1917 [received]

I am glad Theodore got a job and was able to go with training on such short notice. He seems to be obsessed [with] the idea of coming to France. Let him remember that it is easier to get to France than to get back again. He speaks French so well that perhaps they could not do without him, if they ever got him here. When is Foust? I hear Major Jopson is still in town. He said he would be in France with "his Unit" before Base 10 had its supplies bought. He talked much big and caused much amusement for his friends. Heap big Creature, him the Major. ———

Two days later. Not quite so busy now my dear, but still doing something. Censoring letters has been quite a nuisance. Of course, when the hospital is full, and patients begin to improve a little, they write letters, and the mail is very, very heavy. Then on account of recent serious trouble we have some getting past us & being caught by the government, we have all been jacked up and must censor more carefully. It uses the times that I would like to have to write letters. Yesterday, I had a new stunt. I was "Garrison Field Officer." That means, go down to Treport and inspect all of the meat, bread and food stuffs. I looked at 14 000 pounds of bread, tasted a little piece of it and declared it quite satisfactory. Then I inspected the Military Police Station to see if it was in order and if the prisoners (there were none) were properly cared for.

Next the B.E.F. [British Expeditionary Canteen], to see if the stuff was all good, and finally the Y.M.C.A. Hut for the same purpose. This job does not come around often, because it is divided among the Majors & Captains of all the Units here. Today I am Orderly Officer in Waiting, and must attend 4 funerals tomorrow morning at 8 o'clock. Then tomorrow I am Orderly Officer again and will have all kinds of stunts. Everybody is after the Orderly Officer to do something just when he is doing something else. Have not had a French lesson for a week, but will try to get one in this afternoon. Am not very much at French but am going to continue my lessons just the same. Of course, I get no practice, because everything on the "UP" is English. I hear that uniforms are becoming very numerous in our country; that the cities and sea shore resorts are full of them. It is time to sport around in uniforms and I only hope that they that have them will get to work in them.

Please ask Mr. Wall of Ward 9 to send me at once, ½ doz. eye washes – the little wash bottles with a rubber bulb and nozzle that I use to spray or spurt boric acid into eyes. One I think is in my table in the office – that is the kind I want, he will remember. Pay for them or let him contribute as he likes. Also, have Mr. Baer send me 6 doz. eye droppers, the kind I use – straight tips – no ball on end. Just a straight dropper. He will probably let you pay for them, so go ahead and do it, as they are very necessary. Send me yourself the little lamps, etc., which Bob got for me, and which I seem to have left home. Send far more then he got. Lamps for my electric ophthalmoscope; also for my pocket flash light. Mr. Linden of Fruit, Linden & Prospect can get them for you (19th & Chestnut St.). The package which you and Dorothy have referred to has not arrived. Many packages I am mentioning have been sent from St. Mary's Guild – not one has been received. Charlie has been looking for them from himself, but nothing is coming. By the way, are you getting any money or reports from Geisinger? A large number of my best patients are going to him. You should get 1/2 or 2/3 of the receipts. Maybe he will let you have it all, but you are entitled to part of it. Give him my regards and tell him I write to him as soon as possible.

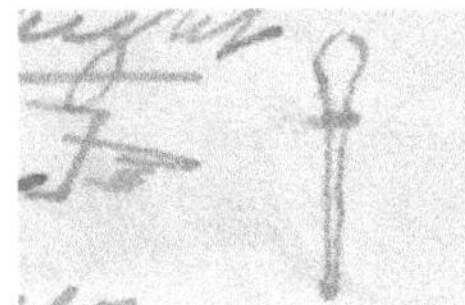

Tommy Holloway has been very kind & thoughtful. I will write to him also I do hope Wallis Johnston will not go to Army, also young Auspark. I want to see them both as soon as I get home. Did you hear anything from Lukens, the boy at Pa. Hosp.?

Now my dear, you have enough. Keep me going as you have been and someday I will be back. Time is passing very quickly. I am very proud of you and your conduct. Indeed everyone tells me I ought to be, just as if I did not know it. I still believe that the war will end this Fall, but no one else does. Everyone almost says another winter, but I can't believe it. Good bye. Much love to you and all.

B

Wm. T. Shoemaker

7/29/17 [Saturday]*
Camp Becket, Becket Mass.
Dear Father:

I can tell you before I start that this will be short because "Colors" just blew and I have fifteen minutes until supper which is not very long. I sent you a "Seen and Heard." The cover was printed by me. Mr. Gibson is giving me half board for doing the printing because they have no good printer this year. A lot of kids are helping me so it is not very hard. A leader at camp Mr. Hussy is thinking of going to France in either hospital or Y.M.C.A. work. He asked me for your address and said he would hunt you up. Try if you can to find him.

I received your postal. Some of it had been censored. The part printed on the top probably telling where it is. The father of a boy at camp Dr. Webster a noted doctor in Massachusetts came to camp and gave free shave and haircuts. I got one and I am a regular convict with a prison shave and a number. Maybe you know Doc Webster.

There goes. Must goodby,

Bob.

Officers of Pennsylvania Hospital Unit, Base Hospital No. 10, U. S. A., in charge of British General Hospital No. 16, B. E. F., Treport, France.

The officers are wearing Sam Browne belts, as ordered by General Pershing. WTS wrote sarcastically about this on September 9, 1917, but he wore it in his formal photo (see cover)

* WTS from his son Robert, ML in pencil, 2pp. on 1p. of Camp Becket letterhead note paper; received 8/21/17.

August 1917

Le Treport – The Third Battle of Ypres

"I am getting so full of war, that you will not love me when I return."

8/2/17 [Thursday]*
Le Treport, France[45]
My dear Mabel,

The Great Drive started yesterday as you know from the papers. We are now in the greatest battle perhaps that the world has ever seen and someone is going to get hurt as I just told my big brothers in a letter thanking them for their kindness in the rent. It is perhaps unnecessary to say that unless you have engaged your accommodations at the "Hotel" far in advance, you are not likely to be well taken care of. More as I told you was in fits and starts; during the last week, I had both a fit and a start. One day before this came, I was officially attending funerals at 8 A.M. – and from then on, I kept going & working until the next day at noon, with the exception of a little nap from 1 to 4 A.M. in Hut 20, waiting for a convoy which came at 4:30. I was Orderly Officer that day, and got hit for a lot. Strange to say, I was not very tired, but sometime toward the end of the day that I went to sleep without rocking the cradle away. I sang "Sing me to Sleep," but it was not necessary. Caught up with 9 hours and feel as fine as a fiddle.

The weather is beastly. What we call a northeaster in all its glory has been performing for 2 days. It's dandy. It came at an unfortunate time, or the drive was started at an unfortunate time, which ever you like. But we are winning anyhow you like it, so it's alright. I could not describe, nor could you realize, what is going on at the front. The work of man in his "inhumanity to man" will make countless thousands undone. You must excuse so much war talk, but since leaving home, nothing has been discussed or scarcely mentioned but the war. It is the day's and the night's topic of conversations. Nothing else is thought about.

I am getting so full of war, that you will not love me when I return. Then I will drive back here! The mails seem to be out of order again. We are not getting letters, and Mary Jack cabled that she had not received a letter for 10 days. That is strange because Charlie writes practically every day. Mary writes every day with the necessary result that Charlie gets his letters 10 or 12 at a time. I would not like that because I can't hope to follow the details of our household hourly. I must now stop and censor my pile of letters. Every man has a pile for which he is responsible. Night is the best time to do it, so that is when we are now doing it. It's a bore sure enough. Good bye.
Much love from your B.
Wm. T. Shoemaker.

August 5, 1917†
Camp Becket, Becket, Mass.
My dear Father –

How are things going with you? We are having a fine time up here although it's been exceptionally hot this year for this part of the country.

The gardens that we planted before camp opened have been coming along fine, especially the corn and "spuds." Also the whole camp has had a mess of radishes, and you can imagine that it takes some radishes to feed two hundred and fifty people. Also the visitors and Mr. Gibson's table have had a mess of peas.

The camera I bought with the five dollars you gave me before you left sure is a peach. I had three rolls of films developed and seventeen of the eighteen pictures came out good. The other was a double exposure. The pictures are

---

* WTS to MWS, ML, 4pp., on 2pp. of plain note paper, double-sided; Censor opened 6064, cancelled ARMY POST OFFICE 3 AU 17. Two U.S. 2¢ postage due stamp cancelled on it. Stamped again at New York Penn Terminal Aug 24, 1917.

† WTS from his son Ted, ML, 4pp. on 2pp. of Camp Becket letterhead paper; postmarked 8/7, received 8/17/17.

so good that three fellows are having sets made of them from the negatives. How are your pictures coming out, or haven't you taken any?

I have to go to Vesper services. I'll continue after, if I get a chance. If not, tomorrow.

August 6, 1917

To-night our tent had supper out on the hill but another fellow and myself took a walk down to Becket and came up in the camp Ford. The Ford didn't start until after six o'clock so we missed our supper.

Last Friday thirty-eight of the fellows who are up for half season left and thirty-eight new ones came. Just think. Camp is over half over and it seems to me as if it just started. But believe me, I'll be glad to get home.

Tonight Bob got a letter from you and also it was opened by the Censor. I guess that it is the one in a thousand that you spoke of. We sure were glad to get a letter from you. Just got a letter from Mother saying she hadn't gotten a letter from you for about a week and a half, but Cousin Mary got a cablegram from Cousin Charlie saying "All well." It was rumored that two mail ships were sunk. I don't know how much truth there is in it.

The stuff you wrote about the Y.M.C.A. will please Gib very much.

There goes "tattoo" so I'll have to stop. Here's hoping you get this letter.

Love,
Ted.

Aug. 7/17 [Tuesday]*
2228 Locust St. [Philadelphia, Penna.][46]
Dear Dr. Shoemaker –

Bonjour Monsieur le Docteur, comment allez vous? Or are you not associated with the Frenchies? By this time, I suppose you are well settled and hard at work. I wonder if you are getting many of the kind of cases we received at first. Imagine you see your cases at the most, two days after injury, which we did not at first. As regards news over here, I have no doubt you are well supplied. Personally, I am still on the shelf, but hope sometime within the next year to be released. I feel like a crab out of water, loafing around with everyone else making sacrifices, and doing his share.

Since I last saw you, I have been in good condition, although I have not done very much work. I have a little piece of work I am tinkering with, out at the University Laboratory. It is along the same line of bugs, which I dabble in over in Dr. Monroe's clinic and may prove interesting. In a case at the Children's Hospital, which looked clinically like a gonorrheal ophthalmia, I found a gram– intracellular diplococcus. The case cured up in three days. I saw another case, similar to that at the University, and three while in Paris.

Our experiments thus far have differentiated this organism from all other diplococci, such as gonococcus, pneumococcus, diplococcus catarrhalis, and diplococ. meningiditis, as well as staphylococcus. It has no action on or in animals. When I get back in the Fall I am going to try some other experiments.

Mrs. Scarlett and I are enjoying the cool breezes of Island Heights, which is a quiet little village, where a cousin of Mrs. Scarlett's loaned us a cottage for the month of August.

Johnnie Carson and Ralph Bromer have been over in your section of the world, and you may possible have seen them. They are back here now, ready to go out with the Episcopal Unit. Rumor had it that Carson caught malaria, and lost 14 lbs, but his mother told me it was a severe cold because of exposure & that altho he had lost a lot of weight, he was well enough to be about.

I hope Dr. Mills will let me start in at the Penn this fall. When do you expect to get home? Hear Dr. Packard is coming this Fall. Wishing you the best of luck. Highest regards to all.

Sincerely yours,
Hunter W. Scarlett

---

* WTS from Hunter W. Scarlett, ML, 4 pp. on 1p of folded watermarked stationery. Cancelled APO 25 AU 17.

8/8/1917 [Thursday]*
Le Treport, France
My dear Mabel,

After about 2 weeks a good big mail arrived and now everyone is happy. I drew 7 letters – one from you; Dorothy, Ted, Barbara, Tom Holloway, one from Dr. Turnbull, and one from Miss Miles. So you see I have much mail from many quarters. Holloway wrote a particularly newsy & gossipy letter. Many things are happening in an Eye line. The fortunes of war are beginning to smile on those who stay at home. Then, Miss Elliott received a letter from Harvey – which she showed me. She always does, but you must not tell. He would be sore, and if "Maud" gets one to him, Miss E. says she won't let him write.

By the way, your cablegram arrived in good time, but nothing has been heard from Drexel as yet. Also, I wrote you a long time ago concerning a money deal, but that has not panned out, so I am out of money – almost out, but it does not matter much, as everybody else has so much that I can't sink. You can't spend a great deal here. $50-$60 a month would be a very liberal amount, so when a man is paid here some $200- a month, he of course has much money. And when is paid in French money, it looks like a million dollars. I was glad to get your cable anyhow. It was a real close communication you know.

I wish you would investigate St. Mary's Guild. Hundreds of things must have been sent through St. Mary. Not a single thing has come through. The Turnbulls sent me something via St. Mary – and you have, but nothing doing. In the future, you had better send everything through St. Mabel. It is much more certain that way. I do sympathize so much with you when you speak of Insurance. For a time you will certainly get hooked, and now is about the time.

I was thinking that it would be a good time to have Heppe "over haul" the Victrola. You might do that if you think well of it. It did not seem to be working just right when I left. I see "by the papers" – 3 weeks old or more, that Forst has signed up with the Jefferson Unit – the one that Mr. & Mrs. Potter advertised regularly during the winter as being ready to sail for France "within a week." This he did before he started to organize it. I presume it is ready now, but I heard somewhere that no more Units were to be sent. Forst is ophthalmologist, which also makes it appear that my friend Bill Sweet worked when the time came. Charlie Nassau is a Major, some Major. Word also reached us today that Isaac Jones was a Major – this created a general laugh and added to the day's pleasures. Little amusing incidents like this lighten our load. The vicissitudes of Majors I suppose and his Unit and the news that he had been sent to Albany to recruit also made light conversation during our resting times. We are all glad that we are in France instead of going to France in a week. Of course, it's nice to be home, but if France is the Order of the Day, let it be France.

I am glad you are putting in a reasonably good summer. Everyone says you look fine and are wonderful. Of course, they need not tell me that, but they like to do it, so I do not mind.

Glad Owen [Shoemaker] & Greta are married again. I think the reason for it is the absence of the little Cherub [Edwin Shoemaker]. It would separate anybody from all that was good in the way of disposition. We are still nearly full up, and as the Great Drive is on or starting, we are receiving a great many convoys. This will probably continue for 6 weeks. I hope this drive finishes the party, but everyone is pessimistic, but yours truly and a very, very few others. If the Boche decides to stick it out another winter, good night! Well, good night anyhow. Much love,

B.

Wm. T. Shoemaker

August 9, 1917 [Thursday]†
[Le Treport, France][47]
My dear Mabel,

Just a shorty to tell you that Morgan Harjes notified me this morning that they had at my disposal 342 Francs. Thank you very much. I wrote you last night saying that no money had reached me. Nothing has happened since then to tell you about except a busy day. Yes, one thing has happened – Dr. Newlin read in the papers here that his

* WTS to MWS, ML, 6pp. on 3p. of plain note paper; opened by Censor 3349; cancelled APO 9 AU 17. Two 2¢ postage due.
† WTS to MWS, ML, 2pp. on 1p. of plain note paper; opened by Censor 6160. Two 2¢ postage due stamps cancelled.

brother's boy, 19-20 yrs. old, had been killed at the front – Ambulance Co. – We had not yet heard that he was here. Dead & buried and Dr. Newlin does not even know where he was. It was not stated in the dispatch. Very sad, but many such tragedies are sure to happen when our men get into the game. I hate to think about what is going to happen. I feel very sorry. I told Newlin that we knew what is lost. Well, I will talk it over with you someday better than I can write about it.

But the wholesale killing of men for no purpose as has been done over here for 3 yrs. is a Crime.

Much love, B.

August 11, 1917 [Saturday]* No. 29 (I think)
Wallingford [Penna.][48]
My dear Billy

As you notice, I am not sure of the number of this letter, as I wrote the last one in town, and put it down on a paper on your desk. That letter was written on the eighth, if I remember rightly.

Well, your pay for June and July has come, but it was only $323.34, which is only lieutenant's pay, and without the 10% for foreign service. Your May pay included foreign service, so I don't understand why they left it out this time. I will have to write to Washington and find out about it.

The principal thing that has happened since I wrote before is that Ralph Straub has enlisted and will go over to Base 10 with the contingent headed by Capt. Austin. It was my letter to Mr. Test that did the trick, evidently, and he and Peg are very much pleased, as otherwise he would probably have to go with the conscription Army, unless he was turned down at the physical examination. One man backed out yesterday morning, and Mr. Test telephoned to Ralph, and he went right in and was accepted. They scarcely examined him at all, did not even make him undress. Owen sent a cable to Ed this morning, which I hope he has received, and indeed Ralph will probably be there himself before this letter reaches you. Poor Peg! She will have to be a war widow with the rest of us.[49]

This is Saturday, so I did not go in town, and as Dorothy is spending the week-end with Olive Alden, she will keep office for me Monday morning, and I will have three consecutive days out here – a good thing, too, for I want to do more canning than I find time for just on Saturdays.

This morning, I canned rhubarb, and Monday, I want to do Swiss chard. Those Greta gives me from the garden, so they don't cost us anything, and we will be mighty glad to have them in the winter.

Did I tell you that Dr. Jones, principal of Penn Charter, died about two weeks ago? He had been ill for several months. Mr. Smith, who was Greek and Latin teacher there for twenty-five years, has been appointed to succeed him. I wrote to Mr. Smith and expressed my regrets on the death of Dr. Jones, and also asked him if it would be possible for Bob to have a scholarship or to make a material reduction in the tuition, owing to the fact that the family income was much reduced as you are serving in France. I wrote a similar letter to Dr. Edgar F. Smith regarding Warren, although even if Warren can get a scholarship, I am not sure but that it would be better for him to continue at work, unless he enlists, as of course he may. I wrote to Warren also to ask him what he was planning to do in the Fall.

Soon it will be time to send you some heavy underwear. If it was cold enough in June so that you needed a warmer uniform, what will it be in December, if you are still at Le Tréport? I was intending to send you those heavy Jaeger's of your father's and went up to the fourth story to get them, and they were nowhere to be found. I believe I gave them to the Emergency Aid last winter, as they were so heavy it seemed unlikely you would ever wear them. Now the emergency has come when you need them yourself and they are gone, so I will have to buy you some new ones, and money is scarce. However, not too scarce for that, and probably new ones will fit you better than those would, even after I made them over.

Barbara left for Ocean City [N.J.] this morning, to spend a few days with Gertrude Russell, going from there to Cape May for the rest of the summer.

---

* MWS to WTS, ML, 10pp. on 5pp. of note paper; postmarked Philadelphia, Aug 13; received 8/29/17. She gave it No. 29, but it is only #7 of the letters that have been preserved. He may not have saved all of hers, and some were doubtless lost in transit.

That seems to be about all the news, as I will go out on the porch with Greta, and possibly add some more to-morrow if anything happens to write about.

Lots and lots of love
From
Mabel

6:30 A.M. Monday, Aug. 13/17

Bright and early in the morning, I want to add a little to your letter, so that Owen can take it with him to town, as I am not going in to-day. The Whitings and Henry Page came out to dinner yesterday, and we had a very nice time. Fresh corn and lima beans seemed to suit our guests, and indeed they were good. In the afternoon, we played croquet, two very even games, Letty and Jack Matthews against Pres and Alice in the first, and Henry and Greta against Owen and me in the second. I forgot to say that Harvey and Alice came out in the Ford in the afternoon. Ann and Marion are at Atlantic City for a week, and Arthur is there over Sunday, so that leaves Alice at home keeping office and house for her father. Next week, Ann will stay at Atlantic and Marion will change places with Alice. Katharine and Margery are still at Cheyney. Arthur is working for J.L.&Co. and he also spent Sunday at Atlantic City.

I don't think I have seen Henry Page since you went away. Nan had been to see me, but not Henry. He is not looking so tired out as I expected to see him. He leaves to-morrow for his vacation. The Lankenau Hospital has not been hit so hard by the war as some of the others. Of course, the Pennsylvania [Hospital] is in the worst way as it is the only one in the city whose Unit has actually gone to France. The extra force mobilizes to-day at nine o'clock. Ralph [Straub] has had no opportunity to purchase a thing or do much in preparation for his absence, as he was only accepted Friday afternoon, every store closed on Saturday, and he has to report for good, early Monday morning. His mother and uncle came down from Millersville and Ralph and Peg hustled with them and did all they could. They are to go to Governor's Island probably to-day, in civilian clothes and send back their suit cases by express after they get uniforms in New York.

I must close now and send the letter, so Bye-bye. Take good care of my dear husband.

Lovingly, your wife,
Mabel

8/15/17 [Wednesday][*]
Sackaley Hall, Ocean City, N.J.[50]
Dearest Dad,

You see now that I am at Ocean City visiting Gertrude Russell. While I was still home, after leaving Camp, I received your letter, enclosing 2 postals. They were very interesting and gave us a better idea of your situation. Last Saturday the 11th I came here. Elizabeth Evans and I were visiting together but Elizabeth just went home to-day. I am going to stay until Friday, when I will go to Cape May for the rest of the summer. The bathing here is wonderful and we're having a fine time. Last Saturday night we went to a dance at the Ocean City Yacht Club – We're going to another one there to-night. We've been to the movies several times.

At Wallingford I had a great time. Several of the girls from Camp lived near, and I went to see one of them – to play tennis. Two others were there, and they both invited me, one to play golf (!?) the other to play tennis and go swimming. I accepted both invites but when the time came to go, I was so busy packing and had such a headache that I gave a late day refusal to each.

Bee belongs to a sort of a swimming club – a lake – or park with a boat house & bath houses – high & low dives, raft & slide – It's a dandy place to swim – but you have to keep swimming, as it is about 15 or 20 ft. deep in its shallowest place! We went over several times. It's "the other side" of Media. One day Bee & I rode over on bicycles – we were going down a hill – Bee in the lead, when she saw a big bump ahead & cut in on me without warning. I was helpless as to stopping, so the wheel just turned over in the road – I scraped my knee quite badly – & my arm, too – the knee was bleeding so we went into a house where Bee knew some people & got it fixed up. I still have a lot of trouble with it, tho' – the minute it gets a scab – off it comes!

---

[*] WTS from his daughter Barbara, ML, 6pp. on 2pp. of folded watermarked stationery; marked received 8/31/17.

I guess Uncle Owen cabled you about Ralph Straub – He was the 37th orderly – & nearly didn't get in at all! It's sort of hard on Peg – but as his draft had just come in and he was to report for examination the next day – it was rather fortunate for him. I'd go in a hospital rather than the regular Army <u>any</u> day in the week!

Dr. Fussell has been made a major – & is at Gettysburg, examining recruits for tuberculosis – pleasant job – but at least he's not likely to be sent abroad! I just saw Isabell to-day – in bathing – She's visiting her brother here.[51]

When you seen [sic] Queen Mary again, remember me to her – oh! yes – and George, too!

Have you heard the new popular piece that the government is thinking of adopting the way England adopted Tipperary? I can't write the tune, but here are the words –

Oh joy, oh boy, where do we go from here
Any place from Harlem to a Jersey City pier.
When Pat would spy a pretty girl, he'd whisper in her ear –
Oh joy, ole boy – where do we go from here!

Oh joy, oh boy, where do we go from here
We'll slip a pill to Kaiser Bill, and make him shed a tear.
And if we see the enemy, we'll shoot them in the rear –
Oh joy, oh boy – when do we go from here!

Cute?

Well, I guess I'll have to stop now as we're going out – Lots & lots of love from

Barbara

Aug. 19, 1917 [Sunday]*
Camp Becket, Becket, Mass.[52]
Dear Pa. –

Bob and I both received your letters about a week ago and we sure were glad to hear from you again. Well, camp is nearly over, only eleven more days before the campers leave, then about a week after that I go. I suppose we'll both be home by the time you get the letter. Mr. Gibson has been getting any number of letters from old Becket fellows who are driving ambulances in hospitals and even fighting at the front in France. Becket seem to be quite well represented. Also, there are a bunch who are trying to decide what service they will go in when they get home. If every body's doing it, I think I ought to, too.

Thursday night we were going to hold our annual circus but as it rained everything except the side show was postponed, that being given indoors the rain didn't interfere. Then the next afternoon, the "mid-way" and "main show" were pulled off. There were a bunch of visitors up from Becket and the whole thing was a great success, and evidently the mid-way had the usual bunch of fakers, because on the whole affair we took in $55.00 most of it being on the mid-way. Bob quite distinguishes himself as an acrobat. He was the top man in all the pyramids and also did some special stunts. The pictures I have taken with the camera I bought with the money you gave me surely have been good. Nearly every picture has been printed over about five times for different fellows. I lend them the negatives and let them have the work done. It's nearly time for chapel now, so I'll stop.

Remember me to Charlie Jack and Ed. Also Hudson Chapman.

Love, Ted.

8/22/1917 [Wednesday]†
Le Treport, France
My dear Mabel,

Several letters came from you yesterday & the day before – two of them were not quite so nice as usual because you said I was "disgruntled." I was not disgruntled but disappointed at the apparently unavoidable mix ups & hold

---

* WTS from his son Ted, ML, 4pp. on 2pp. of Camp Becket letterhead note paper; postmarked AUG 20; received 9/9/17.

† WTS to MWS, ML, 12pp. on 6pp. of plain note paper; opened by Censor 8481; stamped APO 23 AU 17. Two 2¢ postage due stamps cancelled.

ups in the mail. So there now. Also, I notice that you were complaining at not getting letters from me. Many of my letters have not gotten through because I have consistently written about twice a week. I have written you scads of letters. But you do not get them and I can't help it. Nos. 20-21-22-& 23 arrived normally. Please place number also on envelope – perhaps back, if it is more convenient so numbered. I will start this No. 1.

Now, the important thing in your last letter pertained to school matters. I am very much opposed to any change. Unless it positively cannot be done, I want Bob to remain at Penn Charter, and Warren to continue at the University, and Barbara at Miss Hills. If I am at home any time during the winter or Spring, I can take care of it, and I do not feel that we will be here after December. Of course, you never can tell, but I am still for the War ending about November. At all events, please take a chance, and do not upset the children's educational plans. It would be a mistake. Economize in other ways. Of course, if you feel that you must or might follow your plans, go ahead, for you know from your end of the line more about it than I do.

Perhaps, you did not receive my letter telling you that my money would go to the Girard Trust Co., and be placed to your credit, and that unless you so direct, the Girard might not notify you. So it may be in your bank account. If it does not come, write and ask them to send it at once. It's yours and the G.T. has it I am sure.

I am sorry about Martha [Shoemaker], but she has for a husband [Ed] the same proposition that I have told you about for some time. I cannot speak to him about that. I would get nothing in return but a silent smile. The perpetual optimism, graceful ease, quietude, relaxation, polish & courteous manners and the silent smile are with us always and irritate Charlie [Jack] & me but, Edie [Shoemaker] gets away with it and is sure he is handsome, as many think, so there you are. He has lots of money, and purchases suits, shoes, winter coats, rain coats, etc., galore. As I told you once, he would not allow himself to get tired, claiming that he did not come over here to kill himself. He must stop and get in his golf. And like his Dad, when he stops, the war is over until tomorrow. Charlie, on the other hand, is the hardest working man in the Unit. He plays at it every day from 9:30 until 5-6 or 7 o'clock, and often comes in exhausted. The rest of us have gotten after him, and have "instructed" the C.O. to order him to stop work at 4. P.M. This order came yesterday, but Charlie will disregard it, or at least try to for he is particularly resentful of "orders." He likes to argue and object and when doing same he uses his "office manner" and his office tone of voice, with of course the necessary changes in pronunciation to harmonize within the standard professional vernacular of 1533 Locust St. "Goode Mauwn-i-n-g." Very amusing even in France. France to me, Frawnce (way back in the throat) to English officers, C.O.'s, ladies, etc. You are right about Mary. Childish is the word – or pin head.*

Please do not send me just something so as to be devoted; and do not cable unless you have information which you think I ought to get at once. It is almost pathetic the letters, papers, packages, etc., which arrive for Charlie. Mary writes to him every day, so you can see how often we get a mail once a week or sometimes not for two weeks, how these daily letters all come at once. 10-12-14-18 etc. What she can say that she did not say today. I do not know yesterday, with yours to left over, came the usual convoy of bleak edged epistles. Charlie read them & was blue. He wrote at once telling her. She would have to cut out the sadness and sorrow and suffering etc., produced by his absence. She was sure the war would continue for 3 yrs. – and of course, she ever reproaches him for coming. She is a mess. I like you better. You are doing just right. Keep it up. Keep on smiling, if I never come back.

One lovely little anti-climax I must tell you. The Dental Reserve is new, and as luck would have it, Charlie's commission is No. 1. Something to be proud of and feel good about. The devoted one wrote to him saying that she saw in the Dental Journal that he was Commissioned No. 1. Then the lovely sentiment: Why did you do it? O O O O O O O these are tears; little weeps; they take the place of brain cells. Then a cable, then some candy, and we are all ready for the morrow. Repeat about 7 times a week. The little squirt on Franklin was lovely. I have pinned it on our bulletin board. Causes heap big smiles.

I ruined my khaki suit; was etherizing; surgeon opened up the subclavian artery – it behaved like a fire plug; I got my share and looked like a sun set. So far alright, but a nurse told me to use peroxide on the suit to take the blood out. I did, and the suit looked like a very old buffalo robe. One of the nice ones that you can't get any more. I then gave it to my wash lady. It now resembles a cross between a chestnut horse and a garter snake. If I can find a dyer, I will experiment some more.

Please do not publish his letter. I would not like to see it in print.

---

* WTS's brother Owen married Mary Margaret "Greta" Jack; their son Edward "Edie" Shoemaker married Martha Reed; Mary Margaret "Greta" Jack's half-brother Charles Jack is "Charlie." His wife's name was Mary (née Lewis).

Your war news is as good as mine. The French put a good one over at Verdun and the Canadians at our end of the line gave the Boche one of the greatest handouts he has had. And it was not a cold hand out, but a very hot one.

The last five days have been perfect, and the Allies have expended countless tons of ammunition of all kinds into Germany. The Boche has been receiving all kinds of attention and should be satisfied.

I hope you get this letter for you seemed quite "disgruntled." I would not have you that way, if I had to come and cable and send candy every day.

With much love, from B. Wm. T. Shoemaker

August 23, 1917 [Thursday][*]
Wallingford, Pa.
Dear Dad: –

Haven't time for a real letter but here are two things that may amuse you – The little one, Miss Jeannette told me to send you with her compliments but the big one's the funniest. Everyone's going to be, so I must, too – Will write a longer letter soon –

Leave Aug. 25 for Conn. to spent a week with Dot and then home in the Chandler –

Lots of love –
Dorothy

Stringing Up Wilhelm
(With apologies to Rudyard Kipling)

"Wot are the bugles blowing for?" asked Files-on-Parade.
"To celebrate! To celebrate!" the Color Sergeant said.
"Wot mikes you look so glad, so glad?" asked Files-on-Parade.
"I'm joyed at wot I'm going to watch," the Color Sergeant said.
"For they're hanging Bill the Kaiser.
You can 'ear the Dead March play!
The reg'ment's in 'ollow square.
They're 'anging Bill to-day.
They've taken off 'is un'form, an' chucked 'is crown away!
An' they'll string up Bill the Kaiser in the mornin'!"

—*Morrowby Jukes, in New York Evening Sun.*

**I KNOW MOM
AN' THE MINISTER AN'
EVERYBODY IS PRAYIN'
FOR PEACE – BUT I'M
GROWIN' TERRIBLE FAST AN' IF
YOU COULD KEEP HER GOIN'
JES A LITTLE WHILE MORE
'TIL I GIT BIG ENOUGH T'GO
I'D BE MUCH OBLIGED
AMEN**

[*] WTS from his daughter Dorothy, MN, on one page; sent from "The Lindens / Wallingford, Pa.; received 9/9/17.

8/27/1917 [Monday]*
Le Treport, France
My dear Warren,

Received your letter this morning and will answer at once, as I have been trying for some time to write to you. I am glad that you finished your College year so well, but felt sure that you would. Am very anxious to have you return to College next semester if possible, and have so written to your mother. Of course it is up to her as she knows better than I just what can be done.

I do hope, though, that you can find your way clear to return. I think the War will be over this year, in which case, I can come home and make some more money for you. If it continues another year or two, we are up against it like everybody else with red blood, and will take our medicine. I am willing to take a chance however, and have you at last start your course, and if necessary, you can discontinue at any time. However, be guided entirely by what your mother thinks, and I will be satisfied. Am glad that you enjoy your work, for that means that you will do well in it. You have a fine opportunity. Will write again when I get a chance.

Much love,
Your father,
Wm. T. Shoemaker

Aug. 30, 1917 [Thursday]†
Le Treport [France] No. 2
My dear Mabel,

Two letters from you. No 26 – Aug. 6 & No. 29, Aug. 11. You are probably under in your numbering; if not, 24 & 25 are missing. There are many things I want[ed] to tell you that I forgot until the letter is [i.e., was] posted. For instance – knew of Mrs. Andres to need ____ bandages, but I am not using them. I got $60 from Morgan Harjes. If not too late, never mind the *Literary Digest*. Cables twice a week will not materially strengthen my conviction of your devotion. Chocolate, candy & cigarrettes [sic] grow here, however 1000 or 2000 Camel cigarettes would help. Newspapers, magazines, clippings, etc., arrive by the car load. My room mate gets a bushel of *Ledger, Presses, Life, Cosmopolitan, Outlooks, Sat. Evening Posts, Collier's*, etc., about twice a week, and in addition, tons of clippings from other copies of the same. We have little chance to do more than glance through about one in ten. But then, the virtue is there, even if the humor is not.

Clothes:- under ones. I am so glad that Father's Jaegers are lost. I will need some medium weight. Underclothes, but even so, the people here at least know what they are, and there are actually stores where they can be purchased. Of course ones that you would get for me would be nicer, fit better, and have my name on, and I would be pleased as well as comfortable. But fear not, when the time comes I will hie me to Eu or Le Treport & invest.

Christmas:- Please do not bother about me. I do not need any Christmas presents. I am not going to attempt to send any but will bring them with me when I return. For the children and others you must attend to our presents. A van load of stuff will doubtless come to my room, and I can look at that. So I will be perfectly happy & content with a letter. I received a nice little letter from Owen in which he says that you are certainly the cheerful and industrious one, and perhaps the very best sport among the various "soldier's wives." I am awfully glad you are. He speaks about the "weird lamentations" of some of the wives. Very funny expression. The next contingent has not yet arrived. Glad Ralph Straub got aboard. He will not like it, but it beats [is better] than getting shot.

I do not like the situation of Pres & Henry, calmly automobiling and vacationing as usual and just as if there was no war. They are both slackers and I hoped they would not be. But they are so different; Pres would have to be

---

* WTS to his son Warren, ML, 2pp, on both sides of 1p. of note paper. Opened by Censor 3148. Two, 2¢ postage due stamps.
† WTS to MWS, ML, 8pp. on 4pp. of plain note paper; opened by Censor 3189; British censor stamped; APO [date obscured] 17; two 2¢ U.S. postage due stamps cancelled. This letter has unusually poor handwriting, and is one of the few numbered letters.

dragged in and Henry would have to be kicked in. Pres knows he does not want to go in and would fight to stay out. Henry doesn't know what he wants and would go which ever way he was pushed. Geo. Ross will put it all over them. Another mean thing is the way that they men who stayed at home got the high commissions, and are getting the high ones.

The weather here is beastly and in spots very funny. A few nights ago (11 o'clock) we had a bright moon shining, a heavy down pour of rain and a beautiful rainbow – all of course at the same time. Now who ever heard of a rain bow at night, and doesn't the moon generally retire when it rains? It was a lunar rain bow and is said to be very rare. In addition to rain for the past week, we have with us also a gentle breeze of about 60 miles per hour. If you would shut your eyes, and not look at the calendar, you would never know it was August.

Did I tell you that I got a letter from your brother George? I had asked you about this convention you remember some month ago. He saw the letter and at once thought that the *New Church Messenger* might help my war situation. It would really be a shame to burden the mails with that, and never reading [it] at home, this is not a good place to start. I wrote and thanked him, and I think discouraged the proposition in my most masterly style. Now, don't give me away, for he was mighty good to write to me, and I want him to do it again, but I will look over all of the *Messengers* when I get home. I will have lots of time then.

We are not so busy now, although we are doing something. Please do not work too hard yourself. Owen says you are very busy, – don't do it. I don't want you to start the winter all tired out. I hope I can get home in the winter, but you never can tell; it's a hard thing to get out of after you are in. But the war will stop, so that question is settled.

Very, very much love,
from
B.
Wm. T. Shoemaker

September 1917

Le Treport – Reinforcements Arrive from Philadelphia

"If you go to war, make the worst of it and see all the Hell that it in it!!!."

Probably 9/1/1917 [Saturday]*
[Le Treport, France]
[My dear Mabel]
[pages 1-2 missing]

I had better tell you first about the finances – about one week ago I received from ~~Drexel~~ Morgan Harjes Pari[s] that they had some 300 Francs for me (340.50 F). I wrote and asked them to pass it on. It came yesterday. Then this morning Miss Fuhrmann received a cable from home saying that your $75- had been received, so she handed that over. That business action required weeks.

Furthermore, the British Government has decided to give us what is known as Field allowance. This amounts in my case to about $30 per month or perhaps only $20.- but it comes in real money that can be handled and looked at. That will help a lot. Then again, the British are giving us about 1.50 F. per day mess allowance. This we do not see; we just Eat it as it were. All of this being the case, I think you need not send me money again until about November. Of course, things are slow and the paymaster may not come across with the money for a time, but it has been awarded and includes back pay, and is sure to come. My first dose will be about $75.- I am anxious for you to keep for yourself just as much of our money as possible, and I do not think I will be costing much from now on. Will take up the foreign service pay when the time comes.

About the Commission, the most important thing to note is the date, for from that date the pay is due. It will probably be dated May 15, in which case you will get Captain's pay from then. Your telegrams, etc., in re commission did not mention Captain's com. Although I think that is all that it could mean, as they already have my acceptance of the original commission. Do not worry, it will straighten out alright, and all back pay will be forth coming.

I think I wrote you that everybody was ammused [sic] at Isaac Jones – Majority. Captain is good enough for me. When I am discharged, I will receive the rank of Major, which is customary. Everybody moves up one grade when he quits.

Received word through Ed that Ralph Straub has joined Base 10 Enlisted force. He is probably en route
[pages 5-6 missing]

of the Studebaker – right hand side is a valuable battery. That battery must have the three zinc plates always just covered with distilled water. Therefore it must be inspected at least once in 10 days or 2 weeks during hot weather, and if the plates are exposed, distilled water must be added. If this is not done, the battery is sure ruined, and you are out $25-$30-. The battery by the way makes no noise and is apt to be forgotten. I hope you told the Studebaker people what you thought of them. Sorry also that my letters have cost you 4 cts per, but they won't anymore, as Congress has passed a law franking them. All letters from here go that way. We were so instructed. It will be all right in the future.

Now that is a very long letter. Charlie & Ed are OK – send their love. So do I. – very much of it. Good bye

B.

Thanks again for the money & the sugar.
Wm. T. Shoemaker

----

* WTS to MWS, ML, 14pp., on both sides of 7pp. of note paper; opened by Censor 3867; stamped APO 3 SP 17; postmarked New York Penn Station (no date); postage due 4¢. On envelope: "Pages 1-2-5-6 are missing"; probably removed by censor.

September 8th, 1917 [Saturday]*
Bridgeport, Conn.
My dear Father,

Enclosed you will find a very important letter from Uncle Ralph. What do you think of it? Mother decided that it would be best to accept so I am going to return to college in three weeks altho I am sort of "squeamish" about having people pay my way places. However, the way he put it is that it is for the cause for which you are in France seems to make it alright. At any rate it is certainly mighty fine of Uncle Ralph to make that offer.

I expect to stop work two weeks from today and then have the company pay my expenses to Boston as they did last year. I expect to stay up they for a few days and then return in time to register for college on the 27th.

Things have been going along very smoothly in Bridgeport although the company will be pretty hard hit by the draft and by fellows going back to school.

I think that be the end of the summer I will have saved about forty dollars which isn't so bad considering that I paid a premium of twelve dollars on my life insurance policy in favor of the Fraternity, paid Mother twelve dollars that she lent me to cover my fare up here, etc., bought a rain coat and a hat, and am going to buy a pair of shoes.

Over Labor Day a man whom I have met in Bridgeport took me down to New York. We went down Sunday afternoon, had dinner and took a bus ride and then went to his home in East Orange [N.J.] to spend the night. The next day we went sight seeing and I think that we saw just about as much as could be seen in a single day. I am certainly much impressed by the size of the city. I did not realize that it was as big as it is.

I really don't know of anything else that would be of interest to you, so Good Bye,

[s] Warren

[Enclosure]

August 22nd, 1917†
Allerton, Mass.[53]
My dear Mabel: –

Catherine showed me your letter of the 21st and I note you are planning not to have Warren return to college this year, but intend to have him return later. I presume the decision was made on account of the expense which you cannot afford while Will is away. It would be too bad to break into his course by waiting a year and I would be very glad, if agreeable, to be allowed to pay his expenses for the 1917-1918 term. I presume $600.00 should cover his tuition and expense of living at home. I don't mean to loan you the money, but to let it go as a donation to the cause that Will is in France for. I can send you the tuition as needed and $50.00 per month during the term for expenses, and you can give such part as he needs to him.

If this suggestion meets your approval, please let me know at your convenience, and you may tell Warren about it or not, as you think best.

You have doubtless heard that Jesse has been drafted and accepted. I think it (the experience) will do him good, although Margaret feels very badly about it of course.

I am glad to hear that Dorothy got a scholarship and that she will continue her work this year at school. With love to all, I am,

Your affectionate brother,
Ralph L. Warren

----

* WTS from his son Warren, TL [Typed Letter], 1p., enclosing ML to MWS from her brother Ralph, 3pp. on Allerton, Mass., letterhead. [s] = signed

† MWS from her brother Ralph Warren, ML, 3pp. on 1p. of folded personal letter head, enclosed with letter of Sept. 8 to BWS.

9/9/1917 [Sunday]*
Le Treport, France[54]
My dear Mabel,

I am sorry that I must start this letter with another scolding. Word comes today that one of my letters was read at the Women's meeting, whatever you call it, and that Dr. S. says so & so. Now, once again please stop handing my letters around and causing me much embarrassment and trouble. I am misquoted of course, and I told you in the last letter of one of the dangers of your very unwise habit, developed of course in my absence. Geo. Warren & other Warrens are as you know absolutely unsafe and you have by this time cut out that section of correspondents. So be good and keep me out of trouble instead of unauthorizingly getting me into it.

After a week of no mail, yours of Aug 22nd came today. You wrote bitterly of not getting letters. I am sorry, and am afraid that many of my letters have been captured by censors or lost in some way, as you have failed to mentioned many things that I have written about. True, my last letter to you was one week ago today. Several times during the week, I should have written, but thought every morning would bring letters from you, in which case I would know more about it. Of course, I can't do them every day. I think not. I do try to write twice a week.

Who said that Ed was busy? He could be very busy but as I have told you before Ed is Ed. A smile goes a long way. Sorry also that you do not like my letters. It's this way: -- I am not interested in myself, but am intensely interested in my wife and family, so when I start to write, I forget and think of home, etc. Won't do it ever again.

About the middle of the week, Queen Mary's Guild arrived. "Among other things I got," 5 lbs. Domino Sugar, without a card, and several lbs. of Hugler's Sweet Chocolate with a card – "Molly C." Now that was very nice of Molly C. and I sure appreciate it, but I did not receive her address and so cannot write my thanks. Please call up Molly and tell her how pleased I was and that I will write to her just as soon as I know her address. Turnbull I think sent the sugar. If so, thank him. I wrote to him some time ago, thanking him for sending it, but then it had not arrived. My room mate received a complete grocery store, including Telly tea, canned goods of every description, and most everything that is on or is made in the middle of the Atlantic. Poor things. I just knew that when they left here minutes ago, they hoped to be eaten in America. Such a disappointment. Also tobacco, cigars (*bien!*). It was a complete grocery & provision store. What will the Christmas box be?

Oh, I forgot. I must only write about me. Well then, this morn's mail also brought me two nice winter shirts and two nice winter drawers. They are fine and I will look simply beautiful in them. Thank you ever so much.

Speaking about me, our reinforcements arrived Friday night under command of Lt. Harry Wilmer. What in the world we are going to do with 8 or 10 more doctors, I do not know, but here they are. The nurses, as you probably also know, were left in New York. They and the Enlisted men were badly needed. I saw Ralph Straub. He seemed happy and glad he is alive. Working hard. Charlie & I looked him up promptly; Lt. Edwin with his handkerchief up his sleeve & was not very prompt in greeting him. Henry LeBoutellier I have not yet run across; but will find him in a day or so. Ralph S. was to lunch. He would make excellent shooting for the Boche. But with us his chances are better.

The Boche by the way seem to have issued a General Order to bomb hospitals. It will doubtless reach us in time. He is pretty busy and may not get around to us for some time, but he is on the way. Gen. Pershing's order to wear Sam Browne belts, campaign hats, and tailored cams has at last reached us. We have been the only officers over here who have not been wearing Sam Brownes. Our reason was that the Sam Browne was not a part of the American uniform, and we did not want to ape the English. I say we, some of our young men have been dying to wear a Sam Browne, but our colonels said we would not do it until an order came. We saw it coming all right, and it arrived today, so Dr. Ed's, etc, are smiling a little broader. I bought one last week but have not found it yet. I still doubt if it is a part of our uniform, but Gen. P. is boss on this side of the water, and he evidently likes the belt. I do like my cam, not with the blink for it. I don't like my campaign hat, but is on my head for it. Letter from Holloway this morning; he is off active duty and is going on a vacation. From what he says he needs it. Oh, I forgot again; I must not write about things at home. To return to me, I don't like the Army, but I am dandy at doing things I don't like, so am making a most Exemplary Soldier. Everybody says we are staying her for a very long time. By the way, Washington

---

* WTS to MWS, ML, 12pp., on both sides of 6pp. of note paper; opened by Censor 3880; postmarked APO 11 SP 17; postmarked New York Sep 27 1917; postage due 4¢.

has not paid a bit of attention to this since we left. Now our High Commissions are being issued every day but Col. DeLaney cannot get them to give us our promised Commissions. He asked me to write to Furbusch, and say that there was a great deal of dissatisfaction among some of our men, myself included, who were promised Captain's commissions, and are still after 4 mos carried as 1st Lts. and denied pay. He is sore and so are we, but the fellows in Washington won't help us. – I did not write to Furbusch, because I do not want to write, but it is a darn shame. DeLaney has authority to carry us as Captains, but that does not get the pay. The telegram you got probably had nothing to do with a Captain's commission, because I got a letter dated August 6th from the Q.M. [Quarter Master] saying that there was no record in the Adjutant's office that I had ever been promoted. If you see Furbusch, you might tell him that Dr. Harte & Col. DeLaney both think that the Adjutant General might make good the promise to commission certain ones of us as Captains. Of course, we are now here and out of the way, and can't get at them to talk. It's not fair. The fellows at home are Majors. Geo. Müller – a major ordered to stay home and do his own work!!! By this time, I here would think that you were back in what used to be my house, there on 20th. If you know. Of course, I am thinking this over here in France, so it pertains personally to me – of great interest therefore.

Will try to write you in a few days, say about Wednesday. Don't think for one moment that I am not thinking of you just because the mails get banged up. I am just as homesick and Mabel sick as Mabel is Billy sick, but the only thing I can do is to carry on.

Good by – Much love to you and the children.

B.

P.S. Have put picture postals in all of your letters. Do you get them?

Wm. T. Shoemaker
New address = No. 16 (Phila. U.S.A.) / General Hospital / France.

Sept. 12th 1917 [Wednesday]*
Le Treport, France[55]
My dear Mabel,

True to my little promise, I will write you a short letter. Charlie just got a cable saying "Anxious, Wire." What have you heard? We are alright as usual and hope to continue so. The question has the wind up, trying to assure "Mother" that he is comfortable, happy & useful. She is constantly playing her usual part. He has tried hard to shore her up, but no use. Newspaper articles are probably responsible for her worries. Perhaps Geo. Warren has been running his publicity bureau too hard. You know of course what to believe. If you saw an Associated Press dispatch – official – you can I think believe it. But beware of publicity letters from irresponsible persons.

I'm generally against "One Who Knows," and "Who has been there"; "True Americans," etc. They are all like this. We have caught one of these liars through our censorship and will get more of them. They have published all kinds of false information about us and our surroundings & work. You read enough to disregard them no matter how well written & authentic they seem to be. Such rumors and stuff will to be sure make excellent food for the "Lamentations of Mary." [Mary = Charles Jack's wife]

If you go to war, make the worst of it and see all the Hell that it in it!!! It is the only way to be thoroughly unhappy, and you can't properly enjoy the situation and keep a few benefits from it unless you are unhappy. This is not my philosophy you understand, not merely an exercise in sarcasm to keep me from getting rusty. You know, I am afraid sometimes that when I get home, I will be so good and different that you won't love me anymore. So much for that which I am afraid won't interest you.

The War, I can't talk about beyond intimating that it is still in progress. The Boche is still peeved and is throwing all sorts of stuff at us. He seems to have it in for Americans, and especially for American doctors. Why this should be I can't know, for American doctors are surely good doctors. He is mad at hospitals, but really we can't go without them. He has them, why shouldn't we. Most unremarkable function. I think he can't be well. I hope he is not. You of course know about young Mr. Quillen, Fetterbein, Dr. Osler's son and others. I am particularly sorry for

* WTS to MWS, ML, 6pp., on both sides of 3pp. of note paper; opened by Censor 3500; faintly postmarked APO 13 SP.

Mr. Quillen for you remember how he lost his other son 2 yrs. ago in the Bar Harbor express disaster. His troubles this time are not so bad, but still bad enough. Sir Wm. Osler's son was an only one I believe. Sir Wm. dined with us at Cliveden. He is doing his part in uniform like every man in England. Fancy Sir Harvey in uniform trying to do anything not agreeable to Sir Harvey! "I can't. I have to see a patient."

This also does not interest you. I must try again. I have it, listen to this: I broke three pairs of my finest glasses in about a week. I took them to the optician in this place (in France). He sent them to Paris, also in France, and in about 6 weeks, I received them. Now take a deep breath - $25. Ouch! I hadn't $25. That's a month's pay. So different from the way they do at home. Ferguson just says, "that's alright." I was never afraid to break my glasses at home – here, I am scared to death. And $25. is a lot just to look through, and when you haven't it, the feeling is terrible. As somebody has said, I can't recall his name just now, "This is an awful war." Who was it who said that? I don't want to refer to trivial matters at home, but I forgot to congratulate you upon your financing. You are a wonder to pay up and have a balance. I never did it in my life. I am very proud of you.

Did you get any money from Geisinger? You have not said, and it is funny, because he must have had some of my work. Langdon, I last heard was told not to join the Service if he would promise to stay home and examine ears for Major Jones. He promised to stay home all right, and I am sure will keep his promise. Also heard from Wilmer that Whiting got his business at Germantown and was catching it. He & Ross had a bad fight and now scarcely speak. Very interesting. Tell Page, he is a slacker.

Goodbye. Will write very soon. Much love, B.

Sunday, Sept. 16, 1917*
[Le Trepot, France]
My dear Mabel,

Your last letter was very nice because you took back your scolding. Am glad that some of my letters are coming through, but rather think that others have been lost. Ralph [Warren]'s proposition is fine and as usual he is a Prince, but I don't feel like accepting it in view of the fact that I owe him so much money and have not even kept the interest up. I was hoping that I would be able to carry the thing through myself but am not that all sure. While I still stick steadfastly to my personal opinion about the end of this war, it is very difficult to do with everyone against you. Perhaps you might compromise and accept this time only. We certainly should be able to keep Warren, don't you think? I have not received Ralph's letter yet, so have not written him. As you have by this time, I am anxious for him to return to College if possible, and am entirely satisfied with any final decision which you might make.

I am becoming fed up with this place and more and more fixed in my opinions of "the Army." It never was and never will be the life for me. Which of course increases my sacrifice. The powers that be are commencing to make the Army less popular, instead of more popular for the Americans who came across at the Government's first call for assistance. Understand, that I make no criticism of the Army when I say I do not like it. I wonder how anyone can select it as a profession, but am glad that so many can, otherwise we would not have an Army, and first now we need a big one. What surprises me is that more is not done to make it popular.

We are not busy and are not looking for our kind of business for several reasons. When we are busy, we forget our troubles and feel worthwhile. A winter here is going to be very, very tiresome, and winter is coming; – it is at least in sight. We have perhaps two months to run, and then; – let us hope we will not be here. Hammer away at the practice, and keep as much of it in hand has as possible, for from now on, it is very important. During the summer I had a right to be away, and it might have been expected in normal times, but now, I should be at home.

Now I will tell you something funny. Ralph Straub is a chimney sweep, and when sweeping chimneys, certainly looks the part. He does it very well too. I asked him how he was selected; if he had had previous experience, etc. No, the Sergeant asked him what he did in civil life. He was a lawyer. The Sergeant wanted a lawyer about his size to sweep chimneys, so he got the job. He is perfectly well pleased with his job. By the way, I recommended him for a Private 1st Class - $3.00 per month more pay. He will get it too. It is a pitty that Ed. cannot think of anybody but Ed. I do not think he is worried much about his brother-in-law. Johnny LeBoutillier – is in the laboratory. He seems to like it. Charlie is much depressed and chafing under the "Army" as keeping under our C.O.

---

* WTS to MWS, 8pp. on 4pp. of note paper; opened by Censor 3378; APO 17 SP 17; New York [n.d.] 1917, 2¢ postage due.

I have refrained from any criticism of the personnel of our organization and must continue to refrain, but will have some interesting descriptions when I get home.

I am enclosing a letter from Miss Miles. I am much afraid that you did not use your usual tact. At all events, she takes exception to your offers. I have not answered the letter, and probably will not. It is alright. But I do not quite like it, so I will forget it. Went to the usual Sunday afternoon tea given by the Nurses, together with small talk for 20 minutes or so and then went down to Treport to dinner. Always go out Sunday night to dinner as we get here a cold hand out which we have to hand out ourselves. It is a good night to dine out. The *Literary Digest* comes rather regularly. I am very glad to have it. It tells me so much of interest. I think it is awfully good.

Tell Dorothy that she gets the next letter, which will probably come along with this. Speaking of boats, you know of course that we have no acknowledgment of the sailings. If Ed got his letter aboard it was purely accidental. I promised that you also are not told of the sailing dates. If you get a chance, tell Furbush about the way we have been neglected, in regard to commissions. It is a mean bunch in Washington; tell them I said so.

They are boosting up the stay-at-homes & neglecting us because we can't get at them. Well good night. Will write again Wednesday if I can, and I guess I can. Much love to you and the children.

B.

[Letter enclosed]

August 6, 1917 [Monday]*
1912 So. 15th St., Philadelphia, Pa.
My dear Dr. Shoemaker,

I am afraid this is going to be a long story, but you can read it in installments if it is too much to wade through all at one time.

Doctor, I really can't explain adequately about my changing my position as I could if I could talk to you personally. However, I want to tell you all about it. One day, Dr. Peters called me up and asked me if I was to stay with you indefinitely – until you returned; if not, he had a permanent position for me. I told him I believed I was to stay, but that I would ask Mrs. Shoemaker, which I did, explaining, that what if circumstances later on would necessitate my services being dispensed with, that I felt it best to accept Dr. Peters' offer while the opportunity was available, as I would not like to have to look for a position maybe in four or five months from now. Mrs. Shoemaker said she had expected to keep me, but that in September she had sort of planned not to take Bessie back if I would help her with the breakfast dishes and downstairs dusting. I would have liked very much to have stayed and been with you but I do not feel that I want to lose the little I know about stenography and I do not want to earn my living doing maid's work. When you left, I understood you to say Mrs. Shoemaker would keep me busy, but I did not understand it was to be that kind of work. Of course, I realize that there isn't any clerical work to speak of at the office now, and as the days pass, there will be less and I could not expect you to keep me just to do the little bit of office work now necessary for that would be a losing proposition for you financially and for me mentally. Dr. Peter is paying me more money, and that of course if very welcome, but I would not have considered his proposition if you had been home. I feel Mrs. Shoemaker has done the best she could for me, but that under the circumstances there was only this one thing for me to do. I hope you will understand.

I have kept account of everything that has transpired at the office and have told Mrs. Shoemaker that any time I can do anything for you I will be please to do it. I wanted so much to be present when you returned, and I will miss your asking me if I'm thirsty and all the other little things that made it a pleasure for me to come each morning. Today began the reign at Dr. Peters'. It was hard for me. He has me in the room while he examines the patients and dictates the entire history while examining. It was also very hot and we were very busy. At the end of the day, he said, don't get worried, you won't be rushed until about October, as business has fallen off. If this heat keeps up, I'll just be a memory by that time.

I went in to Street Linder & Propert's today to have my spectacles adjusted and Mr. Linder inquired about you and wanted to know how you were. He said to be sure to tell you he sent his regards. I was down to the shore with Florence MacDermott, but it was so unbearably hot we did not enjoy the week, except while bathing.

---

* WTS from [Miss] Ray P. Miles, ML, 8pp. on 4pp. 6.5x5in. notepaper, enclosed with WTS's letter of September 16.

There were just as many people in the water at night during the extremely hot nights as during the day. They have search lights on the beach and it was quite light. Women were sitting on the beach knitting. It appears there isn't any situation in which it is not possible to knit. I trust you are not suffering with ennui by this time, and I hope very much that you are well and safe, and that I may come around to see you when you come home.

With many thanks for all your kindness to me and best wishes, I am,

Very respectfully,

Ray P. Miles.

Sept. 16, 1917 [Sunday]*
Philadelphia, Penna.
Dear Father:

I have not written to you for a long time. I must admit with nothing as an excuse. I finished camp with the rest on the thirtieth of August and went home with Jack, John, and Aunt Greta and Uncle Owen in the Albany boat – it was a fine trip and we had lots of fun.

We are not going to starve this winter, mother has put up a hundred and forty jars most of them quarts and hasn't stopt yet because Ted came from Allen Anderson's with a lot of pears to spice.

There isn't much doing around here at present but two nights about ten o'clock the back of the Aldine Laundry on Samson started to burn but didn't get far enough to do damage. An engine was at Thirtieth and Chestnut, in front of our house and Twentieth and Samson not connected but all ready.

Doctor Ross was to report for duty today and sail either to-day or to-morrow.

I am enclosing a pocket moving picture machine which might give on a laugh. I must close for lack of material (not paper). Your loving son,

Bob.

Sept. 18, 1917 [Tuesday]†
[Le Treport, France][56]
My dear Dorothy,

You have written me several nice letters, and I should of course have written in return, but your Mother keeps me so busy that I could not get around to you.

So far as I can judge, you have had a pretty good Summer for war times, and are now at home getting ready for another Scholarship next year. You will of course find your third year more interesting than the previous two. In a letter days ago you remembered a letter written me about Mrs. Hoffman. He ([Private] William) her husband, was at that time our striker and was in my room at the time I read your letter. I read that part to him and he was much pleased. He is a nice boy, but like all very nice boys needs a spanking. He is no longer our striker, having been court marshaled [sic] for "high crimes," convicted, and punished (not by us). You had better not talk, because he might not want little wifey to hear it. He is still happy as most little darn fools are.

I hunted up a chemist that you spoke of and gave him your regards. I have forgotten his name, but would know him I think by his teeth which are particularly <u>fine</u>. Charles Jack was very much impressed by them. This chap was getting along alright.

Your description of the Judge at Cape May coincides with my opinion of himself. A political gas bag is the best I can hand him. I fancy that a number of your friends will be coming over here as officers and privates. I will not be able to see them unless they come right here, because we are not allowed to travel around. I could go anywhere if I walked and got back the same day, but I am not a fast nor a good walker and of course could not get far. Trains are out of the question. It <u>would</u> probably take a miracle to get a permit to go to Paris for example and then I probably could not get said permit. We are all very hopeful of our American troops here, and look for a good accounting when they get back of the Boche.

---

* WTS from his son Bob, ML, 2pp. on 1p. of notepaper; envelope from 109 South Twentieth St.; received 10/1/17.

† WTS to his daughter Dorothy, ML, 4pp. on 2pp. of note paper; opened by Censor 6079; APO 19 SP 17; postage due 2¢.

I see in today's paper that the Boche has offered a £5. reward for the first American prisoner. It's tough to be thought so cheap, but then £5. these hard times are better than nothing. An American prisoner will be much harder to get than a Hun. They are easy. They just come running over the line with their hands up. They are so glad to be prisoners. I would be too if I were up against this artillery fire of the Allies. Up to here was yesterday – from here on is today. You said something about your rooms being spoiled by rain water coming in from the bay. Then, if I remember, you were going to pay for repairs out of closely hoarded funds. Do not, so don't – I would pay for your rooms if you can only get your mother to hand it over. I think she will. Try it on. Do you remember playing "Long, long trail . . ." Don't forget it, for it is the great song over here with the [British] Tommies. I am very fond of it & will want it when I get back. When I do is becoming very hazy, and indistinct, and begins to look like never. They have a glib way of taking over here, about 2 yrs., 3 yrs., etc. Then of course, it takes a year to get home after the war stops. Very encouraging, but I should worry. Write again when you get time, and don't work too hard at school work.

Much love,

From Dad

Wednesday, Sept. 19th, 1917*
[Le Treport, France]
My dear Mabel,

I forgot to tell you that numbering letters was not a success from this end of the line. I numbered two and then stopped because it is difficult to keep track of the numbers. You also had trouble and were promptly mixed up. So do not worry about the numbers. The letter you get will be the last one to date.

It is a perfectly beautiful warm Sept. day. Nothing to do with it however, but to look at it. The hospital is slacker than it has ever been. Have not had a convoy for 10 days or more, and will not get one for about two weeks.

It is very discouraging to have so little to do. Everybody feels it. I cannot help but think how much more useful I would be at home. But I can't get there so will just have to carry on here and wait for "business." I do not hear much from the front, but that I think that things up there are also very quiet. Comparatively quiet I mean. Strange to say, you get more & quicker news from Flanders than we do. This I know from reading American papers. They describe things we never heard of. But when we open up again, we will probably have more to do then we want.

Am still taking my French lessons, but cannot do much yet. In about a year, I will be quite proficient. You know, I never cared much for French, which makes it harder. However, if I had been in a French community, I think I would have had very little trouble. I take 3 lessons a week, each lesson costing 34 cts., not very expensive.

If you are now at home, don't forget to keep the auto at the Lankenau hospital, and remember to keep it in repair, otherwise it won't be worth much. You can run it a little if you are careful.

Am anxious to hear of your final arrangements á re schools for the children. Hope you can return Bob to Penn Charter. Excuse me for meddling in your affairs at home, but you know I am still interested. I am enclosing a letter to Jimmy, which please forward. You have not said a word about her. You have also not mentioned my retired business man's brother of Newark N.J. Is he just retired?

This letter is short, not more frequent. You know, you said I could write short ones if I did it more often.

Goodbye. Much love. The postal cards are for Jimmy. B.

Tuesday, Sept. 23rd, 1917†
[Le Treport, France]
My dear Mabel,

Letter 33 arrived but 32 of which you spoke as containing so much information has not yet come across. Only one letter last week from home and one very good letter from Creighton. He certainly is faithful to me, and is working hard over my patients. He tells me that he has quite a number of new ones, also a lot of nurses etc. from the German [Hospital] and elsewhere. I am glad that the nurses go where I want them to go and not to some of my stay-at-home confreres like Shumway & Lautler. Of course, Creighton is staying home, but I am glad of that.

---

* WTS to MWS, ML, 4pp. on 2pp. of note paper; passed by G.B. censor; stamped Army Post Office 20 SP 17, 4¢ postage due.
† WTS to MWS, ML, 6pp. on 3pp. of note paper; passed by G.B. censor; opened by Censor 3345; Army Post Office 24 SP 17.

Well at last my Captain's Commission has come, and has been accepted today by cable. DeLaney received it this morning. Taylor's, Packard's and mine. You would get captain's pay, and depending about the date, a certain dividend of back pay, I presume. I am glad it came, because it was annoying not to have it even though "I was going to get it."

Just came back from Treport with Charlie, John Gibbon, Ned Hodge & Bill Cadwallader & myself. I drove. Had a very nice time. We were the guests of Hodge & Cadwallader. Am now quite busy, as you know from watching the Flanders front. I cannot tell you much about it, but it is bad enough. The present operation is said to be the greatest of the entire war. G.B. is prepared, and something is going to break. It won't be G.B. I am still firmly believing in the near future, not as I told you last time I wrote – I got little encouragement. However, I cannot but think that I see the End. I hope so. Flick & Drayton got back from the Casualty Clearing Co. last week. Newlin & Packard are also there but will be here this week. I am very glad, for they were right there this long, and with the team is going with Captain Paul Austin who arrived a few days ago with 30 new nurses. We now have plenty of nurses & Enlisted men, and about now as many doctors as we need. But for war, one week may be heavy.

Your comments on our nephews are interesting. Can't quite understand O'Connor, Linus & Herbert. It must seem that they above all the rest might contribute. What does brother Josh think about it? I suppose he has a very good reason. I read in today's paper of curious things happening in Philadelphia and wondered about this country. The "Daily Mail" has Mayor Smith arrested for murder conspiracy. Now that's a funny one to put over on Mayor Smith, but as he is, I hardly think he would tell anybody. Then it speaks of spies, labor organizers, pro-German papers and the etc. Much trouble brewing, I would say. Byron, the old blather's kid – has joined the belligerents, I believe. It disgusts one to have to read of Byron in the papers over here. If then even deserves daily care, he is it. And for La Follette, etc., they are smelling else that I can't write – XXX.

Monday, 24th. Now a few finishing touches before lunch and before the mail comes. About 2 o'clock. Nothing has happened since. I have been busy in my ward undressing with some interesting & very bad cases.

The weather is wonderful. Such a relief from the days & days of cold rain & high winds that we have been having. I am enclosing a little picture of me that I intended to send to Jane, but I neglected to put it in her letter. [not preserved] She will like it, and won't mind my sending it. Please send it, or another to her. Also, have you thanked Mollie C. for the Hugler's Chocolats.

If you have not gotten my letters, you may not know what I am talking about. I must now stop. One thing more. It worries me a great deal. Do not work too hard. Go out a lot and enjoy yourself. Do not stay home & try to do the work of several. Please remember this, as it is important. When I get home, I want you fresh as a daisy, and of course, glad to see me. Love. Move love to all,

B.

Wm. T. Shoemaker.

Thursday, Sept. 27, 1917*
Le Treport, France[57]
My dear Mabel,

Your letter containing clippings from newspapers came this morning and I am much amazed. I thought I made it clear that my letters were not for publication. You must impress this upon your brother George, and all. What he has done is illegal and is very apt to get me into a lot of trouble. Already here there have been two court martials from similar indiscretions. Mr. Dunham, if that is his name, has no right to do what he is doing, and George knows better than to help him do it.

Please discontinue your circulating library of correspondence before you get into trouble. Remember that we are at war, and that a strict censorship prevails, and that it is your duty and my duty to conform to the material regulations and requirements. Now, having relieved myself, I trust that for the idea, and behave just as you would, were I home to keep you properly informed. Now in your last letter you scolded me for not writing interesting stuff, and I thought that that letter was particularly good. Things have gotten with such a routine here, that it is very difficult to spring anything new. Of course, I do not want to tell you what we have to eat & drink.

---

* WTS to MWS, ML, 6pp. on 3pp. of note paper; opened by Censor 3840; APO 28 SP 1917; postage due 4¢.

Our reinforcements have not yet arrived although they are on the list. I think the article published in the *Bulletin* was an outrage. You may have seen it. Not a single officer here is rewarded but Cousin [Charles] Jack, and he would not be if Ed was worth mentioning. Our nurses are very much over worked, and so are our Enlisted men. The paper spoke of the doctors giving way under the strain of many for so many patients. I do not know what that has to do. Of course, we all do our work and get away with it, but with nothing else to do, it's hard when I am already an officer, my best is generally an officer, I know, but a run of 24-36 hrs. without much done. I strike bad nights as a rule, and am up all through, but then that only happens once in 2 weeks, and I enjoy it. Other fellows draw less days; no convoys or examinations, no deaths, no secondary hemorrhages to Wards, etc. Every time, I have a time or two – or more. They like to die at 2-3-4-5-6 o'clock; and the Orderly Officer has to pronounce them dead and get them sent out at once. The convoy likes to arrive at 3 or 4 A.M. One thing about Army life here is that the help is just as good as directed. We do what is to be done <u>now</u>, and it is of no consequence, what time now happens to be. So you can easily think that our men of Base 10 are working themselves to death.

Report here has it that Harry Wilmer was court marshalled [martialed] <u>for</u> refusing to go to Camp Oglethorpe. Is that true?

The *Literary Digest* arrived same mail as your letter telling of it. I am glad to have it, for I always <u>liked it</u>. I do not believe that we will continue here, but no one knows. It was so cold here last winter that bandages froze to the patients. There was a shortage of coal, and everybody here at that time says Never again!

The enclosed clipping [not preserved] will interest Mr. Harvey. See that he gets it. I caught up with the *Daily Mail* this afternoon. Now for a few closing details – I am going to the Tea given by the Nurses (every Friday afternoon) after which I will walk to Eu – a near-by town, have dinner there, return, climb the 350 steps, because the *tram maritime* stops at 8:30 o'clock; censor letters and go to bed. Get up tomorrow at 7:40 and carry on as usual. Good bye, wish you were with me.

Much love,

<u>B.</u>

<u>Not for publication.</u>

October 1917

Le Treport – Passchendaele Ridge in Flanders

"It would seem that the scheme is to get it at any cost."

Monday, Oct. 1, 1917*
Philadelphia, Penna.[58]
My dear Father –

School started a couple of weeks ago and things are now running in full swing. I've been especially busy this year as every afternoon has been taken up by foot-ball and I've had to spend a good deal of time getting the basketball schedule fixed up, as it wasn't completed before I left school last spring. I don't know what my chances for making the ~~basket~~ foot-ball team are. At present I've been playing with the first team part of the time and the second the rest of the time.

About fifty Highlanders from the trenches are in the city for a couple of days doing recruiting. They have their kilts and the sure do look fine. Also there are a number of English officers here for the same purpose.

Did you know that "Jimmy" Moffatt is in the Army doing some work down at Washington? I believe he rates as a Captain. Dr. Dunn just lately got a promotion and he is now a Commander. Isn't that pretty high for a medical man in the Navy? Harry Dunn and Frank MacFarland are both down at the "Army and Navy Prep. School" in Wash. Harry of course was prepping for the Navy and Mac for West Point. Isn't that pretty good?

Things at home are surely going finely. Mother is doing the cooking and of course we help her all we can. She seems to do an awful lot of work but doesn't get tired out I suppose because she gets to bed early. She surely is wonderful.

Did you get the pack of Camels sent in with the eye droppers? Amos sent them. Did he put his name on them? Today was the first day that seemed at all like fall. Up until this time it has been quite hot. I think I'll quit as it's getting late and I can barely keep my eyes open.

Remember me to Ed and Charlie Jack as well as Hudson Chapman and Frank Chalk if you get a chance.

Love,
Ted.

October 2nd, 1917 [Tuesday]†
[Le Treport, France][59]
My dear Mabel,

No. 16 (Philadelphia U.S.A.) General Hospital. B.E.F. Important notice, as "Orders" say, letters not so directed will not be delivered.

I have been trying for 8 days to write to you, but have been too busy. We are very busy on the Surgical side. My Ward in fact, and I have about 10 that I put outside into tents. So don't blame me. First, I must tell you that the Eye bottles from Wall & Oates, electrical supplies from Linit, Linit & Propert, eye droppers from Baer and Lutereich, bandages from Mrs. Aldren, all arrived last week in sound order. Was very glad to get them as I was running short.

Will write to Mrs. A, W&O, L.L.P. The last Army Guild Box was broken into and much of the contents stolen. Some of our men received empty papers addressed to them. War often times does not seem to have much influence on the meanness of some natural home thieves. There are also plenty of thieves over here. French & Belgian, I fancy – Civil participation.

---

* WTS from his son Ted; ML, 3pp. on 1p. of folded watermarked note paper. Received 27/10/17.

† WTS to MWS, 10pp. on 5pp. of note paper, double-sided; opened by Censor 3159; postage due 2¢.

It is a pity that the mean thieves cannot be collected and put in the front line trenches. Of course, they would not be of any more use than to draw ammunition. Lamenting Mary writes Charlie that you are anxious about our safety. I do not think you need be, so long we are here. Of course, if we move up, it gets more dangerous very quickly, but we are not moving up. A Boche plane went over us yesterday, but he did not drop anything. So don't worry about that. Charlie is getting sick of the daily lamentation. He has spent hours writing cheerful news, and attempting to change Medea's point of view, but what's the use. I have had patients here with more of their brains shot away, than some people have. Sent one to Blighty this morning. About the cable to Rachel; we can't cable any more, unless such message is a military necessity – The privilege was abused, and of course it was naturally stopped. So many people cabling "Howdydo" a couple of time as week, loaded the wires so that it was difficult to send messages of importance.

I don't know how Charlie is going to square himself with Medea. He will be quite angry I am sure. When she got a letter, she would cable that she got it, and when she did not get a letter, she would cable that she did not get it. So, give Rachel my love & blessing and send her our present. Tell her also, that getting married will not excuse service for the great big government. Mr. Matthews might not be able to clean chimneys as well as some others, but surely, he can do something. I am glad Geo. Ross got off as I thought he did. I wonder where he is going? As yet, there has been very little opportunity for Naval service. The boats will probably get in before the Game is finished; they generally do. Geo. has it on Press & Henry alright, and how Mrs. R. will rub it in. Geo. would look swell in a Naval uniform – I can just see his little beady eyes glistening from beneath the visor of his cap. Don't forget to get me his picture.

A note from Martha (Mrs. Edwin) yesterday informed me that Ned Langdon was in a costume which she says probably came from Fran Horns. I understand he was not going in. Speaking of Martha, she told Medea that Harvey was a very mean, tight, white man – or to that effect. He has not let up on the cent or the rent. I always knew he was mean, but that seems too mean. You know, somethings are too mean. Well, I suppose "he can't," which in his case means of course a wonderful credit of a very fine character, because, if he could, just think of all the fine things he would do. Martha is also working hard and must be getting tired of it. It's no easy job to keep Ed.

I am glad Art McGeorge is getting married again. He must have been lonely. It is nice of him to take care of his father who must also be lonely without Minnie.

By the way, the letters you wrote to delinquent patients are all right, but you are outlawed.

Marion Buchard's accent was given to Pam Hamline – ask him about it. And get Paul Hamlin to help you with others if need be. Mrs. Carrick is very nice; she is hard up – paid part of her bill once, and I told her to pay the balance at her convenience. Miss Walsh is a brat, and a bat. You can't reach her. Geo. Clayton is useless – very old account. Heidelbaugh should pay. Adie Reagan is Mr. William's sister-in-law. She pulled off that old trick of paying $10 on a $15 bill, and then sitting perfectly tight; don't bother about her. She is mean – that kind. Emil Müller has shirked me before. He is the gen that Anna Raymond used to work for. Always a hard luck story but no money. I am afraid you tackled a hard proposition. If you can get money from this bunch, you have me beaten.

Your report of Betty O'Connor is interesting. I suppose she is a Romanist. I look for much fuss and feathers and another separation in the family. Dangerous to marry a girl that you just meet with entirely different interests, family associations, etc. It won't last in my opinion, beyond the time when Herbert wants to do one thing and Betty want to do another thing. Then the fire will commence.

I am so glad that you got such excellent terms from Penn Charter & Miss Hill. It is very good of both schools to be so considerate. Received Warren's letter enclosing Ralph's. Have not yet written to Ralph but will do so. You say everybody has been good to you. That is one of my greatest pleasures in being here. If they were not good to you, I would not stay but would come home and be good to you myself. They just must be good to you, so that is all there is about it.

It means so much to me, when people are good to you, that I can't have it any other way.
Some people are about what I would expect – not a profusion of brain cells.

Now I have rambled along trying to make up for lost time. Nothing of special interest, but first chatting. Please do not read to the bereft ones at the guild, to the Laundry Schaubs, to George W. and especially do not publish in any magazine, newspaper or periodical. Very much love to you and all the children. Will write soon to Bob, Barbara, Ted & Warren. Have written to Dorothy.

B.

Oct. 7th, 1917 [Sunday]*
[Le Treport, France][60]
My dear Mabel –

Charlie sends his love as he has every letter I have written. One of the most dreary, dismal days you could think of – cold – raining at times in torrents – and wind running about 40 or 50 miles per hour. But here, we stop not for Sunday, rain, wind or anything else. And just now we are very heavy as you probably know. Passchendaele Ridge is the present objective in Flanders – and it would seem that the scheme is to get it at any cost. The price is great, but I believe they will get it, and when they do, I am told that the Boche will have to retire from Belgium & France. Then, the light ahead shines brightly. Let it be so, not Oh, the Slaughter. You see the destruction and suffering and you wonder what can be the meaning of – "God's will be done." There I trained as I am or have been, trying to think of the "upheaval in the Spiritual World." If we have but a "reflection" what must the real be? One million men were said to have gone over the top one day this week. Think of it, if you can. I believe to think of how a week or two of good weather would finish it forever, but the weather as usual has been bad.

So much for the War. Now of things less serious. Thought of Rachel yesterday at the appointed hour, but as I told you could not cable as it was not a military necessity, I think she has the right fellow that will be alright. She is a lovely girl and couldn't possibly get in worry. Ralph [Straub] is making good and seems to be enjoying the work. He at all [times] exudes the proper spirit. I see him sweeping chimneys and carrying stretchers when a convoy comes in, and he always has a smile and a great salute. Charlie is getting him a better job. He is going to bring him in the Dental Surgery, a clerical job, a much better job. Charlie is very kind and thoughtful. Ralph would sweep chimneys for the duration before his brother-in-law would ever think of bettering his condition. It will be a nice job for Ralph, especially as cold weather is upon us, and walking roofs is fearful in this wind & rain.

I forgot to tell you that I am now a real Captain, and have taken my Oath of Office as such, dating from Sept. 23rd. It came through General Pershing, Paris – So don't worry any more about it. It should have been from the start here, but I can't help that. Will send three vouchers – Nov., Dec., & Jan. to Washington tomorrow, on which you will get Captain's pay. The difference for tail end of September I will collect here. It amounts to some $4-$5. See that you get Captain's pay for October. Our allowance has not yet been paid and I am practically without money, but it is coming probably this or next week, and will carry me through to November as I told you it would. After that it will come regularly every month, and will help.

Speaking of money, you are a wonderful financier, to pay all bills and have a balance. This I told you before, but you are getting some wonderful [receipts], I told you again. I was somewhat disappointed at your receipts from my practice. Sixty dollars a month is not enough. You should get $100 or $200. Perhaps the important accounts will increase it. My calculation was that I was worth about 3000 (2400 + field & foreign pay allowance); my practice about 2000 – my income about 1000. Total $6000 – not a bad war wage! If you can keep up the practice, you can get it.

Monday night: A hard day – almost every surgical bed in the place full. Very bad cases, too. Started at 9:30 this morning operating – it is now 9:30 P.M. I have been on the jump all day long. The weather is fierce. Rain & wind as usual; many tents blown down, and our quarters – quietly rocking when the wind starts to take a sprint. Learned today that our good ship [redacted] has been sunk. It came pretty straight this time so I guess it's true. I am very sorry for she has been a good boat. She was washed, found, and probably had a lot of our mail, so you may not get some letters that I wrote to you.

Will have to close; the lights have gone out, and I have nothing but a candle. The Home Team down to the wire, etc. Do not forget to make the Q.M. give you full pay for October.

Good night – Much love to you and Children.
Am not hearing much from the kids – guess they are busy.

B.

----

* WTS to MWS, ML, 8pp. on 4pp. of note paper; opened by Censor 6331; APO 9 OC 1917; postage due 4¢.

Oct. 7 – 1917 [Sunday][*]
Philadelphia, Pa.
Dearest Dad: –

It's perfectly ages since I've written to you but it seems that whenever I intended to sit down and write, something stops me – I either get too sleepy (which is almost the case now) or start playing the piano and get stuck – Two or three times lately I've come up from dinner thinking I'd write and then I start to play a couple of pieces on the piano and play for a couple of hours. You'll be glad to know, though, that I'm "beating the box" quite considerably of late. Not only popular music but I've been trying to renew my classical music with things I used to play, and new ones.

Warren has gotten quite proficient on the ukulele and Ted of course plays the mandolute a lot, so we're quite a team. Well, school has started again and it's mighty interesting. I'm crazier about it than ever! We have such a dandy bunch down there, the "Krazie Krew" or "Big Six" – We sure do have fun! Mr. Copeland is as dandy as ever.

He wants me to put about four of my little sketches in an exhibition of art students' work at Wanamaker's – So think of your daughter exhibiting "hand painted pictures"! See! Here! It'll be sort of fun to see them in an exhibition though. Wish you could see them, they're not awfully good of course, and are very crude, but you'd be interested.

We have "life" in school this year. You can imagine my surprise when I walked into the room and saw a gentleman of color almost *a'la naturel* standing on the platform – I think it's going to be great, though – very interesting and the teacher is great. A couple of weeks ago today, Hadwin Richardson came down from Allentown and brought a friend along with him – I had Olive in and we had quite a good time – Took them to the zoo in the P.M. – which they seemed to enjoy.

Yesterday was Rachel's wedding – Mother is writing at the same time I'm writing, so I guess you'll get all the details from her. Anyway, it was a lovely wedding and Rachel was sweet. Well, Ted was playing football so couldn't go – He got home before we did and about 5:30, Walter Beadle (remember, he was a leader up at Becket the first time we were up there and then he was here to dinner before the dance) – called up and asked if I were home – Ted, not thinking when I would be home said that we had all gone to a wedding and that he was going out and that he was sorry we wouldn't be able to see him. He said he had his brother with him and wanted to bring him out. Well, at 6:30 we got in. When I heard about it, I was simply furious! There I was, all dressed up and no place to go with a lovely Saturday evening ahead of me, when I might have had something to do. Then Mother had a brilliant thought – She suggested that he had probably called Olive up and so I called her up and sure enough he had – So I went out there and spent the night. It's a good thing she had another girl too, because by the time they arrived they had collected another boy and there were three of them. Herbert Beadle was drafted and is at Camp Dix, Wrightstown, N.J., and Walter Beadle expects to go there very soon – He was drafted, too – He and the other boy goes to Penn and is from Charleston, S.C.

We had a lot of fun with both of them last night and in a crazy mood when they asked us if they could come to our Sunday School class to-day, we said "Sure!" Well, afterwards, when we found that they were really going to take us up, we were petrified! You see they were all six-footers and they said they were going to sit in the little chairs in the front row! Imagine! Well, we were shaking in our boots and darned if they didn't come! Not, Walter, as he had gone to Atlantic City – but the other two! However, they were good sports and didn't embarrass us in any way – They went into Mr. Harvey's class if you please, and didn't pull of anything crazy as we were afraid they'd do.

Then they went to church with us and we invited them here to dinner – they were a perfect scream and insisted on washing the dishes so Mother went up stairs and Olive and the boys and I did it – Then we went out in the park for a walk – It was quite an unexpected week-end and lots of fun.

Yesterday morning I went out to Darby Creek on Westchester Pike sketching. Mr. Copeland was supposed to come but evidently thought it was too cold and never turned up – There were only four girls that came but we had a great time. The autumn coloring is wonderful of course now. I made one little sketch but it wasn't very good. We expect to go every Saturday during October. Of course, it makes a long and hard week, but it will help a lot – Outdoor sketching is fascinating! I just love it.

Did you get a couple of cards tied together with a white ribbon, saying "Dorothy S. Roberts" and "Russell Cuyler Luecleker"? Well, what do you think of my little Dot being engaged? She said she was going to send an

---

[*] WTS from his daughter Dorothy, ML, 8pp. on 4pp. of folded watermarked stationery; received 27/10/17.

announcement to you but you may or may not get it. She's awfully happy and I think it's great but I hardly see a thing of her any more, which isn't quite so nice!

I guess Mother has told you about the kitten. He's perfectly <u>lovely</u>, the prettiest little beast you ever saw and as good as gold – His name is "Chris," for Christopher or Christine as the case may be. Bob is crazy about him and he returns the affection.

Your picture is right over the desk in the sitting room, where I'm writing now. It looks so real, it seems as if you ought to be sitting in the Morris chair right at my left shoulder.

Give my worst to "Crazy Bill" and tell him for goodness sake, why doesn't he admit he's licked!! It's about time you and all the rest of them were coming home! I read "Carry On" by Coningsby Dawson – It certainly is great.

Well, I hope this letter finds you well and as happy as possible – We are all well and happy but would be much happier if you were here –

Lots and lots of love –
Your daughter –
Dorothy

Sunday, October 7th, 1917*
Philadelphia, Pa.
Dear Father –

I'm sorry that I've been such an age writing to you – but you see it to this way – I didn't have anything to say, and it's boring to both end to fill up page after page with nothing (anyway – I think that's what I'll be doing now).

I think the last letter I wrote was from Cape May – well, I've been home now for about 2 ½ weeks but several things of such an ancient date are going into this "epistle." Crosby Boyd, a boy at Cape May had a Ford! Every morning he would bring it down to the beach and teach a bunch of girls how to run it. I'm really quite an expert at Ford driving, and I hope some day to get a whack at the Studebaker. Poor Crosby! He was the best natured thing on earth because he never got a chance to drive his own car an inch! All he did was crank it! Anyway, <u>we</u> had fun & I think driving a car is the best sport in the world. I'll be sixteen this winter and after you come home, I hope I'll take my turn at ours. I know you intended to stop getting chauffeurs – but don't stop until I learn – It won't make any difference about Bob!

I guess you know of Crosby Boyd's father – He was general passenger agent of the Penna. Railroad, aged 70! A couple of days after I came home from Cape May he died. Of course, I felt sorry – but he was so old, and I liked the Ford better than Crosby, anyway

Harriett Wilbur, a girl I met at Cape May, came here before going to New York for the winter. She stayed at the Bellevue with her family for about a week. I went there to lunch one day, and then she asked Connie and me to stay all night with her there. Warren took us over and as there was another boy there, he stayed and we had quite a gay little party. Connie and I didn't know how to act at the Bellevue – it's too sporty for us, and anyway Connie had never spent a night in a hotel at all, and you know how many I've spent! So we were new & inexperienced, – & I'm afraid we acted like country cousins. We discovered a cute little refrigerator arrangement, in the closet, that opened out into the hall. You just put your shoes in it & next morning, they're all beautifully shined. Maybe you who has traveled are acquainted with those things, but it was quite new and novel to us, so of course we tried it, thereby saving one dime ($.10). True economy!

School started about two weeks ago. Worse luck! Of course, Miss Hills' is all right if any school is, but I'm really not in favor of any school! To begin with we had three dry, long books to read during the summer – I started to read them just a week before school began and consequently finished about half a book.

We haven't gone to any "social functions" yet, and Mother declares that in wartimes no one will give even the shadow of a dance, but we're hoping for the best! Last Saturday Louise Harding had about four couples out there, and we had a lot of fun – nothing forward, just a rough house.

I suppose Mother and Dot and everyone told you about the cat. Mother and Dot decided on it before Warren came home, so as Bob always sticks by Mother, Ted & I were the only ones to vote against it. The cat seems to be taking revenge on me now for voting against it, by chasing me from chair to chair in the afternoon until he gets

---

* WTS from his daughter Barbara, ML in pencil; 8pp. on 4pp. of lined notepaper; received 27/10/17.

comfortably settled in the best one. I believe the cat's name is Chris – but Warren calls him "Clarence" (from habit). I call him "Pest (true to life) & Ted calls him "that darned cat" – Bob calls him "Kitty" (sweet & girlish) and Mother & Dot obediently call him Chris. Cats are in one the joys and glooms of life.

To-morrow afternoon I'm going over to Connie Westcott's house to paint the woodwork in her room. Then some afternoon she's going to come here and help me clear Grace's room of my aged trash. Pretty fair trade!

On Wednesday Kitty Smith is taking Connie & me to Keith's – at least I think she's taking us – altho' it may turn out that we're taking ourselves! Yesterday we all went to Rachel's wedding. It was a mighty nice wedding, but then, Harvey balled it up by telling Jack to repeat after him something that shouldn't have been repeated. Dot and Olive figured it out that by that mistake Rachel is married to him. Nice way for Chas. W. to get his wives isn't it? Mrs. LeBoutillier and Howell were there and they brought Mother and Dot and me in their flivver. Howell asked me again to spend a weekend with him. Real coy, wasn't it! Maybe I wasn't peeved!!

Well, it's getting late and Mother is setting a good example by turning in, so I guess I'd better follow suit –
Loads and loads of love
from
Barbara

October 11, 1917 [Thursday]*
[Le Treport, France][61]
My dear Bob,

This time I will write you a short letter, because I have not time for a long one.

Thank you for the Mutt & Jeff, which amused us all very much.

Learned today from your mother that you had returned to school. Very glad indeed that you were able to continue at Penn Charter. It's a dandy school and I am sure you like it. You must however study hard and do well. Charlie Jack saw something this afternoon that was quite thrilling. I was around, but did not see it. An aeroplane came over us quite high. All of a sudden, it seemed to collapse, and it fell head first turning a couple of times. Fortunately, when only a couple of hundred feet from earth, the fellow righted himself and trimmed off unhurt. Of course, it was an accident, but a mighty close call.

We have been having very bad weather. The wind blew down a great many tents. In fact, you have to take off your hat to the tent that did not blow down.

But it is nice and calm today, and that is what we all want, because when it is clear, the boys can fight, and the harder they fight, the sooner the War will be over. When it rains much, the mud makes it impossible for the troops to get their guns etc. up to the advanced line.

Flanders is now a very unhealthy place for the German Army, and getting worse all the time. It is very uncomfortable for everybody, and not particularly heathy for anybody, but is most unhealthy for the Boche.

Give my regards to your brothers & sisters.

Good bye. Love,
Dad

Friday, Oct. 12th, 1917†
[Le Treport, France][62]
My dear Mabel,

We have been right here working for 4 mos. today. It seems like a much shorter time. And in that time a great many sick and wounded have passed through our hands. Many of our patients have "gone west" as the soldier calls it, and many have gone home to spend their remaining days maimed & mutilated, and in a number of cases in Eternal Darkness. I have sent many sick to Blighty. I have yet to hear a complaint or a regret from these unfortunates. Everybody is working for them, and while everybody is tired and sick of the war, the most remote

---

* WTS to his son Bob, ML, 3pp. on 2pp. of note paper; opened by Censor 3349; APO 12 OC 17; postage due 4¢ but not added.
† WTS to MWS, ML, 6pp. on 3pp. of note paper; opened by Censor 6316; postage due 4¢.

thought is that of stopping until the Boche is so completely & permanently beaten that he can never again raise his head.

I am glad you have the "Rhymes upon a Red Cross Man." They are sad, but what you read is actually true and not over drawn for the sake of poetry. The poetic descriptions are every day happenings at a time.

Now I must not talk very much more about the War.

Glad to get Barbara's picture, and would like pictures of the other children and also one of yourself. Barbara wrote about questions to Emline during the winter. You know my ideas on such things and you will have to use your well known tact to get away with the proposition with the least damage to the feeling of those concerned. It's too bad that a father's or a mother's character should affect a child's life, but tell Barbara if I were colored, she would be part colored anyhow.

By the way, if you ever want to send me something, send it direct as an individual, and not through the Guild. I say ever, because just now I feel as if I would be her forever. Queen Mary's Guild does not work out just right. Last time the boxes were robbed, and in any event, it takes a very long time for the boxes to arrive. Anything you send individually comes right across just like a letter. We are wearing Sam Browne belts and campaign hats, and our nice little caps are taboo. We do not like the hats, but Gen. Pershing does, so there you are.

My little cam hangs only on the wall.

Winter is coming upon us; the wind and the rain are never weary and the days are mostly dark & dreary. Little rain you never saw and the merry way it keeps a coming down & up and sideways is very annoying. But I am getting used to it, and with the exception of gum boots am fairly well prepared for it. My sweater is a good piece of work. Would like winter pajamas, most say of flannel instead of the thin summer ones which I have, but then can sleep in my under clothes if necessary, so am not worrying. If I only did not have to get up in the morning.

That little introduction to the day's labor is not as pleasant as it might be. It seems to be regulation for all we do it, but it seems to me that some more kindly way of starting the day might be introduced. When I am Orderly Officer, it is not so hard, because I do not take my clothes off, and am up most of the night anyhow, but I can't be O.O. every day.

Well this will do for the present. Good night,

Much love,

B.

Oct. 19, 1917 [Friday]*
Philadelphia, Pa.
Dear Father:

I am now back in college, indeed have been back for three weeks, and I am working pretty hard. There are a great many more fellows back this year than I expected to see, but the juniors and seniors are greatly depleted. There are only eight of last year's Dekes [Delta Kappa Epsilon, DKE] back, but we are running the house and expect to take in some freshmen in several weeks. Rushing season has been moved up to Nov. 1st this year.

At present I am trying out for coxswain of the Penn crew. I have no hope of making the varsity this year at least, but it gives me something to do and it's a lot of fun. I am rather surprised but athletics seem to be running along almost the same as usual in all the colleges. Football games go on just the same as if three were no war – indeed everything over here is that way. It really seems hard to realize because life in this country is practically the same it has always been. Of course, there is "Red-Cross," "Emergency Aid" and all that, but they are the only reminders. I suppose [ran out of ink]

I guess you have heard of the second edition of Liberty Bonds with the U.S. has just issued. There is a very large campaign on at the University for the sale of these bonds. Most everyone seems to be buying them. It makes us all sore as the deuce to see lots of young and ignorant looking doctors wandering around town with Major's insignia on their uniforms, and then to think that all of you, real doctors, are over in France working hard and giving up so much when you are only Lieutenants and Captains. It' pretty darn rotten, but I think it will "all come out in the wash" after the war's over when the men who really did something will be recognized and the military slackers who get in some safe and easy branch of the service and stay at home to pull strings for high ranks will get what's coming

---

* WTS from his son Warren, ML, 5pp. on 3pp. of unlined note stationery; date of receipt not shown.

to them. Believe me, when you come home, patients will just come flocking to you and leave the stay-at-homes in the lerch. Well, I don't know much else, so good-bye,

Warren

Oct. 20th, 1917 [Saturday]*
[Le Treport, France]
My dear Mabel,

I do not remember when I wrote to you last, but believe it was one week ago. Since then, I have been too busy to write – so I hope you won't mind. Love you just as much and think of you all the time.

My work has been trying and continues, and I am beginning to get nervous, and more like myself than I have been since I left home. Of course, when I am like myself, I am not nice as you understand, and for 5 mos. I have been quite unlike myself in that I remained good natured and could not get Cross. Now, I am often on edge, so that I can explode. I don't want to explode because when I do, I do it pretty hard. I want the party to End. We have been doing great things up in the line in this last week or ten days, and the weather is again fine, so I hope the good work may be done.

The mails have gone wrong again; nothing for over a week, and even then, for the last month about, I have received just one letter a week and that from you. The children have evidently stopped writing. Have not heard from them for a long time. Given, they are busy.

I am sorry to say that I have cut your income $5 a month. The President through Gen. Pershing "wants" the Army to buy Liberty Bonds. I resisted a long time because I have no money to invest, but they made such a fuss about it, and made it appear so mean, disloyal, unpatriotic, etc., etc., not to "show confidence" in the Government and the "President wants it," you know, that I said alright, put me down for a bond and my wife will have to pay for it. So you will have $5.00 per month dropped off your pay to please the President. I hope he will be pleased, but my personal opinion is that it is a characteristic Democratic hold up. However, my dear, I will give you the Bond, so there now. Don't cry. In the course of time it will be delivered. Look it up in our letter box.

I finally got my Field allowance, and it will be just about what I consider one. But you had better send me a little money in November. Of course, it was balled up. After sending me my voucher for $94.00 they sent an S.O.P. [Standard Operating Procedure] "Do not cash it yet because we paid you too much." In other words, they paid me a Captain's allowance since June 15th and then discovered that my own Government only considered me a Captain from sometime in Sept. The difference between Capt. & Lieut. is not a great deal, but I have to get less, and I do not like less.

Elephants are still coming my way. Four wary ones have been given to me so far. Got a nice one today from my nurse who went today to Dieppe & yesterday another nurse gave me one some time ago, and Ed gave me two, so my collection is growing.

But the time the Colonel ran an indelible pencil in his eye a few day ago and has kept me on the jump since. He is some patient. If I had him in the States, it would not be for long. We had quite an eye "at that." He is getting better, "chez Cherie." Some day when I get home after this War, I will tell you all about Colonels; I don't think much of them. Captains, Lieutenants & some Majors are fine; Colonels are all p_ _ _ ks.

Dear old Molly C. sent me a box of Chocolates, last week, and Dr. Turnbull a box of Domino Sugar which my nurse was very glad to get. Of course, you must not tell Dr. T. what I do with the sugar, for he is awfully proud to think of me, now that sugar here is rationed next to American items in quantity; then some jams, tea, like raisins, onions, and potatoes. A ton of coal would be gratefully received. Coal is scarce, and while we can steal some it is hard to steal enough. The blamed Huns [Germans] are over in the coal fields you know and the French asked them to get out, but they won't do it.

That reminds me, in speaking of pajamas of different kinds, since as some are Jersey ones, and mine are flannels ones, the one for summer and the other for winter, I neglected to say that in event that I should someday have the nice ones of flannel, or a flannel variety, I would look much nicer and better in night dress, if I had also, two tiny slumber socks, one for each foot. That would be nice. We all have dandy sleeping bags – soft as silk; I think they are

---

* WTS to MWS, ML, 12pp. on 6pp. of note paper; opened by Censor 6064; APO 32 [sic] OC 17; postage due 4¢.

Jaeger. Mrs. Whitelaw Reid presented one to every nurse and officer in our original Unit. They are beautiful, and I think they belong to us. The Army gives us certain things, but anytime the motive strikes, all said Army just send around and takes them back. No motive ever said – "Government property." I think that Mrs. Reid will [not] be so mean. I don't like Government Property because it is so difficult to pinch.

I can't think of any more to say just now, but come back to a little more business, Good night.

Tuesday, 21st. A very beautiful but slow & tiresome day. Nothing has happened except no mail arrived. I guess it has been sunk. I see by the paper that quite a number of boats are getting caught. Most of them are caught going Eastward. This I think is because they are not Censored much going that way. After sufficient time, I presume they will be, but certainly not until a great many have been lost.

Notice. Miss Furhman came wants to send $30 home. She asked me today if I would do as I did once before for her. Of course, I will, so please send to Mr. J. W. Rutan, 1305 Arch St., your check for $30. immediately upon receipt of this. If you have already sent me some money, I am sorry, but I will be able to go longer next time. If you have not yet sent me my "allowance," you can count the $30 in it. I thought the account was not large enough to hurt you much one way or the other.

Charlie is quite peeved at me tonight because he wanted to have his way, as he is quite used to having with his mother and Medea. So, like a little child he has begun to pout. He will get over it. Ed is well and getting fat. Straub is a 1st class private and has the appropriate do dab on his arm. He is still sweeping chimneys and sweeping them very well for a lawyer. He has the darndest teeth I have seen on an American in France. They are like English teeth.

Now I have properly mentioned all of your family, and told you all important news, so I must stop.

Thank you for discontinuing your press campaign.

Good bye, much love,

From B.

Oct. 26th, 1917 [Friday]*
[Le Treport] France[63]
My dear Mabel,

Such a rainy dreary day you have never seen; cold also. Charlie has gone to Boulogne for a week to study yesterday, so I had to make a fire all myself. I am very proud of it too, for my first fire. You know I am not much on fires and everyone was much alarmed when Charlie went away because I must have no one to call for the fire. So Cadwallader who is Mess Captain ordered a striker to make me a fire every eight o'clock. Very nice & consistent – but today I wanted it in the afternoon – have my experiences & great pride.

What do you think of the French? For a people "bled white" I think they accomplished a wonderful feat, and I do hope that when the End comes, they will have the honor & pleasure of delivering the final blow, the Boche out. Indeed, one or two more like this last one would come close to being final, I think. May they continue their good work. With their bombastic talk about what they will do & what they won't do, the Boches are a pretty sick crowd, and with all their clevernesses they are in ways wonderfully stupid. When they see American troops & officers, they think they are English troops dressed in U.S. uniforms to fool them. That is what they are told, and they smile and say you can't fool them. We have a lot of Boche prisoners working here now. They look at us with great interest. They are a very happy looking lot, and I don't believe you could make one of them run away, even though he wouldn't get far if he tried. They are all well treated, much better than are our prisoners in Germany.

Received a few days ago, a dandy box of peanuts from you. Thank you ever so much. They are the best things I have had since leaving home; not at all stale. Also, some knit bandages. I can't place the sender, but you must thank her for me. No mail has reached us for almost 3 weeks, so I have no word what you are doing and how you all are. I can only hope that everything is alright. Of course, we do not know what has happened to our letters so can only speculate and wait impatiently.

Just did something to my fire – hope I did not put it out. It is some fire, but it does not quite understand me. When I push and pull the various little things and make little passes here and there, I try to inspire confidence, by acting just as if I know how, but the old fire seems to be a little suspicious and doubtful.

---

* WTS to MWS, ML, 8pp., on 4pp. of note paper; opened by Censor 3711; passed by G.B. censor; postage due 4¢.

We are not so busy now – having evacuated a great many patients to England. We are expecting though, and can get busy in a remarkably short time. Excuse me a moment. The fire is calling me. – It makes very nervous to have to nurse a fire. Bill Drayton was just here, and I consulted him about it. He said it would last an hour if left alone, so that's alright. I can then go to bed.

I have permission to go to Boulogne anytime I want, but I can't get away. That is one disadvantage of being the only eye man in the outfit. You know Boulogne is the great hospital center; some 60,000 beds. I am told; – spread out of course in the town and "neighborhood." It is the eye center, and I have been invited down by Col. Lister the Consulting Ophthalmologist of France, to see what they are doing. I don't want to go alone, and hope to be able to get away with Charlie, but could not make it, so if not ordered, I will probably not go. We have not been bombed yet, and I guess we won't be. I doubt if Fritz has enough ammunition to waste on just us – a hospital. He generally likes to pot stores of ammunition, railroads, supplies, etc. and incidentally of course anything that is alive and more or less human, but I don't think he will bother with us. He certainly would not try such a thing if he knew what Mary Medea would do & say. I think she could stop the war is she tried. If Fritz knew Mary Medea and knew that she wanted the war to stop so that CHAUWLES could come home, I believe he would quit. He would be awfully foolish not to. I wish M.M. would stop this rain first, just in the way of practice.

Maybe my dear, I have written you nothing but nonsense, <u>but</u> that is because I have nothing to tell you that would get through. I am very anxious to get that belated mail. <u>Then</u> maybe I can do better. So good night and God bless you.

Much love to you & the kids,

<u>B</u>.

#16 (Philadelphia U.S.A.) General Hospital. <u>B.E.F.</u> France

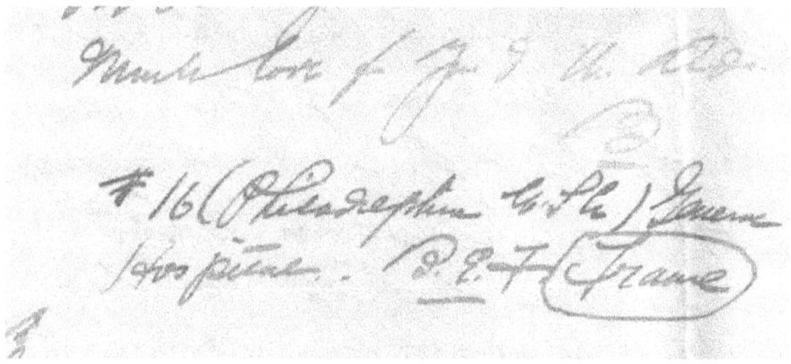

Oct. 31, 1917 [Wednesday][*]
[Philadelphia, Pa.][64]
Dear Pa -

I'm a bum. I haven't written to you for nearly a month. Every night when I get finished studying it's too late to start a letter, so consequently I've been putting it off. It's now quarter after eleven but I'm tired putting this letter off. Tonight is "Halloween" [and] we had a very exciting time of it. Bob, Barbara and Dot went out to a party given by the Sunday School at Mr. Harvey's while Mother, Warren and I stayed home. We didn't even hear a horn. About five years ago, that was one of the big days of the year for Warren, Wilkins Thatcher, John Arnold and myself. When we'd go out and just be general nuisances.

By the way, Mrs. or rather Dr. Arnold got married again. I don't know any of the particulars, not even the man's name. The only thing I know is that he isn't a doctor.

I've been very busy ever since school opened as I told you in the last letter. Things are about the same. Our football season is more than half over and has been quite successful so far. But I sure will be glad when it's over however. What's the difference, the same week the foot-ball season closes the basket-ball opens and as I'm manager of that, I'll still have my afternoons taken up.

I've been getting along pretty well with my school work. The first month I got a clear report. Another report is due quite soon, and all though I'm not sure what I'll get, I feel quite certain it won't be bad.

Friday night Barbara, Warren and myself are going to a small dance given by Dorothy Healy out in Roxborough. Speaking of dance reminds me – the Tuxedo that I got a couple of yrs. ago is so small I can't wear it, so I thought I would have to go without one, which didn't make much difference as there won't be many parties this winter, I don't think. But it just happened Aunt Lulu was cleaning out her house and found a Tuxedo which just fits me. So I'm fixed as far as that goes.

---

[*] WTS from his son Ted, ML, 3pp. on 1p. of heavy note stationery; received 25/11/17

Things at home are going finely. Mother surely is doing wonderfully. She does quite a lot of work but somehow doesn't get tired out. And she looks fine.

As it's getting quite late, I think I had better close. Remember me to Charlie Jack, Ed, Frank Chalk, Hudson Chapman and Ralph Straub.

Lots of love,

Ted.

November 1917

Le Treport – Hopes Rise at Passchendaele and then Fade at Cambrai

"I am far out here, and the war is not over and nothing is nice."

Nov. 1, 1917 [Thursday]*
Philadelphia
Dearest Father –

It's quite a while since I've written to you but really 9 A.M. to 4 P.M. keeps me terribly on the jump – at school for the last two weeks we've been working frantically on a competition given by the Overland Auto Co. for color scheme and top design for their Country Club remodel. Well, then, Mr. Copeland wanted us to try, first for the other experience of going into a thing of that kind, so we all did. It was some job! We did two views (profile and bird's eye), traced from photographs. – We had to do them in oil – altogether it made a "hand-painted picture" about 16 x 22 inches. Mr. Copeland says that whoever wins, it will just be luck, as technique, etc., do not enter in and it will first be what happens to hit the judge's fancy. However, we have as good a chance as any one! Five art schools are competing. We sent our entries (I guess you call them that!) off a couple of days ago. Don't know when we'll hear – There are three prizes, $150, $75, and $50. Now don't start in and mentally spend it for me because I have about ½ a chance in 10,000!

I also sent four of my little sketches that I did last June to a competition for Art students at Wanamaker's. Mr. Copeland picked out the ones to send and I had them framed. There again, I have practically no chance at a prize, because you can be an "art student" until you've had something hung in a big exhibition like the Academy – and that may take a long time; so down home those hopes that have been rising while you've been reading this!!!

Had a letter from Walter Rodman a little while ago. He is in France now. He didn't attempt to tell me any news at all until very near the end and then when he did, the censor thought it was too interesting and applied scissors! My but I was peeved!

On Monday night, Marjory Eastlake had a Halloween party at her home in Swarthmore – all the "Krazie Krew" (the bunch at school, otherwise known as the "Big Six") were all there. It was a dance and we certainly did have a peach of a time.

I went as a harem girl in a dandy costume that Mother and I concocted out of things borrowed from everybody "such as" – Aunt Florie, Mrs. Lewis – Olive, etc. It really did look great – One of the boys that was there told me that if I'd go out on the street in it in the city, he'd take me to see "The Boomerang" which is a peach of a play I've heard. I wish I had taken him up – It wouldn't have been bad to go out in on Halloween night. However, I thought it would have been a trifle fresh!

My room is at last curtained and couch covered! It looks great but a bit startling and rather daring! The curtains are pumpkin color and the couch covers and my little chair are cretonne (a new one, not denim, stockinette or divinity!). The cretonne is black, peacock blue and green with a little orange to match the curtain. I like it ever so much. You were a big dear to tell me you'd pay for having it papered, so I still have your five dollar gold piece to do something else with.

The Rosses have one of those red flags with a white centre and a blue star to show that a man from that house is in France – You see loads of them in stores & office buildings, etc., with lots of stars, but I don't like it much for individuals. If many people do it, you can bet we would!

---

* WTS from daughter Dorothy; ML, 6pp. on 2pp. of notepaper stationery; received 25/11/17 (WTS replied 27 Nov 17).

We're going to have a play up at the church some time in December, which I'm to be in, I think. That will mean a lot more work. We are all well and Mother certainly is a wonder – She does an awful lot but seems to enjoy it and we're getting along finely – Mother seems very well – She also seems to have plenty of time to do outside things as usual. You can't imagine the scrambles to hear your letters when they come – They sure are great – Write again soon to me! Lots and lots of love –

Dorothy

Nov. 4th, 1917 [Sunday]*
[Le Treport] France[65]
My dear Mabel,

At last the spell was broken and two nice mails arrived, bringing me letters from you and all of the children. Our children write very interesting letters. All are good. Barbara particularly so; Dorothy likewise – and also Warren, Theodore & Bob. You all gave me so much more that I feel good – up to date. I gather from one of your letters that mine do not contain enough local color. True. I have not told you much of little happenings. It is difficult for me to do that. However, tonight Charlie & I dined with Dr. & Mrs. Krumbhaar. Krumbhaar you know brought his wife. She works in the laboratory with him and they have a tiny 2x4 flat in Treport. I enjoyed it very much. With Charlie, it is different. He writes a description of every day from start to finish and gets same in return. That would be tiresome and just impossible for me. I don't care much what you do at 3:10 P.M. Tuesday 3 weeks ago, and you surely don't care what I did.

Here there are no amusements. About the only "thing" that we can do is to go down town to dinner. This we do practically every Sunday & sometimes during the week. The mess is good to be sure, but I get so tired of it that I would often like to jump out of my skin. You spoke of Charlie's letter written the same time as mine arriving first. Your explanation is wrong. The mail is put into a box and that box is taken to the post office every day at 1:30 o'clock. There is one mail each way per day, <u>and</u> that's all. Your letter goes today or if not tomorrow. So it is not a question of an hour or so. And we can't post in letter boxes around town. There aren't any.

Received a letter from Miss Tilge – a nice letter. She told me of her visit to you and how nice you were. She remembered what I had told her about you and agreed with me. She is a very nice good woman, and if you are willing to listen, you will find her very interesting. She might well be called a scream.

I am sorry about the automobile. But remember that it is yours to command, and I want you to have it and use it just as much or as little as you want to. There again, I am a "thinking" that my poor little Studebaker will be worth 30 cts. when Louis Herbert gets through with it. They are his regs. But if you want it, tell Uncle Josh. With care & attention it should not have cost much money during the summer. Of course, if you don't understand it, you can get in quite deep financially, and I guess they did not understand it.

The news this week has been rather depressing. The Italian situation is not nice, but I have a theory. It will be a boomerang. I believe it is the start of the finish. You notice that our front is all to the good, and that we have about gotten Passchendaele Ridge. Once that is gotten, bang goes Mr. Boche out of Belgium. The Allied front to Flanders is the strongest thing in the world. So keep your eye on Sir Douglas Haig, and don't worry about Italy. Kind of sad though, Yes?

By the way, an "interesting" book for the family is "Over the Top." You can read it in a short time. It will tell you all the fighting stories. Get it and read it to the children. You spoke of one of your letters of the Warriors of Hell [Sikhs]. We treat lots of them here. They are probably the greatest fighters in the war. At all events, the Germans dread them most. When they go over the top, Hell breaks loose in all its frightfulness. Almost on a par are the Australians and the New Zealanders – the <u>Anzacs</u>. They would rather go over the top than eat. Then the Canadians are wonderful fighters and have made for themselves a great name. Finally comes the English Tommy who has them all beaten in his way. His great trick is to hold on and strike. When he gets a new objective, he can't be driven out, he just naturally won't get out. So you see they are all good, and I am anxious to see what the striking characteristics

* WTS to MWS, ML, 10pp. on 5pp. of note paper; opened by Censor 3678; postmarked APO 6 NO 17; with original paperclip.

of our boys will be. One thing sure in the works, the Anzacs, the Tommies (Imperials), the Canadians & the U.S.A. will make a team that will take care of the Huns. Read "Over the Top" and see how it's done.

Monday, Nov. 5 – A little local color. Went to Dieppe today with the Quartermaster. He was after money, so we brought back some 10 or 12 000 dollars to pay our men & nurses. I enjoyed it very much. It is an interesting old town, and I had not been there for many years. Had dinner, did some shopping, bought an ordnance rain coat and came home. Very exciting when you consider that I have not been 2 miles from this place since I came, except one day last summer when I took gas instructions at Abbeville. Dieppe is only 20 miles – we went in a Ford [ambulance].

Last week a camouflaged ship came into Treport. It was very interesting, and very well done. It was difficult to determine just where it started or stopped or whether it was a ship or a row of houses, or a wharf or what not. Of course, you have read much of camouflage. Everything is fixed that way at the front.

Your letter with the new address arrived about 3 days after its fellow. But it may not bad address, letters are handled now in any old way. The new address is official and came out in orders which will eventually reach you.

I think this is a very cozy letter and you should be much pleased if you can find time to read it.

I almost forgot two questions. 1st, the picture for Jane. Of course, any way you like, so she gets one. Send her a good one. Christmas as you intended.

2nd. The Red Cross brassard. While a part of our uniform it has never been worn by us, nor have I ever seen one worn by anybody over here. I wish I had not put it in the picture, but it makes no difference one way or the other. It belongs on the left arm. When we declared war, so to speak, we did not know much about dressing soldiers, as we scarcely had enough to learn on. Now this is really the end. Good night. Much love properly distributed.

B.

Wm. T. Shoemaker.

Nov. 6th, 1917 [Tuesday]*
[Le Treport] France
My dear Bobby,

I saw a very fine camouflaged ship last week that would have interested you. It was particularly well done, and was fixed up in great style to fool the Boche. You know sometimes they paint the bow like white waves & spray & at a distance the boat looks as if it were going very fast. Then Mr. Submarine shoots far ahead and is surprised at his bad marksmanship. Another nice way is to paint the ship blue to match the spray & water, and on the deck paint a smaller ship going like thunder in the opposite direction. Then when looked at through a periscope, it is almost impossible to detect the fraud. Everything up at the front is done the same way. Cannons, wagons, mules, the tin helmets, etc. It is very important not to be seen, and they have become very expert in blending in with the surroundings. All of which the painter learned from the lower animals. Zebras, hyenas, bugs, snakes and a whole lot of animals look so much like the surroundings that their enemies can't see them.

Your mother tells me you are a very good & useful boy, as well as a fine athlete & scholar. This is very gratifying to me, and very good for you, because if you were not, I might have to come home to see about it. Then how would you feel? Of course, you and I would not like that. OK now!

Well, as you would say, here comes the end of the letter as I have nothing more to say.

"Remember me to all the folks," including your mother.

Lots of love from

Your Father

----

* WTS to son Bob, ML, 3pp. on 2pp. of notepaper; opened by Censor 3340; postmarked APO 7 NO 17; with original paperclip.

Nov. 11 – 1917 [Sunday]*
[Le Treport] France[66]
My dear Mabel,

Sent to you today a little Christmas box – not to be opened until Christmas Eve – When you will all sit in the nice little dining room (without me) and commence the festivities, which I hope you will have in good measure. The box contains some little things but a big thought for all of you. You may and probably will have to pay duty, which I trust you won't mind, as it is impossible for me to pay it in advance. But that makes it necessary for me to list the contents, and even give the price, which is such a vulgar thing to do you know. However, you understand and will not tell the children anything about it until Christmas. I have purposely sent this package very early so as to not get caught up in the Christmas mails which this year will be overwhelming. Now you have the whole story. Get your little package, keep it for Christmas, and think that I am doing the best I can with my limited means and faculties.

The package

1 tin box which contained about 5 lbs. of chocolates sent to me by Mollie C. The said chocolates are retained by me, but that is nothing here nor there. Within tin box = One lace scarf for my best girl. One lace scarf for my big girl. The same for my little girl (my Barbara). One pipe for Warren, a wallet for Theodore, and a very excellent knife for little Bobbly – also a handkerchief for Jane and one for Mollie C. – which kindly deliver. The scarf for my best girl is Dutchess [sic] lace-applique. The best part of the applique is that if so desired, she can rip it off and sew it on the edge of her panties. It is said to be very pretty, but of course you know what I know about it. The scarfs for my big and little girls are alike, but that makes no difference. They are Maltese lace which I was always very fond of. The handkerchiefs are embroidered, and I think will stand a reasonably bad cold in the head, if careful. The pipe, pocket box & knife need no explanation. I am sorry to have to tell you all about it, but I feel that you should know, in case the package is tampered with or does not arrive in good condition. It is registered – so when you get your notice in case it is held for customs duty – take money with you when you claim it. See? So you see, all my Christmas shopping is done early as it should be. I am sorry I could not to do more, but will try to next year.

It has been rather a slow week. We are not busy just now. The Russians and Italians make us sick, but you notice that we have Passchendeale. That's a lot. Your letter last week was quite interesting. I hope you had not sent Mr. Winder one of your dunning letters. He is a very decent sort of chap, and as he told you, I have known him many years. The Jones-Langdon reputations are interesting and amusing. They do not carry much weight here however, and the tussels still continue. A reputation once gotten is very difficult to change. Isaac has a rep. for being light and no matter what he does it stinks. His "great" week is the work of Barany of Sienna, and he has worked & played – hard & long on this particular stunt, and very likely his work has been good and he may make a first class Major. Langdon has a cushy job and is doubtless up to it. No harm in his being a Captain, for Captains are not much of a much. You must not worry about R. & C. The money is of course important. When I get out of the Army, they can have all of their ranks and go with them. I am first last & all the time a doctor and not an Army man, or a Navy man. The A. & the N. are fine for those who like them, but not for me. Just a matter of taste, you know. I am very glad I am in it just now however, and want to do my little part. Will be very glad also when that little part is done.

Today (Monday) is a wonderful day. Clear as a bell & like Spring or early Fall. We got a big convoy last night, but I received nothing very bad. I think I will take a walk this afternoon. The trouble is it always gets dark at 4:30 so we can't walk far. Charlie has had rather a bad week – homesick, mostly, but also some sciatica – cold, etc. Don't tell Medea. No indeed.

Good bye – Will write very soon again. Hope my little package gets across ok. Much love to you & the kids.

B.

Wm. T. Shoemaker

----

* WTS to MWS, ML, 8pp. on 4pp. of note paper; opened by Censor 6417; postmarked APO 12 NO 17; passed by G.B. censor.

Monday, Nov. 12th, 1917*
[Le Treport, France]
My dear Mabel,

I neglected to enclose list in your letter so send it here with. You may not have to use it, but if you should it is best to have it. See? – B

[Enclosed]

	Francs		Francs
1 Dutchess lace scarf	35.	1 Handkerchief	6.
1 Maltese " "	22.	1 Handkerchief	6.
1 Maltese " "	22.		137.
1 Pipe	12.		
1 Wallet	22.		137÷5.7 = $24.03

Sunday, Nov. 11, 1917†
Philadelphia[67]
Dear Father

We just this minute got in – Aunt Lulu invited us all to her house for dinner – Aunt Florie & Betty O'Connor were there – Betty you know is Herbert's fiancé. The baby was there, too - & she is the cutest ever! We're very popular, not something new – but we're being engaged ahead for dinners – Aunt Caroline is going to have us all out there for Thanksgiving and Aunt Ann engaged us a month ago for Christmas dinner. Say that isn't economy! We've been having quite a gay time lately – About two weeks ago, Ted & I went to a dance in Germantown – When we started it was clear as crystal – but when we got there, oh my! The most terrible thunderstorm I have ever been in! During the dance the lightning was so bad that all the electricity went off & we danced in darkness for over a half hour. It wasn't really very dark because the lightning was mighty bright & plentiful!

Dorothy Keely gave a dance the next week at Roxborough – That afternoon our school played Devon Manor in hockey & Gertrude Russell was taking a bunch of girls out in her auto – eight girls & six suit cases. Well, on our way out on Walnut St. bridge, we had a puncture. Three boys came along & fixed it for us, & we got out there in good time – But on our home, late in the afternoon & after dark, we had another blow out! As luck would have it, we were on Lancaster Pike (which by the way has no toll gates, anymore), and directly in front of Mrs. Boericke's place. The girls were dressed in all sorts of simple ways – half of them had kept on their bloomers & middies. I went into Boericke's & phoned for Elizabeth Evans' car – Their chauffeur fixed us up & we started home again. I didn't get home till half past six - & the train for Roxy left at half past seven – Of course we couldn't make that, but the next one was half past eight, so we didn't get to the dance till half past nine! It was a slick dance tho & every one had a wonderful time.

One afternoon Connie Westcott, two other girls & myself had a little peanut party at the Berad. Ruth Chatterton was there in "Come Out of the Kitchen." It was a terribly cute play, & Ruth Chatterton is sweet! If it gets to Paris, I would go see it.

Last night we had one sick cat! When Mother came home in the late afternoon, she found the cat asleep in a candy box. No one had been home all afternoon – Mother picked him out of the box & just put him down cellar. He meowed in the most frightful way, & when he walked his legs were perfectly stiff. He seemed to be in terrible pain. After supper we phoned to a veterinary – Dr. Glass, & he said the cat had had a fit, & to give him olive oil. So Mother and Bob went down to give him the oil, & Mother took him by the back of the neck & when he opened his mouth to yell she poured the oil in – Well the cat resisted & then first fell back stiff as a board with his feet up in the air. Bob yelled "He's dead" & Mother & Dot & I just went limp. The old cat wasn't dead tho, just having another fit. But he stayed perfectly stiff with his feet jiggling & kicking in the air – This morning he was all right – but he looked like the last rose of summer.

---

* WTS to MWS, ML, 1p., with enclosure; opened by Censor 6238; postmarked APO 12 NO 17; passed by G.B. censor.
† WTS from daughter Barbara, ML, 8pp. on 3pp. of folded stationery; received 3/12/17.

Last night Dot & I went to the League Dancing Class – Some class! Anyone & everyone belongs – all ages – simply girls! Not a soul would dance – at least only a very few. Herbert Swing couldn't dance an inch – perfectly rank he was – well, I got him a Paul Jones & naturally couldn't dance with him – no one could! But the conceit of it – when I said, Oh I can't do this, or something of the sort – says he, "I'll take you on to Miss Schelty (the teacher) & let her teach you" Sore – well, I thought I might do something wild! But when he said – "After you've danced for about a year, you'll probably catch on to it," – I was furious! Can you blame me – there he was, couldn't dance a step & blaming it all on me!

Guess what I have now – Lumbago! Yes – full-fledged, honest to goodness lumbago & it isn't one inch of free, either. Last night when I tried to get into bed, every time I moved it nearly killed me - & to make it worse, I couldn't do a thing but giggle! It may have been funny – but I didn't see it, & it was most humiliating to just have to giggle! I wouldn't care if I took after you in some things – nice ones, of course – but when it comes to picking out lumbago – Oh, gee!!

Yesterday I saw for the first time – a foot-ball game! Allen Anderson took Ted & Margaret Curtis & me out to Haverford in this car – It was the Haverford-Hill game – not colleges – but anyway it was foot-ball - & it was something very new to me – a part of my education that had never been completed – It isn't completed yet – but at least I'm not quite so ignorant!

Well – I'm just about exhausted – with news – I mean –

So good-by –

Loads of love,

Barbara

Sunday, Nov. 18th, 1917*
[Le Treport], France[68]
My dear Mabel,

No letters this week – everybody disappointed. Have about completed our first 6 mos. of service, tomorrow being the day. Things do not look particularly good, and I am afraid that we commence another 6 mos. As you know, I hoped to be home [at] Christmas, but that particular hope has now been abandoned, and I will spend my Christmas in France, and preparations for the festivities have started. Plans are under way for our celebration. One little plan for this time is that the nurses are going to sing carols to all the wards Christmas morning. For Thanksgiving, which is next week, the Q.M. has bought turkey for every living soul in the Army, and plenty of it too. Then the New York group sent us though their "fund" 200 000 cigarettes which have arrived and are being issued. Cigars are coming too. So you see, we are well looked after even if we can't get home.

Having been here so long, the British Government has now started on our leave. We get two weeks. When my time comes I suppose I will have to go to Paris. You see, we can only go [to] certain places. Our way is paid and we are supposed to take our leave when offered. I don't want much to go to Paris, but where else can I go. Would like to go to Edinburgh, but it is rather difficult to get permission on account of limited accommodations on the Channel boats. They are so full up with troops & supplies. Would like to go to Madrid Spain, or to Switzerland, but can't go to a neutral country, so there you are. Can go to Nice or Monte Carlo and many do it, but I am not very interested, so will probably land in Paris. Of course, my time may not come for several months, and I am going to try to have Charlie [Jack] get his at the same time. It is no fun for me to go alone. So if I can't have it my way, I may pass it up and stay right here. I can't get very enthusiastic. It is getting deadly slow here, and we are all getting more or less fed up. So much of the same thing is getting tiresome and when the glooms talk "duration" – 2 years if it was 3 y, etc., it does not help one's spirits a great deal. However, we do not complain – yet. When I have been out one year, you are going to hear from me. My willingness & good nature are of one year's duration. Anything less than that I will accept, anything more than that I risk and you can risk, too. But what is the use of discussing this now. Come what so ever.

---

* WTS to MWS, ML, 8pp. on 4pp. of note paper; opened by Censor 6316; not postmarked by APO; passed by G.B. censor.

Received from Drexel or rather Morgan Harjes notice that they had 5 fr. for me. Asked them to produce it and report it every mail. I think you need send me no more until February 1st. Surely not unless I let you know. I am expecting money from the British Government because I am a specialist. I will get 2/6 pay in addition to field allowance. The claim has just recently been made and if it goes through, which I trust it will, I will get all the back pay amounting to some $107.00. If it does not go through, I may have to tap you in January. With my specialist allowance, the B.G. will give me in cash about $45. per month. Now, that is not bad. If I were with the American Forces, I would get my little salary & no more. I am glad I am with the B.E.F. But I may not connect with the specialist allowance, so not be too cheerful. I am waiting for it patiently. My officer allowance I get regularly - $25. per month. You do not tell me about your "Emblem" from the Pa. Hospital. Did you get it, is it nice, do you wear it, and all about it? ( ) I understand you have to pay a tax on everything, including the automobile. About $25. for the auto I think. If Josh has appropriated it permanently – surely he should pay it – Please? I will stop now until tomorrow, when I will write some more.

Monday. Very cold, bleak, dull, cheerless, uncomfortable, etc. One of the melancholy days that my mother used to tell me about. One interesting rumor this morning from the Camp – entirely without foundation of course says we are to go to Egypt. I would not mind that, for it is not cold over there. Of course, the British can put us where they please. I have thought that in view of the Italian troubles, we might be sent to Italy, but that was just a "think" of my own. Personally, I don't want to move until I move Westward to the neighborhood of 109. But you never can tell what an Army will do in time of war.

By the way, tell Holloway that I have seen Geo. Darby several times. He was here last week, and I will see him again when I go to Boulogne. He is now at Boulogne. He likes the Army about as well as I do, but has not had as good work. We have the crack place alright here. Well good bye, with the usual love *ratiner* [?],

B.

Wm. T. Shoemaker

Saturday, Nov. 24, 1917*
[Le Treport, France][69]
My dear Mabel,

Saturday night and I ought to be at home; the war ought to be over and everything ought to be nice & in good order again. But here I am far out here, and the war is not over and nothing is nice. No letters from home for more than 2 weeks, and when they come, they will be more than a month old, if indeed they ever come. I fancy they are on transports and transports take their time and necessarily travel in curious ways. It would not do for a transport to travel on schedule time. Only one good thing happened this week, and you know all about that. The victory up around Cambrai was a hummer. It offsets the Italian disaster, and means a great deal more. To break the Hindenburg Line was some break. Mr. H. thought and said it could not be done, now it has been done, and will be done again. And that is the only thing worth while for a week or two weeks. Did you know the Prince was in Paris? I mean the Philadelphia U.S.A. Prince. He and Major Jones arrived about a week ago. He has not been here, but hopes to come. Dr. Harte saw him in Paris early in this week. So far as I can gather, he is looking things over and will go home & make recommendations to the Government. We will then I suppose establish hospitals & centers in France for our Army. I only hope that I do not get hung up in that game. It would be a nice enough game, but unless I could play it home, I would not like it. Morgan Harjes have not yet sent me my money, and I am very poor, but I look to that again, so about it will soon be forth coming.

The hospital is pretty full just now, but the character of the cases is different. The casualties do not have to be so bad, and while my ward is full up, it takes me very little time to attend to it.

Have just read another interesting book that I think you would enjoy – get it and read it. It is *The German Spy in America* by John Price Jones – published by Hutchinson & Co. – London. The account of the Lusitania murder is startling – it makes you cry. And the whole thing is undoubtedly true. If there is anything more treacherous and destructive than the German Government it has yet to be created. I wish the civilized based God spent in the total

---

* WTS to MWS, ML, 10pp. on 5pp. of note paper, with 2pp. enclosed ML written on both sides of 1p. of note paper; opened by Censor 6259; APO postmark 28 NO 17; passed by G.B. censor.

ambivalence of it. I want to see the Kaiser and his crowd a thousand times dead and damned. That's the way I feel about that. That is my little "Song of Hate."

The Army has issued a new round of cheap insurance which is compulsory for English men. I am at present automatically insured for $4000 – until February, when I am able to have as much or little as I like – of $1,000 or $10,000. There will be no pensions after this war, but a minimal payment for 240 mos. corresponding to the amount of insurance carried. It is very cheap – for my age, about $10. per 1000 annual premium. Then you may die anyway you like, and your beneficiary lives happily ever afterward. In 5 years, you may drop it as the war will be nearly over then, or convert it with some other form of insurance. I think I will take $5 or $6,000 worth, just to try it. What do you think of it? I can't write – more later. Hospital business. Good night –

Tuesday – 27th. You see I have your trouble. When I try to write, I am disturbed & then can't get back to it. Well, Sunday, a fine mail arrived, so everybody is again more or less happy.

Three letters from you – very nice ones; one from Dorothy, one from Ted, one from Jane; a nice letter which I shall enclose, so that you may read it too; one from Dr. Turnbull, Mrs. Wall, Women's Medical College, and others that you forwarded. A pretty good haul for me. The children are all high in their praises of their mother – they say she is wonderful. Well, looks fine and all sorts of things. I am so glad that you are well and have learned to Carry On. You get the cake of Carry On very well. It is so much better than to fret & worry about me. The only thing to do, the time will pass much more quickly. You are doing your part wonderfully well, and I am of course very proud of you. The children all seen to be doing very well. It will be very nice when Dorothy gets all of those prizes.

By the way, in one of your letters some time ago, you said Mr. Burnham phoned for Warren's office address – you did not remember ever hearing it, etc. Now, just what did you mean – Our Warren & his office, and the address thereof? I did not know that he had an office. Have learned that Arthur – he of the Harvey & Arthur variety has gone, or is going west on a ranch. Just where is he going, and what is the idea? I presume the ranch belong to "one of Daddy's patients" – and also the where withal. Ed told me that fires had not yet been started at 2011 and that Martha was cold. All of which I found natural. Is the winter residence to be occupied this year, or is it closed until *apres la guerre*? In England, most people with "Estates" have closed them and are living in the Lodge. 2011 is quite a clever little Lodge, and the place will doubtless be adopted in the States. Much Holiday preparation has started, and we are to have a most wonderful Christmas. 52 cases of X-mas presents have already come across for our Unit, and more to follow. Thursday is Thanksgiving – we are to have much turkey and all pertaining there to. It's all very nice, but I know what would be nicer. Charlie send his love to you and all your children. The boys for a Shoemaker, your husband believes girls – well. Oh, I forgot – Sorry Horme Smith left the Church, but knew he would do so often Mrs. Matthews died. He of course changed his Church not his religion. Good bye,

Much love,

B.

[Enclosure]

Dear Willie

I cannot thank you and Mabel for thinking about me. Was surprised and pleased so far away to often think what good times I had with you all but I am happy and so glad to hear from you but my I did not think you would have time to write to me you have so many to write to I never thought you would have been [in] the army but we never know what is laid out but I did not think when you and the rest of the boys were little that I would end my days in the South it is a beautiful country. I hope you will not be away long I know they all miss you very much.

Thanking you again. With much love,

Jane

Our address Boute 97 Bose 422

THE FIRST DEATH
In November our first sorrow came in the sickness and death of Kenneth Hay.[70]

Private Kenneth B. Hay died November 29, 1917

The funeral of Private Kenneth Hay.

December 1917

Le Treport – Wartime Christmas

" A Zeppelin just went over us. He did not drop anything, or has not yet."

Saturday, Dec 1st, 1917*
[Le Treport, France]
My dear Mabel,

This has been a more interesting week but with sadness as well as with gladness. We lost our first man by death Thanksgiving morning. Pvt. Hay, a very nice orderly, died of appendicitis following an operation. Fortunately in a way, he was an orphan, so a father and mother did not have to spend a wretchedly sad Thanksgiving day. He was buried yesterday, and we had our first Garrison funeral. At a Garrison funeral everybody who possibly can is in attendance. I was not there, because I went to Rouen yesterday. Will tell you about that later. But I saw the procession and it was very impressive. Almost all of the Officers, a great many of the Nurses, and about 150 Enlisted men. Will send you a picture of it when I get one. It was of course our first American funeral, and the American flag was held, and "taps" were sounded instead of the "last post," which is sounded at all British funerals.

Thanksgiving day was properly celebrated. A foot ball game was played between the Surgical & Medicine staffs of the Hospital (Enlisted men). The Medicals beat us, and it was an interesting game. We have with us some star players. It did not seem right to call it off on account of Hay, so we went ahead with it. Then a wonderful Turkey dinner; just as good as anybody in Philadelphia had. A dandy dinner and this evening a reception at the Nurses' mess – good music, singing, etc. So having our one misfortune, we had a very enjoyable Thanksgiving day. Oh, I forgot, at 3 o'clock we all went to church. Padre Jeffreys gave us a very good short service. We are going to have a very big time Christmas.

Yesterday, I went to Rouen. I took a patient to a hospital there in the ambulance. Ambulances can carry 7 people. We had two chauffereines [ambulance drivers], Charlie Jack, Les Norris, Ned Krumbhaar, myself and the patient. It was a beautiful ride, through the best country I have seen in France. We dumped our patient about 2 o'clock, then had lunch, and spent about 2 hrs. looking over the town. Rouen is a very old & interesting town. The Cathedral which I have seen many times from the train, is a beauty. Rouen is between Dieppe & Paris, and I have always gone that way, but never stopped there. Joan of Arc was burned at the stake in Rouen, if you remember.

It is quite a city – (over 100 000) and very quaint & interesting. Then we went to Dieppe, where we got dinner after which we came home, arriving about 10:15 P.M. I had a very nice day and it was a change, as you can imagine. Right here for 6 mos. and can't get away – is tiresome. I told you I think, that I expect to go to Boulogne for about a week, but I do not know when I shall go. I am not pleased with that, that is, not <u>very</u> pleased.

I have two Christmas presents. One from Aunt Caroline, which I have not opened, and which will be a "surprise" when Christmas comes, and one from Henry Page & Nan. They sent me 200 Fatima cigarettes, 6 packages of Wilbur's Chocolates, <u>and a package of toilet paper</u>. The card said smokes and other "useful articles." XXXXXX

<u>Sunday</u>, Dec. 2nd.
Hospital very full. Convoy after convoy has convoyed on us. 700 patients in last 24 hrs. and more to follow. Bad times you know around Cambrai. Also a mail arrived this morning. One letter from you No. 50. The one from Barbara, 2 from Gleninagle & a "copy" for Ralph. Two more Christmas presents which I must [not?] open until Christmas. A very interesting looking package from Jack Phillips and one also from Mrs. O. Shoemaker. I tell you we are going to have some Christmas. The socks from Florence have not been received, but may be here; also the wooly pajamas have not come. Both socks and pajamas are needed. Some 50 cases of goods from America are over in the G.M. stores unopened. Perhaps my things are among them. I hope my little package has arrived safely. It is not very much to be sure, but I want you to get it of course.

---

* WTS to MWS, ML, 12pp. on 6pp. of note paper; not marked opened by censor; APO postmark 3 DE 17; passed by G.B. censor.

I am sorry you are having trouble with your pay. I would simply write to the Depot Q.M. and tell him that I have a perfectly good Captain's Commission dating from the date I gave you. That I am entitled to pay accordingly, and that if he does not know it, it is his business to find it out. My vouchers are all perfectly made out. Nov. Dec. & Jan. are marked out Captain. Previous ones are made out Lt. However, I did collect here the difference in pay for one month, so I will have to teach myself and straighten the thing out. Oh temptation. XXXXXX English kisses.

Tuesday night –

Have just been talking with Kidwell, our Q.M. The situation is thus. I became a Captain Sept. 23rd. The additional pay of Captain – for balance of September amounting to about $70 – I collect here. Vouchers for Nov. Dec. & Jan have gone in as Captain, and you should have no trouble with them. The only month in question then is October. The Q.M. cannot pay you the excess without authority from me. Therefore, I will send him that authority in the form of a voucher tomorrow. The amount in question is about $36.37. This will come to you and with another month's pay, see? – So do not worry about it; do not write to the Q.M. & do not worry Furbush. You are only shy $36. And that will come along. Now, about the Commission dating from Sept 23rd instead of May 15th. You cannot help that, nor can Furbush, nor any one else. So forget it. The good Government has simply done me out of about $160 – not intentionally of course, but though the customary inefficiency that pervades officialdom during a crisis. And you do know, everybody has been very busy and thousands of new & inexperienced clerks had to be employed. But no power under Heaven could change it now. So carry on and do not fret. If you do not get full pay for Nov. raise H___. Received yesterday £5 from Morgan Harjes – I am alright now until February, I think.

Monday morning Dec. 3rd.

I seem to get disturbed whenever I try to write. However, such a letter as this covers several days and comes first as some. Have sent to Washington D.C. this morning voucher for $36.67 additional pay for October, you will get it in time. It is getting cold here – real December weather. I do not like it cold.

Can't think of any more to say just now. And this is probably the last letter you will receive before Christmas. I must wish you all in it a Very Merry Christmas & Happy New Year, etc. Have the best Holiday season ever, and next year we will have a better one. Just learned yesterday that Jim McMichael died. Sorry, because, now the Fares, for whom I have no respect or use will be out in their fight. Oh well!

Good bye. Lots of love to you and all the children.

B.

Wm. T. Shoemaker.

Dec. 6th, 1917 [Thursday]*

[Le Treport], France[71]

My dear Dorothy,

Just a little shorty, because I feel like writing to you. So that shows that I am thinking of you. You have written me quite a number of very interesting letters, and theoretically all should have been pretty soon answered, but practically, it is very difficult to do such things as write letters. I am delighted with the work you are doing at the Art school and so glad that you copped both prizes about which you told me. It shows that you are all to the good & doing very well. I forgot to tell you to be careful in your art studies not to get artistic in temperament.

I think that she of the artistic temper^a^ment is sad. Paint the pictures you know, but do it without getting sloppy & "peculiar." Do not join the class which is described as being a "law unto itself." I hate that kind – both males & females. Am also very glad that you are continuing your music, because I like that. So much for the lecture.

Now as to things here. Cold as the devil, and with no wooly pajamas and no tiny slumber socks. One of my non-artistic temperament dislikes much the getting from the warm bed into the cold damp clothes. Which needs happen the once in the 24 hrs.

All of your friends here are doing well. I do not remember now just who they are, but it's a safe bet that they are alright. Our men have to work very hard. All are up carrying stretchers for every convoy & every evacuation, and convoys are generally at night or at the wee hours in the morning – 2-3-4-5-6 o'clock. So sometimes they get but little sleep. Just at the present time, the Boche is behaving very badly around Cambrai. He is awfully mad to start

---

* WTS to daughter Dorothy, ML, 4pp. on 2pp. of note paper; no censor number; APO postmark 7 DE 17; passed by G.B. censor.

with and he is awfully numerous. He is throwing everything over but the kitchen stove! But, he is getting awfully much killed, and I am glad of that. He evidently likes to get killed, because he comes right along in mass formation making a wonderful target for our machine guns. It is all dreadful business and it is difficult to figure out how it lasts so long.

I still stick to my original belief that the end is close at hand. I sure do hope so. It seems a very long time since I left home, and 7 mos. is quite a long time for some things. But your mother tells me how good you all are, and you all tell me how good she is, so I suppose it's all right. Of course, it's not good for the eye business, but that's only a side show in war times. The stay-at-homes and the profiteers are welcome. That's all – Good night – Much love –

Your Father

Merry Christmas, Happy New Year, Joyous Easter. Congratulations, love & best wishes to Dorothy and the others. Anything else you can think of that I have mentioned.

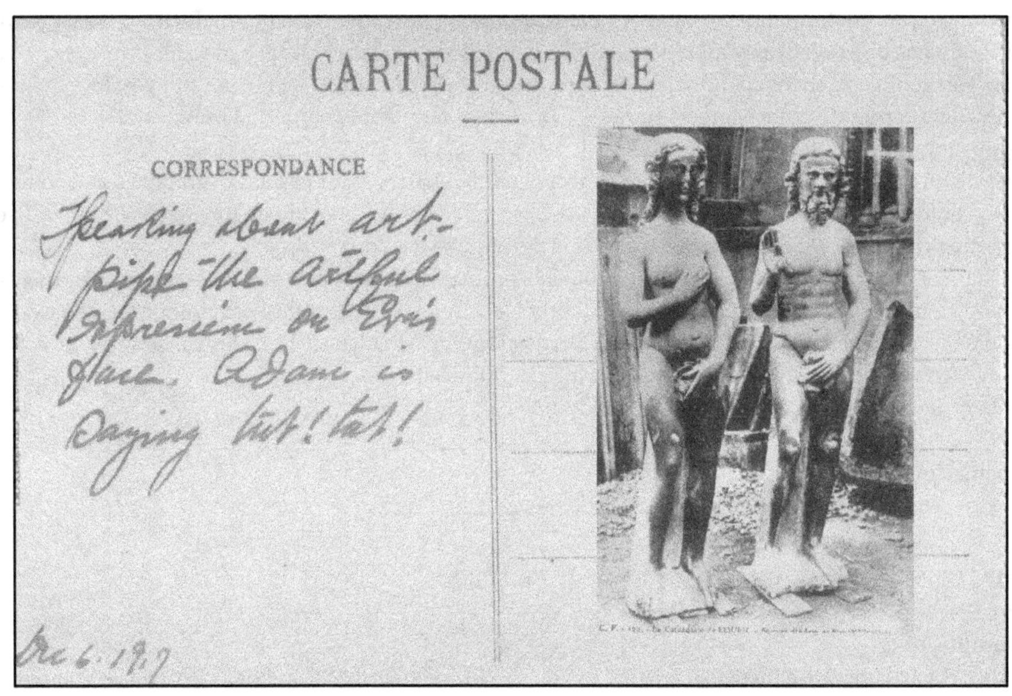

WTS wrote to BWS from Rouen, France, on Dec. 6, 1917,
"Thinking about art - pipe the artful expression on Eve's face.
Adam is saying tut! tut!"

Sunday, Dec. 9, 1917[*]
[Le Treport, France][72]
My dear Mabel,

Another week and I guess no nearer the end. Wooly pajamas arrived this morning – also a package from you marked – "Christmas mail" which I will not open until Christmas. The package from Strawbrige and Clothier has not come. But many other things have. Cigarettes & tobacco. My tobacco from Jack Philipps. Box of candy, an elephant & an enamel tray from Harvey & Nancy, a box containing Checkers, Chess, Card games of various kinds, –

[*] WTS to MWS, ML, 8pp. on 4pp. of note paper; opened by Censor 3231; APO postmark 10 DE 17; passed by G.B. censor.

a dozen pencils, etc., etc., from Phillips as Secretary. I presume it comes from the ____. Very nice of them. Some things, you see, I have to open at once. Others, I am keeping for Christmas. Mrs. Wall sent me some more eye washes. Cards are coming, – a very pretty one from Olive Alden and one from Charlie Judson. Our friends are very kind & thoughtful but I'd rather be at home. One good mail last week – 2 letters from you 49-50. – also letters from the children. Congratulate you on your play. I hear it was very good and that you were as good as any of them. Of course you were. You frighten me when you lose so much weight. Please stop it. I don't want you to get thin you know. You might get sick. Everybody tells me you are very well. I wonder if they write the truth.

We have read of the frightful calamity that has befallen Halifax, and are wondering if Carral Jack is safe. It looks like a Hun trick to me; it is their kind of work. They love to murder women & children, and unless investigation shows another cause, I would hand it to them. Hymn of Hate ~~XXXX~~ [English kisses crossed out]

You have spoken several times of Norman Gear. Look in my case records and I think you will find that he was a patient of mine. Sent to me by Vincent Lyon. He was a long time paying $5- and I think only came once. But that is the chap – Norman Gear. A good looking fellow the dark hair. He came from Washington. Look it up & let me know. Am not altogether surprised at what you say concerning ____. My job was to please my patients with safe & careful hands, and I was not so much interested in sending them to doctors that they would like as much as they liked me. I want some back some day. ___ is alright but now a bromide, and with such a wobbly spine. I hope conscription gets him, and some others in Phila. Well, I am glad Nancy did not go to Shumway. Griscom is a Quaker and a friend of Mrs. Chambers, but Mrs. C. will come back, I think. Have a letter from Stuart Smith. He is in London. The Lurie Science contributes, Ross, Stuart Smith, Moffat & myself. Not bad, do you think. Jack Phillips tells me that several others are wavering. I know which ones are not wavering, but can't quite seize on the ones that are. Some of my friends are very strong for staying home for [the] duration. Stuart Smith said he could not practice law while things were as they are. He is a fine fellow, do you know him? I am very fond of him.

Ralph just called. He brings me his letter to Censor & frequently calls anyhow. He never calls on Ed. In fact, I do not think that Lt. Edwin knows him. Sometime in the future, I think that Ralph at home will perhaps remember having seen Lt. Edwin Shoemaker in France.

I can't think of any more to write tonight, so I will stop for the present. XXXXXXXX. [English kisses resume]

Oh, all of our men have returned from the Casualty Clearing, and no more will go up until next Spring or Summer. Thank Heaven, nobody was hurt. I guess I will get up year after next. That's the kind of talk I have to listen to. Next year, next Fall, year after next. 1920 – 1921, etc. Very encouraging XXXXXXX.

Monday 12th. Cold & dreary. News not good. The hospital is quite full but we have trouble getting the patients evacuated to England, so that many of them are just carrying on, and do not require so much attention. Few operations at present. I will send you this week your birthday present – for January 5th. It will come in about 3 registered envelopes, possibly only two. Hope you will like it. It is "just a little remembrance" don't you know. By the way, did I tell you that my nurse, Miss Fuhrmann took care of Alice Jenkins Fuller when she had her last baby! She knows her very well & likes her, and also feels sorry for her. She agrees that Alice married a patch of lemons.

Tell Creighton that I have written him – 2 or 3 times, but will do it again. Possibly the Censor destroyed my first letter. I might have mentioned too many details of our place here. Remember me to Buchanan, and tell her I have been on the point of writing to her many times, but it's hard to write.

Good bye – I must stop. Please be good to yourself. Expect to be home in 4 or 5 years, according to the glooms over here. Will write again however before coming. Much love as usual.

B.

Dec. 11, 1917 [Tuesday]*
[Le Treport,] France[73]
My dear Mabel,

Enclosed is one third of your birthday present. I wish you many happy returns of the day, but sincerely hope that next time I can do it all verbally. I hope you will like these little handkerchiefs, but if not, I will exchange them for something else *apres la guerre*. Of course, I know nothing about it, but they are said to be very nice linen, and the

---

* WTS to MWS, ML, 4pp. on 3pp. of note paper; Registered; APO postmark 12 DE 17; passed by G.B. censor, with 2d stamp.

embroidery was done by my little French teacher last summer. Am sending three envelopes, each containing two. Hope you get them all, but this one especially, or won't know anything about it. See?

I know you will see the news yesterday, so this is only a little love note wishing you a happy birthday & many of them. Charlie Jack sends his love & wishes you the same. He is not my kind. Good night XXXXXXXXX

B. Wm. T. Shoemaker

France, Dec. 11, 1917

My dear Mabel,

Enclosed is another 1/3 of your birthday presents. All that I said about the first 1/3 pertains to this envelope. Am still loving you, only just a little more

Your B.

Wm. T. Shoemaker

France, Dec. 11, 1917

My dear Mabel again,

This is the last 1/3 of your birthday present. Now that you have all of it, you can blow yourself. This section contains just a little more love then the other love. Good night from

B.

Wm. T. Shoemaker

Dec. 15, 1917, Saturday night*

[Le Treport,] France

My dear Mabel,

I will start your letter and go until I am too tired, then will finish it later. It does not seem a long time since I wrote to you, but it is, and so one more is put behind us. But I wish a week of that kind was a month, and a month a year, i.e., looking backward. War news is not encouraging and things are beginning to come out not my way. I am sorry for my way was a nice way. Received some 2nd class mail during the week.

A very pretty calendar from Benj. W. Morris, but I have not the slightest idea who Benjamin is. Have tried to place him, but can't. He was very kind to think of me, and I will write to tell him so. In the meantime, can you describe him, and tell me who referred him, etc. Glasses & lenses from Joe Ferguson today. But your package from S. & C. [Strawbridge & Clothier] has not come. The one I have was wrapped & addressed by you; is about 14 inches long, 9 inches thick & 11 inches wide; it leaks excelsior, and is marked Christmas mail. Should not open it as I told you. You see, if your other package does not come, I will soon have something to open from you. Forgot to tell you that I also have a little package from Warren, not opened. I wish I could have sent you big nice expensive things, as many here sent their wives & families, but I had no money big enough, so you will see some grand laces, etc., around Christmas time, but they won't mean any more love. See? Then, I would have to give some c.p.'s [Christmas presents] here – Charlie, Ed, Ralph [Straub], Miss [Amina] Fuhrmann, Miss [Bertha] Elliott, Miss [Bessie] Metz, & maybe more. The nurses, except Elliott, are the ones on my ward.

You know how I get Christmas presents – my double [i.e., Mabel] does it. Well, I can't do it any better here than I can at home. The next time I join the Army, which by the way is not likely to happen, I am going to bring my wife with me. It's very much against the law, but we just have to try to find that in some way. Did I tell you that Frank [Isaac B.] Roberts is now a Captain? He gave a party down in Treport, The Regina – nice party.

Jud [J. E.] Sweet is a Major and I think Charlie is going to be one soon. Don't say anything about it, because it might not come through. But the Colonel recommended him I think. But the Dental Corps is getting well organized in Commissions corresponding to those of the Medical Corps are on the way. You remember Charlie holds commission No. 1 in the Dental Reserve and Ed is No. 2. Ed I guess will get a Captain.

Entered Ben [Wiliams B.] Cadwallader for a chat XXXX.

Tuesday, 16th. This is the most awful day yet. Cold & snowing for the first time. The discomfort comes however from the fact that we have no coal or very little, and the water has been shut off for a couple of days. Well a day like

* WTS to MWS, ML, 8pp. on 4pp. of note paper; opened by Censor 8678; APO postmark, no date; passed by G.B. censor.

this without coal is depressing. I have been frozen all day. Speaking of coal, how are you fixed? I read of a shortage, on account of labor and very high prices, also of blizzards in New York, and I suppose also in Phila. If I were you, I would buy some oil stores and have them in the service room you have. Read in Phila. papers today, that Con. Beran's young wife was killed in an auto accident. That is very sad. Did you ever meet Con – you remember his sister at our dancing series. He is a dizzy fellow. He was my resident in Pennsylvania [Hospital] and a good one. I am very sorry. There are more or less persistent rumors that we are going to move from here. One, rather authentic has it in May; another has it in February. For heaven's sake, I don't want to stay here until next May. Of course, both rumors put us with our forces for the next couple of years. The moving from here would be alright, but I would suggest a better place to move, than back of our own forces. And then, I do not like the careless way they have of making time afterwards. They hand out over two or three years in a very careless and thoughtless manner. It has to be said very fast to get by me.

Of course, I enjoy thinking, including the hope that there can be but the ending to this war – a complete victory for civilization & humanity, but why should it take so long. And think of the period of return to normal. You & I will scarcely live to see good normal prosperous times again. France is making money. Did you read that the British had to pay rent for the trenches? Also, large rentals for hospitals & ground such as we are occupying. It seems very funny to charge rent for trenches, but it's fact. Then all of the Allied forces have to pay for everything they get, so you see it makes a very big business. You spoke of Warren & Ted getting hit – the show. If so, I hope they get into some unit like this, for if they enter after their grandfather & father, trench warfare will not come easy to them. Then they might enter the Navy for navies do not sleep in tents.

I have exhausted all the news. Hope to hear from you tomorrow, or next day, or someday.

Love all –

B.

Wm. T. Shoemaker

Saturday, Dec. 22, 1917*
[Le Treport, France]
My dear Mabel,

Almost Christmas, think of it. Christmas without the one essential – Peace on Earth & Good Will toward Men. There is no peace on Earth and the good will toward men is limited. I am dreadfully disappointed as I was perfectly sure and confident that [in] this war would not there [be] a 4th Christmas. But it is about to take the 4th of many like any number more. You remember that I used to tell you that we could not escape our trials – that we were staying out but would have to get in for our freedom. Well we have started in and to get our full share may take a very long time. I read in the papers that disagreeable things are commencing at home. No lights, no food, no heat, etc. in spots. That is the day it is over here. Everybody leads a life of more or less discomfort as compared with life in normal times. Not actual discomfort perhaps, but just lack of freedom, much espionage, etc. In other words, all of the things that we don't like, never having been trained to them.

I am more worried about you at home than about us here. What our thieving politicians in Phila. got with the war game, they may make it very disagreeable for you. Speaking of our politicians, I see that our Mayor was acquitted of the murder conspiracy charge, but is indicted for some minor crime, with a view of course to impeachment. Of course, I hope they get him, but then he will only be traded for one of a similar kind. I guess the Fares are now on top with McMichael dead. All of which does not interest you from me over here. Here, very cold, 5 or 5 inches of snow – ice & icicles. Very prickly of course, but who cares for appearances. Preparations are about, simpler for Christmas. Our presents via Q.M.G. [Quartermaster General] have not arrived. Some 21 boxes. I am sorry, because the presents for the men & nurses were supposed to be in them.

Great excitement! A Zeppelin just went over us. He did not drop anything, or has not yet. There is a beautiful moon a-shinin', so that he could see us alright. We all put our lights out & hustled outside to see him, but could only hear him. Hope he does not get back for Christmas. Zepp excitement over – Dispersed caller XXXXXXX.

Tuesday 23rd. All ready for Christmas. Shopping done, etc. Most of the Wards have been decorated by the nurses & patients. Decoration consists of colored papers. Much use of mistletoe, flags, Kris Kringles, etc. etc. They

---

* WTS to MWS, ML, 8pp. on 4pp. of note paper; opened by Censor 3685; APO postmark 24 DE 17; passed by G.B. censor.

look very gay & festive. All patients have been put on chicken diet for Christmas; all unnecessary work has been ordered stopped. Invitations to teas & replies have been issued & received from all the hospitals on the top – for different days of course, and everything is lovely except that this is no place to spend Christmas.

Many church services are scheduled, commencing with Holy Communion Ch[ristmas]E[ve] Midnight Tuesday morning – then another at 6 o'clock – 11 o'clock, etc. The Catholics will also carry on. The Officers' program is – Christmas dinner tomorrow (Monday) night – 7 o'clock. Big party for entire Unit in one of the Huts at 9 o'clock. Tuesday – no special eats. We entertain the Nurses at our Mess. That's all – at other times we feast around and look happy. We never have much for patients – nobody is professionally busy.

I know just what you are doing. All of your shopping was finished last week except for the two wreaths for my office windows – those you will get tomorrow. The kids are home from school & busy with expectations. You all went to church this morning, and now have nothing to do but await the great day. How I wish I could be with you. Well, if thinking of you will help any, I'll do a lot of it. By the way – no American mail has come through for 16 days. So we know no nothing of your doings. You are to take your Christmas dinner with Harvey & his. In which of his residences? The winter palace on 34th St., the summer place at 2011 or the chateau at Cheney? So much for Christmas, until we know more about it. –

One more point – last year – a convoy arrived just as they were sitting down to Christmas dinner. That may happen to us. Have intended asking you if the Union League & the University Club remitted my dues. Did you receive a bill from either club? Also, do you sometimes use the League restaurant? I wish you would. You can you know use the Ladies Café just as if I were home. So go ahead. Also, what has been of my poor auto? Watch him carefully for I can't get another. Maybe [I'll] have to get a Ford – Ugh!! Will write again during the week & tell you all about it, but your letter would come first. Good bye. Much love to all & a Merry Christmas. B.

Wm. T. Shoemaker.

Saturday, Dec. 26, 1917*
[Le Treport, France][74]
My dear Mabel,

Christmas is over thank goodness. It is the third I have spent in this side of the Atlantic far away from you, and the worst. One in Buda Pest, one in Jerusalem, and one in Le Treport. A very thick & heavy gloom pervades the entire outfit and while everything was done to make us cheerful, and while we had every possible creature comfort necessary for a good time, I doubt if anyone really had it. What a difference it would have meant, had we received our mail, but it did not come through, and it is under within one day of three weeks since we have heard from home. Three weeks is a long time. We all wanted our letters so much and looked for them but in vain. Also our Christmas boxes from Q.M.'s General did not arrive. Which of course caused general disappointment. But our letters would have fixed us alright. You will be interested in what we did. Monday night the Officers had their Christmas dinner, and a good one, too good for war times, and the entire day might have been better than the folks at home had.

While we were at dinner the Singers – our Nurses (ours & an English Sister) and some men, serenaded us with the usual Christmas songs that we enjoy so much. "Come hither ye faithful," "Noel," etc. It was very pretty and as Ned Hodge said – "If they sing "Holy Night, I'll cry." These same singers started out at 5 o'clock Christmas morning with lanterns, and went through the Ward singing. It was very beautiful. I was awakened by the singing of "Come hither you faithful," outside my window for they also visited the Officers' Mess. After dinner Monday night for a party – as I told you – for everybody – Singing & Eats. The boys enjoyed it Christmas day. We entertained the Nurses at our Mess. Very fine here, but I thought I would jump out of my skin, and others felt likewise. Monday night, first at the time you were at supper and opening your things – I opened mine. It was quarter to 12 here – quarter to 4 with you. Good people had sent me gifts – a number of remembrances & good wishes.

The package from you contained the picture in good condition. It's very good and I am delighted & so glad to have it, but it is not complete, and I look in vain for Billie, and I feel so bad. I have not tried to mention Billie but not a day has passed since leaving home that I have not thought of him, and I have always thought I felt that he was over

* WTS to MWS, ML, 8pp. on 4pp. of note paper; opened by Censor 6316; APO postmark illegible; passed by G.B. censor.

here with me. Well to come back to the picture, you are all good, but you are too thin, you know. I don't want you too thin. I have the picture hanging beside me and I am very proud of you all.

Greta sent me a beautiful pair of gloves – dandies. Some lime, some cute little shaving things and soap. Warren some nice cigarettes; Miss Furhmann gave me a silver nose inhaler, Dr. Gibbon a box of cigars, the Germantown Hospital – Chocolates, cigarettes, pipe tobacco, chewing gum, and a card of greeting with the names of all the Nurses attached since our duty there. Very nice. Next Hodge – cigarettes. So you see, I had quite a Christmas for a soldier – Also, Molly C. 200 Camel Cigarettes. Molly C.'s present was well timed – marked "Don't open until Christmas," and I received it Monday afternoon – pretty good calculation, don't you think? Oh, I forgot, at our dinner Monday night, Santa Claus left a lot of toys for us, which Dr. Gibbon presented in one of the happy series. Mine was fuzzy bear performing on a horizontal bar. But the patients had a real nice Christmas – they had Eats & tea & smokes galore.

Well after the reception at our Mess, we felt so depressed that Ed invited me to dine with him in his place. He, Bill Drayton, Johnnie Flick & I went, but it was of no use. We scarcely said a word or cracked a smile, so I came back and at 20 minutes to 11 was in my downy trying to dream of better times after the war. And all because we did not get our letters from home – Wasn't it too bad. But we carried on & looked happy.

Thank you ever so much for the picture. Now that Christmas is over, I feel that a large part of our time (maybe ½) here is passed, and I like to think that we are on the home stretch. Of course, I must keep away from the glooms. The biggest gloom is Dr. [Richard H.] Harte [Director (Executive Officer) of B.H. No. 10]. He seems to have an obsession that the war will never be over in our time, and that none of us will ever get home and he loves to explain how nice & grand it would be to occupy "unmarked graves in France." He is welcome to one, but not for me. Whenever he makes a speech, he delivers just about the above. You know, he started at our banquet at the Bellevue-Stratford before leaving. I did not tell you but his speech there was a classic – he had us all in unmarked graves. There are many phases of senile deterioration; perhaps he exhibits one. Good night my dear, God bless you. Happy New Year. Love to all, B.

Dec. 29, 1917 [Saturday]*
[Le Treport,] France
My dear Dorothy,

Just a few lines now to thank you for the fine sweater, which arrived a few days after Christmas. It's a very nice one, and is the first sample of your knitting, keep interior decorating and think to pearls. It is very cold now, and sweaters are much in order. My Mabel – your mother , tells me you are having quite a gay time this winter. That's the stuff, I wish I were. Please watch your mother carefully & see that she does not work too hard – In the picture, which is fine (all but your right hand), she looks pretty thin to me. She looks nice though & very Mabely. I can't tell you anything, because I spilt all the Christmas news in my letter to your Mother. This is just a little thank you for the sweater. Good bye – Much love,

Dad.

Dec. 29, 1917 [Saturday]†
[Le Treport,] France
My dear Bobby,

The picture puzzles arrived safely. You are quite – an artist at this I think. Thank you ever so much. I will put them, in fact have put some of them in my ward for my patients to work with. They enjoy such things very much, for you know in a war hospital there is not much to do. While everybody is doing something for the poor soldiers, there are so many of them that it is very difficult to supply sufficient amounts. So you see, you could not have sent anything better – I get very good reports of you from home, as I think I told you in my last letter, but I am glad that they continue good.

It is not nice over here now. When I get out of bed in the morning, I find the water in my pitcher & basin frozen solid. Now you know how I enjoy that. Then my fingers get so cold that I can't button my blouse & snap my belt on.

---

* WTS to Dorothy, ML, 2pp. on 1p. of note paper; not opened by censor; APO postmark 30 DE 17; passed by G.B. censor.
† WTS to son Bob, ML, 2pp. on 1p. of note paper; not opened by censor; APO postmark 30 DE 17; passed by G.B. censor.

I want instead to have the thermometer – 55° or 60°. Now it stays around 28°-30° & I can't start anybody. So Bobby, I don't like it. My next war will have to be in the tropics.

Good bye, Bobby, thank you again,

Much love,

Dad.

Wm. T. Shoemaker.

Dec. 31, 1917 [Monday]*
[Le Treport,] France[75]
My dear Mabel,

This will be your last letter from me this year. Next year you will doubtless get many. At last, after more than 3 weeks, the mail arrived. Saturday & Sunday, we were very glad to get it, I can tell you. Packages also came. Your funny slumber socks, Dorothy's sweater, Florence's hand warmers. The slumber socks are alright but too heavy to start with. So I keep them in bed, and along toward morning when it is very cold, and when I always awaken, I reach down put them on. That is really a very good way. Thank you very much for them.

Then I got cards from the Westcotts & Mr. & Mrs. Aldrich. Was very much pleased to hear from them and hope you will tell them so as I surely will not be able to write. Also, a nice card from Miss Garrett, thank her for me, and Dr. Burnham, all acting under your permission wrote me a nice long newsy letter. Was awfully glad to hear from him. Your letters are up to Thanksgiving day. It seems like a very long time. All time is so long now that I am getting greatly deranged. It also seems to me that things at home are now made unnecessarily hard. We talk one moment about our tremendous & unlimited resources and the war sentiments [and then] publish food costs [and] propose great difficulties in getting ordinary things to eat. We have no trouble getting food in France. Why should we have trouble at home before we start! Don't you think there is a lot of Bull [expletive] somewhere? You ask about Dr. [1st Lt. J. Howard] Cloud – He is a lovely fellow and sits next to me at the table. I am very fond of him. True, they call me Bad Bill, Old Bad Bill, Uncle Bill, & Bill. I think [1st Lt. Arthur] Gerhard introduced the bad part of it, but as you suggest it is not as bad as it sounds.

Am sorry you are not getting your full pay – or rather did not for the moment as I figure it out. Please remember that hereafter there will be a slight reduction for the Liberty Bond, and also a reduction for insurance which I ought to take. I think I will take $6000 or 5000. I do not like the way it is arranged and indeed under such arrangements, it does not amount to much. You see, the beneficiary gets for 20 yrs. an equal monthly installment. That makes it never enough to amount to anything. Each month 1/240 of the sum total. However, it will just keep you in scruffy little white caps such as ladies wear whose husbands get mixed up in the last war.

Your remarks about Dorothy and her friends are interesting but I would be more interested if she would travel along American lines & cut out the Boches. Dorothy told me that Edwin B. was crazy to enlist and has been having much trouble with his Aunts on account of it. I hope he gets in the service and tell his Aunts to go plumb straight to _____ [Hell]. Barbara must be as tall as I am. Please do not let her get any taller, because while these great long-legged girls may be fine, you know I prefer than just your height or Dorothy's.

It's not quite so cold now. Last week, I had to wash in cracked ice, and think I shook from noon until night. A little of that goes a long way with me. You know better than I can tell you how I stand that kind of a racket. I would not mind it so much if I thought the war would ever be over, but when our wise guys at home tell us that we are just starting, and that it will take a century or so to lick the Boche, I am not sure.

Well my dear, I will try to write you more cheerful letters next year. A very happy New Year & Birthday to you. Good night. Lots & lots of love,

From

B.

---

* WTS to MWS, ML, 6pp. on 3pp. of note paper; not opened by censor; APO postmark 1 JA 18; passed by G.B. censor.

January 1918

Le Treport - 7 months

"I am so sick of uniforms it spoils my pleasure to see them around."

Jan. 5, 1918 [Saturday]*
[Le Treport] France[76]
My dear Mabel,

Your birthday; may you have many of them, and in the future, may I be on hand to celebrate them with you. Two nice letters from you this week – one particularly nice but in it you said two funny things – "be good" & "keep loving me." Now what do you mean by be good. Do you think I am not being good – and as for keep loving you, what else could I do? I am being as good as I can be, and I am loving you more & more if possible. So you need not worry on those scores. You will be pleased and interested in the enclosed slip of paper written by yourself last July [not preserved]. It's nice to know that your gift to a soldier in France actually got to him. He was given the pajamas, and found this paper in the pocket. He gave it to the nurse who gave it to Miss [Margaret A.] Dunlop who in turn brought it straight to me. So I went over to Ward 28 where he was to see him and the pajamas which you made. There he was. With them on [I am] warm as toast and proud as punch. Now isn't that a nice story? Incidentally, you make very nice looking pajamas. You ask about your letters. Very few are missing. Before me I have 49-56, with but one missing. No. 50, and it seems to me that No. 50 is in my trunk. I am almost certain that I received it. In fact, I doubt if any have been lost, for a long time. You do not please me any too much when you speak of Dorothy having a "steady." Please do not encourage it. I have very definite reasons for not favoring either of the ones you mentioned. I can't of course discuss them at this distance, but they are real reasons. Naturally, there is nothing to it, but don't let anything get to it. See?

From what you say, you must be the busy lady. Tell me more of the work at the Lankenau Hospital. Is it Red Cross work? How many go and who are they. Are the Pages and Whitings doing any thing at all for the war except being comfortable and undisturbed and making much money? I received a most interesting letter from the little Deaf Children at Bald. Will bring it home with me; it's a scream. About a dozen of them with their personal ideas, etc. It was very cute. One said, "I saw so many kittens yesterday – much love," etc. Then another: "I am so happy because it snowed." I must write them a letter and send them picture postcards. They have a service flag with a star for every doctor who has gone into active service. I have a star in it. It is nice but yet pathetic that they should be so happy with such afflictions. XXXXXXXX

Sunday, Jan. 6th. Weeks keep coming around, but it takes a great many to get far ahead. Here 7 months next Saturday and 8 also from home. Perhaps my time is more than half over. I don't want another 8 mos. As usual, no one thinks this war will ever stop, so I will just have to come home before it is over. It is so much easier to say than to do. However, it does no harm to say it fast & try to think it. It is still cold here, had to smash up ice in my basin again this morning. What did you do about my income tax; did you make a return? Of course, I had not enough to pay on, but the return is a nuisance. Received the ballots from the Union League. Very sorry they made Sproul president, because, first he is an ardent politician, slated to be our next Governor, and second, he is not a resident of Philadelphia, but has always lived in Chester. So I think the League has a bum president, but I don't care much just now. Well, news is very scarce. Nothing happens; we are not busy, and are all pretty well fed up.

Please take care of yourself, in case I ever get home, and incidentally for the sake of those at home, and incidentally again for your dear self.

Very much love to you & the children.

Love, B.

---

* WTS to MWS, 6pp., on 3pp. note paper; not opened by censor; passed by G.B. censor; postmarked APO 7 JA 18.

Jan. 12, 1918 [Saturday]*
[Le Treport] France
My dear Mabel,

The further I go in this affair the more I am impressed with your wonderfulness. You have such a fine spirit and are doing so much for everybody, that I marvel at your helpfulness. That is right old girl, keep it up and you won't be sorry. Was not much pleased when I learned that you were going to take Christmas dinner with H. & A. because I too am not in sympathy. So when I learned of your change of plans, I was more pleased. H. has contributed less for others – both in peace times & war times, than any man I know. He is indeed a cipher, and A. is an adventuress, who has been up to the present time more or less successful in her way. But such a selfish and irresponsible life as they live will waste itself in time.

Sorry to hear that Aunt Jane is getting so old. Hope she lasts until Charlie gets back. He has a notion she won't and frets quite a lot about it. He is terribly fed up and homesick – same as I am, but more so. If she should go West, while he is here, I don't know what he would do. Of course, he gets no help from Medea. And with her for a wife, it would be a pleasure for him to stay here indefinitely. There is great big difference in wives, both in personal appearance and in suitability. His sister is a very nice one.

This week's mail brought me 2 lbs. of dandy candy from Miss Jane and several lbs. Domino sugar from Dr. Turnbull. Wasn't it nice of Miss Jane to think of me? Everybody is awfully kind but the Germans, and they won't be liked. By the way, no German songs were sung at the Christmas Exercises at the Lankenau. That is, I do not think Horace Smith would allow it, and furthermore, I do not think anybody would sing them. You just find out; I bet there was not a word of German. German there is the thing of the past.

Before I forget, let me tell you about pay vouchers. I have sent them to Washington for Feb., March & April. You will receive from them $220. less $5. for Liberty Bond & less $6.48 for <u>insurance</u> or <u>$208.52.</u> I am sorry to cut you down, but it's alright. I have to take the Liberty Bond, and should certainly take the insurance. I took $6,000, which you will get when I die in 240 monthly installments. It won't be much, but when I am gone, every little [bit] will help. Also, if I become permanently disabled, I will get it, and give it to you. The Liberty Bond $5.00 of course only calls for 10 mos. or until $50 has been paid. In 3 or 4 yrs. I may be a Major, in which case I can send you quite a little more money. And in 14 yrs. or when I am 63, I will be <u>retired</u>, and expect then to come home. So you see it's not such a terrible war after all.

Now that's something in the nature of the gloom's hand out. A little exaggerated perhaps, but keep smiling – the glooms may be wrong. Everybody has been so far in this affair. Was very much shaken & sorry to learn of Dr. Dean's death. Strange – he was apparently healthy, young & strong, and seemed to take good care of himself. It's too bad. I hope Henry will not quit the Navy. You see, if he went into the business or professions, he could have a home & take care of his mother. But in the Navy – no home. To my mind the Navy or the Army would be the last to consider. Dr. Dean himself did not like the Navy. In fact, I do not think anybody in the Army or Navy would go in again. You see, they look attractive from the outside to the young man. However, I am speaking mostly of the medical branch, and Henry as I remember was going [to] Annapolis. That might be different.

I think you did very well with your Red Cross flag & your service flag. But I can't remember for Sheets & Pritchard had so many others. They must be surprising some servants. Your story about the clothes, etc., of your English friend is quite interesting. I am glad that Mrs. Roberts & Mrs. Bird Grubb thought they knew me, but I do not know them – or rather I do not think so. Don't tell them though, for it was very kind of them to help you out.

Now about the war sermons, tracts, etc. Send me one, but do not expect me to distribute such things, because you see about half the people in this war are Padres, and they attend to that department. It would be quite as unethical for me to distribute tracts as it would be for Padre [Edward M.] Jeffreys to distribute medicines to my patients. That would be very unprofessional. It would also be very unprofessional for me to dabble in the religious department. That's not my business.

<u>Monday</u>, Jan 14th. Now just a few words and will get this letter on its way.

If you will, I wish you would call on Mrs. Sweet. You know her, she was in our dancing "series." Dr. Just says she takes this whole thing very seriously and knows very few of the wives, etc., and feels rather blue. She is a nice

---

* WTS to MWS, 10pp., on 5pp. note paper; opened by Censor 6274; passed by G.B. censor; postmarked APO 14 JA 18.

little girl as I remember her, so cheer her up. I am not going on leave. I can't afford it. Hope you will send me some money soon (Feb. 1st) as I haven't any. I am just standing pat & waiting. But leave would be an unnecessary expense and I do not feel that I either need it, or care for it. If I could get away from my uniform, and the uniforms of others, I might enjoy it. But I am so sick of uniforms it spoils my pleasure to see them around. I would love to get in a blue suit & a sporty neck tie.

You might send me some batteries for the pocket flash light that Street, Harder & Crepart sent me. They will know the kind. I am out and have no flash light at present. Flash lights are very essential things over here. They sent me two and I had the smaller one, but without batteries they are not worth much.

This is a long rambling – disjointed letter. It will probably bore you to death – but I can't help it. Will do better some other time.

Goodbye. Please take very careful care of yourself. Much love to you & the kids.

B.

Jan. 26th, 1918, Saturday night*
[Le Treport] France[77]
My dear Mabel,

This week your letters 59-60-61-62 arrived and very nice letters they were. I am so glad that you all had a nice Christmas and from the letters everyone seemed to be thoroughly happy in spite of war times. You deserve great credit for running the show so successfully. You will be pleased to know that your letter wishing me a happy birthday arrived Jan. 22nd, right on the dot. And with it the other nice letters. I have them for a very happy birthday. Accidentally, I mentioned the day before to a nurse who said the 21st was her birthday and the 20th some other nurse's birthday, that the 22nd was mine. It was thoughtless of me, for she flew straight away and told my friends, so when I appeared in my ward next morning I was properly greeted. Miss [Amina] Fuhrmann, Miss [Bessie A.] Metz (the little one as I call her) and Miss Gault (Jessie) gave me a beautiful pair of fleece lined gloves, and Miss Fidler (Sallie) gave me a plant with beautiful flowers on it. Miss [Bertha] Elliott was a little sore because she did not know anything about it. Wasn't it nice of "my girls" to remember me? They are four darn nice girls, and they certainly are good to me. I have a good many girls here who will fight for me when I want them to, and you bet I will fight for them. You need not be jealous, because I am old enough to be the father of any one of them. Then at night Charlie & Ed invited me to dinner down town. We had a nice time, and altogether the 49th B.D. was quite a success. We will repeat it next year!

The day after, I received a package from Charlie, Lulu & Ethyl. Lulu sent me a dozen of the dandiest handkerchiefs that you ever saw. 6 were embroidered silk, and 6 white ones, all nicely marked. They are fine and I was much pleased. Also, a box of candy – very acceptable, and Charlie 200 Camel cigarettes, "*bien*"! I suppose they were intended for Christmas, but's fine to get them "after the rush."

I am now quite busy as I have been given a medicine ward of 40 patients in addition to my eye ward. So, I am an internist as well as a specialist. I would rather have a surgical ward, but in the Army, one does anything, and takes what is handed to him, and says Thank You. At present both wards are almost full. The weather is perfect, almost an Indian summer, and if it continues, the much "talked of drive" may be pulled off and the show ended. Wouldn't that be nice? When Mr. Boche tackles this Western front of ours, he is going to get licked, so the sooner the better, say I.

Crazy little bulbs are actually poking out of the ground; the poor seeds think it is Spring. I am afraid that along about February they will realize their mistake. The weather you have had must be a special war kind. You may be thankful that we took such care in rebuilding our house. You got through very well, but it must have been uncomfortable. It makes me sick, to see our great country fall down before they get really started in war. Where are our boasted wonderful resources, and our wonderful railroads, and general efficiency, when we can't even handle a cold snap? Unlimited coal and can't get it. Unlimited food, and can't get salt & sugar. What will it be like when we have been in the place 3 yrs.? Something must be wrong. We have coal, sugar & salt; perhaps France & England can

---

* WTS to MWS, 12pp., on 6pp. note paper; opened by Censor 6129; passed by G.B. censor; postmarked APO 28 JA 18.

lend the U.S. some until it gets warm. You at home are having a much harder time than we are over here. That does not seem right, at least not so early in the game.

I am surprised at your successful financing. I know now that our income this year would be what you made it, and you did not go into your personal income. Well I said I could stand it for a year, but if it goes another year, look out. It will drop to one half. You have done very well, and should be thankful. I am sorry you sent Norman Gear a bill; it was not worth it for $5.00. When I found that he would not pay, I dropped him, as I dropped many others. You know what a bunch of that kind I have. They are what I call "people." I can't understand Mr. Butler, unless he is hard up. If you want it, ask Geo. Müller about him. He seemed to be a pretty decent fellow.

About the tax: I heard that the Government was going to make the soldiers pay it, and that we would shortly be served with "shot at sunrise" papers to be made out, and that any amount they thought they wanted would be deducted from our pay. You can trust the Government to get us on taxes, so don't worry about it. If you should have to make it out, which I do not think you can legally do, watch your enemy's lines carefully.

Oh, I nearly forgot – Your birthday present is fine. Just the thing; how did you think of it – Tom Holloway! It has not come, but I will be mighty glad to have it when it does. Also, you can save $4.00 on the *N.Y. Medicine* if you can square up & find the date of subscription to me.

I have received from A. G. Elliott, I think, the bill is $5.00, but I refused to pay more than $4.00, which they were glad to have. I think I told you not to renew Malpractice Insurance. The agent of that is a friend of Henry Page – and is in business with Harry Lewis. He will renew it and send you a bill, if you don't stop him. Save $15.00!!! I cannot recall his name, but Harry Lewis will tell you all about it. My Accident Insurance I presume is cancelled. Watch out for it. If Campion sends you a bill for it $29.00, be sure that it is in force and just to what extent. My insurance will bother you more than anything else as does me. If I die though, you will be so pleased with it that you will wish I had more. Unfortunately, I have not had much, as it is very expensive.

Mr. Harvey will be pleased to know that an English officer here the other night, speaking of actors, told me that Martin Harvey was simply wonderful. That he was the best actor England had now. I told him that Martin's brother was a very good minister also.

Will write soon to the children. Ted spoke of "Camels" – I did not get them. Also, if Barbara sent anything, it did not get over. Warren however, sounded funny.

Tuesday, Jan 27. This is the Kaiser's birthday – May the Crazy Hun never have another. That is pretty near the world's wishes for him. No returns of the day. This was a Spring day – and the night is for gloves, with a moon just like that

Speaking of Majors, I read that Burton Chance went to a training camp for 4 weeks, and they kept him 6 weeks. He put up in a hotel because he could not "make both ends meet" – got released and returned home a Major. That's going some.

Dearest, are you really well? I hear you look thin & tired. Please don't do it. It's not worth while. Let them go first, because your health is most important. Now be good & mind me (me only).

Good night. Much love,

and please miss me.

B.

February 1918

Le Treport – Expectations on the Western Front

"There is no <u>leisure</u> in the gas business. You have but a few seconds between life and death."

Feb. 2nd, 1918 [Saturday night][*]
[Le Treport] France[78]
My dear Mabel,

Saturday night again, but I don't know what I am going to tell you, because nothing has happened. You have been reading a great deal about the Western front. Well, it's still there, and everybody is ready for a great drive. The Hun is very active just now, and very nervous. I feel that he is also a little doubtful of his ability to break through or come out of a mix up right side up. He sure won't, so let him try it soon. It was amusing how he had to bomb Paris by way of "reprisal." The "trash" actually dropped bombs on some of his cities. To think that anyone would do that was too much for him, and his vultures. Of course, he was over London at the same time, but that was different. If you want to read a good and interesting book, read *The Land of Deepening Shadow: German at War* – by D. Thomas Curtin. He writes splendidly and is most interesting. It will give you a good line on Germany & her methods, which by the way are now becoming very well known. "They always were a sneaky people." Ask Dorothy if she did not understand it, too. I think things are working around behind the End.

Read today that Mr. Wilson said in a speech that he expected a victorious peace this year. That sounds better to me than Mr. Taft's talk about 6 yrs., and Leonard Wood spoke about many years. That kind of talk is very discouraging, and it is refreshing to have someone speak in terms a bit reasonable.

Today, I increased my insurance to the limit - $10,000. It is such a bully & inexpensive proposition that I could not afford not to. I will carry the payments on the additional 4,000 here for the next 3 mons. – as my vouchers for Feb., March & April are already in Washington with the premium instead. At the end of 5 yrs. after the war, it must be converted or dropped. Or any part of it can be converted into one of the several kinds of insurance without medical re-examination and at a very much lower rate than it could be gotten for at that time. The monthly payment in case of death or disability is $7 for 20 years – not so bad. And the cost for the 10,000 is only about $11- per month. If you will examine some of the insurance which you have been paying you will see how very cheap this is. My largest policy is 5,000, and others at 4 – 3 – 2,500, etc. But you know by this time what they cost. So it's a good thing. Our nurses took over half a million dollars' worth.

No mail this week, so I can't think of a thing to say. Good night. XXXXXXXXXX

<u>Sunday night</u> – Two letters today – one from Creighton and your #63. Glad you spoke of insurance to limit. – I saw after taking $6,000 worth that it was worth going the limit, so was glad that Charlie thought so too. Then I received two Christmas presents, much belated. Frick, London & Prospect sent me a dandy pair of skin gloves lined – a sweater and 100 cigarettes. Wasn't that nice of them? The gloves I can't wear because first I can't get my hands into them, and second, they are a very light gray – almost white – But that makes no difference. I am just as much pleased. I have 5 pairs of good skin gloves, and 2 pairs of woolen gloves, so I am really not suffering in that way.

Now about the picture [gas masks, not preserved]. I was afraid to say much about the masks, and also, I thought that you people at home knew all about them. The complicated one is a box respirator and is the one worn in action. If it leaks or is in the way or what not, one must change in 4 seconds to the other one called P.H. Mask. In other words, the P.H. is an emergency one. You cannot live in it long. The respirator can be worn for 5 or 6 hrs. – and some have slept in them all night. Just how, I don't know, for they are very uncomfortable. About 20 minutes would be enough for me. When gas is reported, a signal is given for "gas alert." The respirator is then gotten out, ready to be pulled out of the bag and put on. Then would be the signal to put on and it must be in place in 3 or 4 seconds, or you are dead. As I said, if it leaks or goes wrong, you make a lightning change to the P.H. and beat it.

---

[*] WTS to MWS, 12pp., on 6pp. note paper; not opened by censor; passed by G.B. censor; postmarked APO 4 FE 18.

You wear also the tin hat. That is worn always at the front and has saved thousands of lives. I have not lost much weight, but have done away with superfluous fat, which accumulates from inactivity. I am a perfectly beautiful figure as it were!!! So my dear, your guess on gas masks was only partially correct. There is no leisure in the gas business. You have but a few seconds between life and death. When we are being instructed, if a fellow could not come through in time, the instructor would say to him, "You're dead." In fact, I think I died, if I remember correctly.

Now one word about prescriptions:

OD means right eye
OS means left eye
O2 means two years or both eyes.
If both eyes are alike, I write O2 – See?

You are a little confused on the life insurance. Most of the premiums I pay semi-annually. That makes them not come so often. The N.E. Maritime works the notice in the form of a note, but it is just the same really as any other notice. You have 30 days without interest with the N.E. Mer. I often use it, and you should if you want. Just do not let any of them lapse. Then unfortunately you get bills for interest. They cover other losses, not ours. You will have a rest now for about 3 mos., or a separation rest. Watch also the dividends and use them to pay the premiums. The Provident & Northern Trust will give good big dividends.

Heard from Ed. that the Pope [Joshua Shoemaker] had finally left the Vatican (2011) for good. Now that is interesting, but you did not tell Anne. Martha says she has a 3 yr. lease, but supposes that that will not bother the Pope much. She calls him the Pope you know. Now are you not glad that I waited until tonight to finish? Couldn't touch a thing last night. Good night; be very good until I get home – years hence.

Much love – love,

B.

P.S. I would pay the County Medical dues. After a certain date they add 10% and still later drop you for payment. The County membership carries with also the State Society and the A.M.A. I think they are kind of mean, but what's the difference. Lots of people are.

Feb. 5th, 1918 [Saturday night]*
[Le Treport] France
My dear Ted,

It is a long time since I have written to you, but you have been very good writing to me so that evens mother up. You seem to be doing very well in school, and having your share of honors. You must be a natural born President or manager. I congratulate you, and of course am very proud to know you, to say nothing of being your father. Am also very proud of your mother, not only because she is your mother, but for many other reasons also. You say you are in Class C. – C.H.S [Central High School]. I remember when I was, and how big I felt as I walked around with the older boys and looked upon the younger ones with patronizing assurance. When in Class C, I spelled just a trifle better than you have been spelling in your letters to me. You would enjoy getting with the games & goals over here. We have had base ball, foot ball, and at present are playing hockey. Do not understand that I am playing hockey – Oh no, we are. Tuesday nights – there are boxing bouts in the Y.M.C.A. – Then of course there are concerts, cinemas, theatricals, etc. The soldiers are entertained as well as they can be, and it is a good thing, because they have a deuce of time of it up the line. If you ever get to be the Emperor of Germany, think twice before you start a war. Aside from being a poor game, you would be very unpopular. And like the present Crazy holder of that title, you might not be able to stop it.

You spoke in your letter of Camels cigarettes, "fags" as the Tommy calls them. If you sent me some, I regret to say that they missed. But I thank you just the same. Also, remember me to Amos (Baer) and thank him for the package of Camels which he sent me ages ago. Well, I guess this will do for the present.

Take good care of Mabel, my Mabel you understand. Tell Barbara she comes next; then Dorothy.

Love to all, Your

Dad.

---

* WTS to son Ted, 4pp., on 2pp. note paper; opened by Censor 3341; passed by G.B. censor; postmarked APO 6 FE 18.

Feb. 10, 1918 [Sunday night]*
[Le Treport] In France as usual[79]
My dear Mabel,

Two nice letters from you and a nice one from Greta today. And yesterday – peanuts and two lovely pairs of woolen socks from Dorothy. She sure is industrious, and if she is to graduate this year, I am beginning to think that she is working too hard. Do not let her do that. Now the things not received are – Dr. Searfoss' package – Barbara's chocolate – Ted's cigarettes, the two stockings, etc. I told you I think that one Sunday some weeks ago. Everybody but me received a Christmas stocking. I thought it was alright because they came by Queen Mary's Guild and I had told you not to bother with the Guild. But I was the only bare footed one that day, so some one must have been wearing my foot wear. I am however just as grateful as I am sorry, and therefore thank our children just the same. The peanuts are great – best I ever tasted. By the way, before I forget, send me tooth picks – preferably Japanese, not any wooden kind. They are quite necessary and do not grow here so far as I know.

Now about the automobile – First, you are responsible for everything I own, including my wife and my children. As for the auto, I will give it to you if you want it. So, you told Josh just the proper thing. I am sorry that you sent him $81.00 – because you do not owe it to him. Also, he received the car with 5 tires, and should so return it. If you have not already done so, do not dispose of the tires. If you do it will cost me $100.00 to move the car when I return. An automobile without tires is a stationary. If the car is out of use, the tires should be deflated, and the car jacked. Then with a little attention, they will last a long time. If the car is really junk, which I do not believe, it never was – perhaps, you had better have Mr. Williams sell it. If however you sell it, do not eat the money, but keep it for a car which I will have to have when I return. That is, put it in a special fund, behind a new and less expensive car like a Dodge or Overland – anything but a Ford. It might be well to have Mr. Williams look it over for you and determine just where we get off. It can't be junk, but it may need a new battery, which Uncle Josh never thought about, and I think cost me $20 - $30 – but even if we have been to war, we can't resume our work *apres la guerre* without a car.

The 2011 proposition is very interesting, but why hesitate to ask questions. Sail right in and find out all you can. The callous disregard for the tenant is strange, but characteristic. I think I see a woman in it, and if I do not mistake, my "esteemed" brother [Josh] will have many examples of what a woman can do before he is though. Of course, "one of Daddy's patients" may be back of it. I wonder which one! The episode will certainly go down in history as one of the results of the Great War. Attributes – wartime morale – women – ambivalence – victory. I will someday write a lovely novel on the subject. I will start in the South & carry the tale to the North. It will be a best seller, I am sure. I suppose the Estate will make an apartment house out of the remains – a small one. Temporarily, our income will be reduced a little, but after a while, it will be a mint.

From news coming to this side, it would seem that the ravages of war are much more terrible with you than with the expeditionary forces. I cannot quite follow Mr. [Augustus] Garfield. If I remember correctly, he is a school teacher or rather a professor & college president. That has excused a great deal in the past of certain dignitaries, but even a professor should have a heart. He must be very uncomfortable with the country's industries stopped, and a million people put in a forced holiday unable to buy anything, or do anything. If however, that is going to win the war, I hope no one in America ever does another stroke of work, or eats another thing.

Were we not after all very lucky with our convoy? Of course, certain losses are known to occur, but think what might have happened. I am glad it was no worse, but sorry that it had to be at all. Hope none of our former patients are among the missing. The list has not yet been published. When it comes to troops, the Huns have gotten very few – not one in thousands. I think the submarine business is about played out. Of course, they can always get something if they keep at it, but they can't win the war in a hundred years. But the air plane can & will win the war. So make air planes *tout suite,* and send them right over here. They will give the Boche a terrible head ache.

Well, this is all I can think of just now. Will hold over until tomorrow and perhaps add another line. Good night, my dear.

Much love, B.

---

* WTS to MWS, 13pp., on 7pp. note paper; opened by Censor 3905; passed by G.B. censor; postmarked APO 11 FE 18.

Monday morning,

Just a word my dear to finish up yesterday's letter. Nothing has happened since of course. Forgot to tell you that I have finished the *Veil of Life*. I do not think a whole lot of it, because so much of it seems deficient to me. The author was not to my mind a clever writer. Then there are several extraordinary statements. For his time, in one place, he says "that one going to the other world would not know or recognize his father or of course his other relatives." Then we have always heard so much of seeing & remembering those who have gone before. So you see, I would not know or recognize my parents or Billie. That is not what they tell us in Church. There is another place – it says that all animals – horses, dogs, cats, birds – are plentiful there. If that is true, then I would be induced to think that animals have souls the same as people. Why not? Well, ministers, fathers, etc., were always very particular to explain that they had not. But, I guess it is alright if you know how to read it. However, I always thought it was gotten up for those who did not know how to read it. Well, I must now stop and get this letter in the box.

Good bye again,

With just a little more "love" than yesterday, ever increasing, from B.

Feb. 10, 1918 [Sunday]*
[Le Treport] In France[80]
My dear Barbara,

Thank you ever so much for the Chocolate – just as if I had received it and eaten it. Somebody ate it I guess. Much of our mail has been stolen, under the caption of lost. I understand you have had a very active & pleasant "social season." All reports of you have been favorable and have pleased me very much. By the time I get home, I suppose you will have your hair up and your dresses down, but you will have to reverse it for me or I might not know you.

There is a beautiful little French girl about 5 or 6 yrs. old who spends most of her time at Martha's house. Martha was my French teacher and lace "depot" you know. Also, my "do anything to help you I can" girl. Well, I always take the little girl Genevieve – on my knee – and Christmas I gave her a nice doll. She named the doll Philadelphia and thinks Capt. Shoemaker is some fine Santa Claus. Now when I go around, she promptly climbs on me and kisses me. She is a great little kid. I am very fond of her and quite often stop in to see her. So you see, I have a girl here as well as two or three at home. I do not see any girls here just your size. I guess French girls don't come in that size. They are either little kids or "working women." Are you studying French? You ought to, for it is going to be a very popular language. It is quite popular here already. You need not bother with German. Your grandchildren can study that as a dead language – just for mental training, you know. It will be an elective in the Classical courses. Barbara 3rd can take either Greek, Latin, or German.

Excuse the brevity of this letter. It is late, and I have just written to your mother, so will have to stop. Good bye. Much love from

Dad

Feb. 16th, 1918, Saturday†
[Le Treport] France
My dear Mabel,

Writing time again, which means that another one of those things called weeks has been disposed of and if this war with its duration was estimated in weeks instead of years it might be more encouraging. You are doubtless expecting as we are the big push. It was supposed to start yesterday, but nothing doing yet. If it comes it will be big and desperate and may end the war.

There is a feeling of great confidence on the part of the Allies, and I think I told you last week that it was doubtful if the Huns enjoyed the same amount of confidence. We have a great many men & guns massed on the Western front, but he has not more than we have, and our men [are] better than his 3 & 4 to 1. Russia's collapse

---

* WTS to Barbara, 4pp., on 2pp. note paper; opened by Censor 3905; passed by G.B. censor; postmarked APO 11 FE 18.
† WTS to MWS, 13pp., on 7pp. note paper; opened by Censor 3905; passed by G.B. censor; postmarked APO 11 FE 18.

certainly prolonged the war, but it cannot influence the outcome. The weather has been wonderful and everybody is all fixed up for the showdown. Wouldn't it be great if the war would end this winter or spring? Then we might get home during the summer – a fine time to come home. Well, there is not much use speculating about it.

This week, we had a visit from Bob Le Conte – Head of the Naval unit to which Ross belong. I was sorry Ross could not come with him. Le Conte looks fine, and reports his outfit – well & happy. Of course, they are not doing any war surgery – not a single battle casualty. But then, they are organizing and carrying on. Le Conte was the first Naval man to visit us – the first we had seen. His uniform was pretty because it was blue, and I am so tired of our brown stuff. You know, I love blue, and the only blue that the Army will let me wear is in my two eyes. I suppose they will someday issue one order – no blue eyes allowed. They issue orders of that kind in the Army you know.

But to return to uniforms. I think the Army has it on the Navy. The Army is smarter and more business-like you know. Oh, another little patch of blue, I sometime wear is my disposition. When I get home, I am going to wear blue socks, a blue neck tie, and carry a blue pencil. Also, on the subject of clothes, I will just have to get a new suit. My old one is done as you remember; can't be worn anyhow except occasionally in July & August. My O.D. [Olive Drab] was declared by General Pershing out of bounds, and I am living in my very nice but very light & cool "best suit" which we purchased some years ago for service on the Mexican border. I was keeping it carefully to go home in, but alas if I wear it continually here – not a chance.

Geo. Norris told me tonight that I looked like a Storm King in my summer tops; and Miss Fuhrmann told me when I took to the suit several weeks ago that I looked like an Easter parade. When cold, as it now is – about 30°, I feel like September morn and wonder if anyone is looking. The feeling is a little like that of one telephoning when in the bath tub. You know, how he instinctively wonders if it is perhaps possible to see over the telephone. So war or no war – Army or no Army – I am going to Eu tomorrow and have my carcass measured for a suit suitable for a soldier such as me, situated in sunny France, not in December, so I sure swear. Clothes like airplanes have gone up. The best I can do in Eu is about $40. Not too bad. But in London & Paris and I understand at home - $60. – I just won't pay $60.00 for a boom-a-laddy's suit of clothes. Just as soon as this little French tailor catches on to the advanced price of clothing, he will jump too, so I must go *tout de suite.* I am going to take Martha (French teacher) with me so that I can impress upon him the great necessity of making me what I want instead of what he thinks I ought to have. That is very important when dealing with English & French tailors. They like to dress you the way they think you ought to be dressed. If they dressed you, you look like the devil. However – leave it to me; we will have some outfit.

Now from Morgan Harjes carried the joyful news that they had 425 francs for me. I have written for it and it is momentarily expected. Thank you very much, considering the aforementioned clothes and insurance which I wrote to you about. I think your next issue will have to be about April 1st. But I will let you know. Tomorrow, I expect two nice letters from you, after which I will continue this letter. Good night, my dear XXXXXXXXXX

Tuesday – 17th.

Well my dear, your letters did not come – no mail today, so I have nothing to write about except the suit of clothes. We went to Eu this afternoon – Miss Fuhrman, Martha, Charlie & Ed, and I selected the cloth that I wanted & got off for $34 – not so bad. Others as I told you are paying $60. They are rich like Ed, etc. So I am quite happy & well pleased with my economy. Then of course we had tea with the usual trimmings of toast, butter, jam & cakes. We had a real nice time.

I forgot to tell you that "Ladie" (Miss Aldie) – druggist – Gen. Hosp. sent me a nice little elephant. I guess elephants will come to me where ever I am – war or no war. I have quite a collection in my trunk. When I get to New York, I may be taken for a dealer. Ladie you remember gave me the two big fellows on the office table.

Well, I wish that your letters came so that I could comment on the interesting things you wrote in them, but will have to wait until next time for that.

Good night and be a good girl. Please miss me a lot because I would not like it if you did not miss me.

Much love as usual,

Yours,

B.

Feb. 17th, 1918 [Sunday]*
[Le Treport] France[81]
My dear Dorothy,

I cannot remember whether or not I thanked you for the lovely sweater. It is very nice and you have become some knitter. I understand that you successfully warmed the entire family with sweaters about Christmas, and considering the weather you have had, you were certainly far-sighted in your efforts. Also, two pairs of soft woolly socks parked at my door along about the anniversary of my most fortunate birth. Also, some artist are you on socks. Glad you got a mention on your designs submitted in competition. Of course, you will eventually get the prize, because the prize is always given for the best work, and that's it.

Did you ever receive the pictures of Adam & Eve that I sent you some months ago? They are said by those who knew A. & E. to be very good likenesses. How they would have enjoyed sweaters such as you made for me, but if I remember correctly the man who invented the sweater was not born until <u>after</u> Adam & Eve. Of course, I may be wrong on this – look it up.

I am not quite sure about the advisability of doing two years' work in one. Unless you are in a hurry to finish, I would take the four years. I have just counted my money and find that I have just enough to keep you another year at school on the usual scholarship – or even without it. Of course, after the war, you will have to keep your mother & me because we will then be old and decrepit. But you will have plenty of time to prepare for that. I am so glad that you saw the Music Master as played by Warfield. It is indeed fine acting under most artistic settings. Now, when you see Warfield in the Grand Army Man or The Return of Peter Grim or in something else, you will be just a little disappointed, although you still will find him good. Both your mother and I enjoy most plays that we can think about and enjoy for a long time afterward. With me that is particularly true of the Music Master, for I still think of it and enjoy it – although I saw it some years ago. Another great play for me was The Little Minister – Maude Adams – and still others – several by Willard – especially the melodramas. Also, the Witching Hour – John Mason, and last but not least, The Man Who Married a Dumb Wife – Shoemaker – mother and daughters. The elder Shoemaker was particularly good and I often think of her.

While "on the Stage" – last week, a British officer stopped in for Tea, which we have every day [at] 4 P.M. – rain or shine – a nice-looking fellow, and a nice fellow just like the rest of us, carrying on in this damnable war. He was Brandon (?) leading man in "Peg O' my Heart" which played in Phila. for 3 mos. He played mostly the part of the villain. But you notice, he left a large salary, fine engagements – was an actor and probably temperamental – and became an officer in the British Army – interested in & devoted to but one purpose – that constantly in the minds of the millions in uniform – to lick the Boche and set the world straight again. And mind you – as he came to us, and as we saw him, he might have been John Triste of Oshkosh. He was a soldier, not an actor. I would not be out of this show for anything, but I would not protest hard if they would end it. But when it is over, to have been in it, will be well worth any sacrifices which we have made yet, or probably will make. I read today that Vernon Castle was killed. He is said to have dropped a salary of $10,000 – per week, for an Army commission.

Now that shows some stuff. And yet, I know a bunch of fellows [at] home who are satisfied to sit tight and use the war situation to their own advantage. They will be rich, but they are welcome to their gains.

Well, my dear, thanks again for [the] sweater, socks & such. Remember me to your mother should you see her.
Much love,
from
<u>Dad</u>.

----

* WTS to Dorothy, 6pp., on 3pp. note paper; enclosed in envelope with letter mailed on February 17 to MWS.

Feb. 23rd, [19]18, Saturday*
[Le Treport] France
My dear Mabel,

Here we are again, one week nearer infinity, which according to mathematicians is no nearer the End. However, winter is leaving us, the days are much longer and the customary early Spring is coming. In a recent letter to Dorothy, I spoke much, or wrote much of colors. Today, for Tea we had no less than a one than E. H. Sothern, your friend of Lord Chummelly, "my friend" of the sunrise Bell!! He is in the Y.M.C.A., attached to the Entertainment Department, naturally enough. In the Y.M.C.A. uniform and looking like a soldier, he was quite interesting. He is playing tonight to the Y.M.C.A. Hut in Abbeveille. He was with Mr. Ames, Dr. Harte's brother-in-law – a theatrical manager & promoter – also in the Service. Mr. Ames is in service as a Soldier, and with the Chief of the American Ambulance business – for the trenches, etc. It does you good to see such men, doing their bit, just like anybody else. There are men you know who think they are too big to stay out, and men who think they are too big & important to come in & leave home. It requires less bigness to stay home, but those who do it can't understand how things at home could move without them, because of their importance. They are too big to come!

Have just returned from a show in our Y.M.C.A. arena given by our nurses. It was very good. All musical, and we have some dandy voices, for solo, duet, quartet, and chorus. We also had a Madam Faller or something like that – who is from the Opera Co. in Paris. She played the violin wonderfully.

Another important happening this week was last night when I have a small dinner in payment of a bet made last summer, and I am happy to say lost. It was on the question of moving from here. Bill Cadwallader bet a dinner for Ned Hodge, Charlie & myself that we would be here Washington's birthday. I understood that we were to move in February, so I told him so. Well, I am very glad that we are still here, and we understand now that we will remain here for [the] duration or indefinitely. That is fine, for [all] four of us [don't] want to go with our own forces away from here. We had a nice time and enjoyed the menu which I enclose [not preserved]. For France, after 9 mos. it was fine, but it would not have amounted to much in good old Philly.

A nice big mail came Friday. You, Warren, Dorothy & Ted were in it. You were twice and that was fine. Very nice mail. Charlie got his usual 20 letters. Do you know that Medea still writes him countless pages every day? They all depress him, for she is eternally beefing about "the outrage" and her sorrow. He sighs and say, "Well I'll read another blue letter." And yet he is insanely devoted to her. It must be a form of insanity. Gosh, if I were not insane when I took that on, I soon would be, believe me. But his mother's letters are fine. All cheerful and free of gratitude and satisfaction with his work over here. She would not have him home under present conditions. It is kind of pathetic to see Charlie's devotion to Medea. She can do, think, act, love, or do no wrong. It's a crime. XXXXXXX

Sunday, 24th. By the way, Charlie is quite miserable with his teeth. Has had trouble since the start and has lost 3 good molars. The one lost today leaves him without anything to eat with but front teeth, and he must get a plate. It's hard luck to have Shoemaker teeth, and he sure has them. I have had trouble, but am better now, and have not lost any more yet. There were times though when the dental fraternity wanted to extract. But I am still in good form & will be until the next outbreak. You must not let Medea know about Charles. Oh no, the war is bad enough.

Now about money. Last week I received from Paris $75. which you kindly sent early this month. But today, I received notice from Morgan Harjes that they had for me 568 F. sent by you. This I do not understand. I have plenty of money for my needs and you cannot afford to send me so much. Of course, I keep pretty poor but that seems perfectly natural to me, and I don't mind it. And I guess it is better, because you know that money burns my pocket. You have been very good and liberal, but I am worrying all the time for fear you have not enough at your end. That is far more important than this end. This month has been a more expensive one because of insurance and clothes, and then my mess bill was also bigger. I do not understand the additional 100, so will let it rest in Paris until I find out about it. I have decided to borrow the 200 F. for my suit from Charlie and pay him back as convenient. He has plenty and the amount is only about $35 so that's alright. But you surely could not afford to send me so much all at once, and I do hope you have not put a crimp in your home accounts. However, it is perfectly safe with Morgan Harjes, and I won't touch it until I know more about it. In fact, I think I will keep it to come home with. Wouldn't that be nice? They say money doubles itself in 11 years so I will have a fine lot to get me home - $200 - $300 –

---

* WTS to MWS, 14pp., on 7pp. note paper; not opened by censor; passed by G.B. censor; postmarked APO 25 FE 18.

Have just reread your two letters and find so many things to talk about that I guess I had better not try. Am glad Josh returned the money, for you surely did not owe him a cent. Now have Mr. Williams look over the car and recommend necessary repairs. Then be sure and do what is necessary. Then tell him or Dr. Page that the Hospital can use the car anytime it wants it; then send up one of your chauffeurs and use it yourself any time you want, being sure of course to pay for your gasoline and oil, which you get from the Hospital. Any moderate "fixing" you can probably get Mike & Frank to do for you much cheaper than it could be done elsewhere. In other words, do as I used to do, and go see Mr. Williams and learn how to do it. See? I would like you to get some use out of your car, and if you are careful, it will not cost you much. Would also like the Hospital to feel free to use it. They have and are helping me a great deal, so why should I not help them?

Commissions: Charlie & Ed are to be made Majors in the Dental Corps. You need not worry about your Captain. He has not a chance for promotion. All recommendations must come from the C.O. and our C.O. is not fond of me. So there is nothing doing for me. XXXXXXX

Ralph Straub stopped by today.

Monday – 25th.

Well, my dear, I will try to finish this letter. I was speaking of Majors. Would like the money, but otherwise do not care a snap for them.

Captains are very nice,
Majors may be alright,
and Colonels are rotten.

This really is a very long letter, and this is not my time for writing (daytime) so I will not try to write more just now. Much love, etc.

B.

P.S. Got a dandy box of candy from Molly C. last week.

March 1918

Le Treport – The German Army Attacks on the Western Front

"The German offensive has started and we are indeed busy."

March 3rd, Sunday*
[Le Treport] France as usual.
My dear Mabel,

Your letters explain the $100 & other things arrived Friday. It was now very good of you to send the money, but you must not worry about me. I have at the present time no notion of taking it. However, I feel much better with some money now at hand for a leave or an emergency if necessary, so I wrote to Morgan Harjes and asked them to keep it until such time as I advise them. Thank you very much. At present, I do not need a leave, and do not care for one. Now, if you think better, I can draw on this money instead of calling on you in the usual way. I will need some early in April, and if you send me some, about $25., which I will ask you to send upon receipt of this letter to Mr. Rutan for Miss Fuhrmann. Now do not forget. Send Mr. Rutan $25. *tout de suite.* She – Miss F. – will give me a like amount April 1st or 2nd or 3rd or 4th.

You asked about Gerald [J. Sullivan, Pvt.] & [Capt. J. Paul] Austin. I have purposely not discussed the doings of individual members of our Unit, because such things come right back here, in just the time it takes a boat to cross the Atlantic Ocean twice. And then of course, they gain in weight, strength & appearance on the sea cliff. However, there can be no harm in telling you now that Gerald is where you think, about 2 miles from here, and while not detached from our Unit, has been loaned or detailed for service there. [1st Lt. John B.] Flick is also there now.

Austin's story is quite different. He has from the start had a raw deal. You know he & 5 or 6 others came as reinforcements, sometime in July. Austin was in command enroute and I believe did not make good, to the satisfaction of Washington. After a short time at U[nit], he was detached to 3- on the top – about 2 squares [kilometers] from here. That was preliminary. He was shortly permanently detached from our Unit, and the British forthwith sent him up the line. He kept going until he was finally in command of stretcher bearers in the main land. That's where the dead Boches are you know – and the dead Allies. You of course follow the troops over the top. Then he got P.&.O. and was in a Hospital and then I think he was with a Field Ambulance. Some experience for a nice fellow like Paul Austin. But, the worst part of it was that just as that time his wife was so ill, and she naturally did not tell him of her vicissitudes. And I did not want to run the risk of spreading the information. I was very much worried because I was fond of him, and especially fond of his wife. You know she was Margaret Thomas, and was on my special list of "dears"! You remember that I have such a list; well she has been on it for years. She is a lovely girl. Then if a fellow gets in the front line, as Paul was, he is very apt to get gassed and killed. Well, I hoped so hard that Margaret would come through alright, and that Paul would not get killed or much hurt, that I guess I saved both of their lives. But, it was a raw deal. Then the usual Army side of it was amusing – because they did pick a good man for the job after all. They put Austin over the Top!!

You know by this time the sad news that came by cable yesterday. Mrs. Harte is very ill and from this, I hear it looks bad. I hope she pulls through, and I hope Dr. [Maj. Richard J.] Harte will go home where he belongs. He is such a stubborn dick, to put it mildly, that I am afraid he won't. I think I told you once that he does not consider that a fellow's duty was done over here unless he is buried here. He has a very well refined and set notion about "unmarked graves." I really believe that he hopes to fill one, and would like as many of us as possible to do likewise. But it's tough luck and I do hope that Mrs. H. will recover.

This week, the boys have been giving their play, "What happened to Jones." It was simply fine – as good as any professional show in London. I saw it twice. It is very funny – on the order of "Charlie's Aunt," I must say. Costumes from Paris and everything else made here. It was a great success. They are going to take it to Dieppe tomorrow. You have no idea what some of these boys can do. If you ever hear of the war ending, kindly let me

---

* WTS to MWS, 14pp., on 7pp. note paper; not opened by censor; passed by G.B. censor; postmarked APO 25 FE 18.

know, for I do not want to miss it. Such a thing is not considered possible over here, so it might of course happen without our knowing it. Now I will stop for a while and perhaps write just a little more tonight. XXXXXXXXXX

Having finished my cold Sunday night's supper in the cold mess hall, I will return to you. Speaking of cold reminds me that I have not yet discussed the weather. Well, it was fine until March 1st, and how it knew that it should change on that date, I do not know, but it did. It changed to the real March weather with wind, snow, sleet, cold, etc. And so it continues.

When is Ann going to have the doctor's office? Or is she going to retire him. It's too bad that he will have to live with his family, but wars always cause great suffering for the rich ones as well as for the proletariat. I received a Valentine containing 400 cigarettes. Very nice and I think I have ordered the send, but must await developments. There was a written verse in it and I will in time identify the sender.

Ted tells me that he is still taking on honors. Good work for Ted. Congratulations. Bob says he attended at "Mascerade" – alright for Bob. Give my congratulations and best wishes to Dorothy Roberts. Now I have finished everything. Don't you get tired of my letters? Some of them I am afraid are rather stupid. And, can you read them? I cannot. From here on is just to love – much for you of an only kind. Much for the children of very fine quality.
Good night my dear,

B.

Sat., March 9th, 1918*
[Le Treport] Same old France[82]
My dear Mabel,

Would you believe it, Spring has come and beautiful daffydils – just as big and yellow as Mr. P____'s are growing wild in fields. You can pick a dollar's worth in a minute – Philadelphia dollar's worth. Little Miss Sally is making up its mind and should be along in a few days. We were very fortunate in our winter – which was mild. Last year they had one more like the one you have been enjoying. I do not know what kind of a winter we will have next year, but fancy I will be here to find out. The war seems to me to be settling down with a double defensive proposition as it were. Well a defensive battle seldom wins so I presume a double defensive will never win. In other words, the beligerants [sic] can sit and look at each other for 4-5-6-7-8-9-10-50 years. It costs much less and it does not kill you so much, don't you know.

And both sides are getting tired of being killed so much. It's not so bad to be killed just a little. Well, to come back to the beautiful Spring time, I took an interesting walk with [Maj. Joshua Edward] Ted Sweet this afternoon and in addition to seeing the various Spring time I saw two things that I never saw before, even though I am now an experienced soldier. I saw a tank – many tanks, and even had a short ride in one, and I saw four or five air planes start from the ground. It's funny, you will think that I had never seen a tank, and as for never having seen an air plane leave the ground, well you will think that is ridiculous. I had been to the tank camp, but at night, and while I heard them playing around, could not see them. But today, they were playing all around me and close by was an aerodrome, from which the little birds issued. Air planes to me are fascinating. They will win the war. Tanks are nice too in the way that ~~Hippotomas~~ Hippopotamuses (gee, that's a hard word to write) are nice. Tanks look and act just like Hippos – walk like them, and are just as dirty. The Tank Corps men love them, but give me the air plane. You have noticed of course, the Allies' superiority in the air lately. They have it all over the Boche and if the Archbishop of Canterbury did not object so much to reprisals, G.B. could make those beautiful German towns and cities look like Camden, N.J. Then you can bet London & Paris would not be bothered. You know the Huns invented the game of bombing open towns, but strenuously maintain that there are no open towns outside of Germany. Here they can visit London & Paris every night, but only barbarians would bomb Berlin or Cologne. That would be against all international law & treaties. Damn the Huns.

Mrs. Harte is still very sick and the Dr. requested leave to go home. It was turned down flat. He is only 62 yrs. old – Came in as a volunteer, has given liberally of his time and money, and has helped his country in every way that he could. He gets his reward. Nice Army. How I do love the War and Army. How I wish I were younger, so that I could give to the Army more of my time. How sorry I will be when the war is over, to have to get out of the Army –

---

* WTS to MWS, 16pp., on 8pp. note paper; opened by Censor 3923; passed by G.B. censor; postmark illegible; received April 5.

and sever my connection with the wonderful bunch of men running it. I feel sorry for Dr. H. but think he may still get home. He is going to G.H.Q. tomorrow and I hope will succeed in getting what he wants. You see the Army is very strict. For example, it would be quite impossible for a man to go home just because his wife died, even though he had 22 dependent children and a couple of widowed mothers. But he might easily go home to read a paper before a medical society on "What I don't know about war surgery"; or "Cross-eyed men, I have seen in the Army – How can they shoot?" Now you see the supreme wisdom of the military – a cross-eyed man might occasionally hit a Boche, but a widowed mother or 22 dependent kids, never. So my dear, please don't die, because I hate to read papers, and naturally enough, I would want to go home to read it, under those conditions.

Now the next important part of news is that I have a new suit, and it's a *bon.* If it were not a soldier's suit, I would like it very much. It's a dandy camouflage, because anybody seeing me in it would think that I was a real soldier. They could not possibly guess that I was only a poor doctor.

Oh, I almost forgot some very interesting news. Col. Dr. [Matthey A.] DeLaney leaves us Monday. He has been ordered to London as a Liaison Officer. Everybody is just tickled to death – the whole camp – Dr. D's Nurses, and Men are on top with glee. He is so popular you know that that's the way they show it. Funny, isn't it? If they should weep, it would be because "tears cannot restore thee." Major Harte will be C.O. so long as he is here, and when not here, [Major] George Norris. Norris will be a good C.O. Major Harte will be an awfully good Dr. Harte. Did you ever happen to know a Dr. Harte – they are fine, but are a complete education on running things on the bum. The characters of a Dr. H. and a Major H. are just the same but more so. By the way, speaking thusly as above, reminds me that the other afternoon at Tea, I was speaking of daffydils and Easter. Harte said, "Shoemaker doesn't know anything about Easter – he is a Swedenborgian." I shot back, "Don't worry, we have more Resurrections than you have." He laughed – He's a Unitarian. But, I was surprised that he knew or remembered what I was, because I had always though he knew and remembered nothing. You never can tell in war times.

Now some family gossip. Charlie's recommendation for a Majority went through and the Commission will arrive in course. Ed was turned down for a Majority, but will get a Captaincy. The headquarters A.E.F. told DeLaney – "H---, he only graduated in 1912 – nothing doing – too young." Ed is just as much pleased, and he is really too young. Medea will now be greatly pleased with Chawles, and I hope will let him alone for a space. Returning now to Capt. Edwin – he is up in Isolation with German Measles. Please don't spill this until you hear it from some other source. He is fine as a fiddle because you know German Measles do not hurt. You can have a whole lot of them and be perfectly comfortable.

Received today the package from Dr. Stauffer about which you wrote some time ago. What it has been doing I do not know, but it arrived in good condition. Everything O.K. Fine. Am expecting letters from you tomorrow. Will add to this or write another if you tell me anything nice. You certainly will be glad to reach the end of this letter, for it is too long. Fancy Medea writing this much to Chawles every day and then having them accumulate in transit and arriving here 15 or 20 strong. Goodnight my dear, with kindest regards,

Yours sincerely,

Wm. T. Shoemaker

How do you like that? Now you think I don't love you. Ha! Ha!

Tuesday, 10th – No mail, so I just won't write another word. A reasonable amount of Love.

FINANCE DIVISION*
OFFICE OF THE DEPOT QUARTERMASTER
WASHINGTON, D.C.

January 30, 1918
From: Depot Quartermaster.
Captain Wm. T. Shoemaker, M.C., U.S.R.,
To: Thru Chief Q.M., Amex. Force.
Subject: Pay accounts [s] B.H. #10

1. Your pay accounts for the months of February, March, and April, 1918 forwarded under the provisions of Army Regulations 1259, as amended, have been received for payment at maturity.
2. Your vouchers will be paid in the sum of $208.52, War Risk Insurance in the amount of $6.48, and $5.00 for allotment to the Liberty Loan, having been deducted therefrom.
3. The 10% increase for foreign service is included in these vouchers.

By direction of the Depot Quartermaster:
P. L. SMITH
Lt. Col. Q. M. Corps.

[Stamped] MAR 2 18 HEADQUARTERS AMERICAN EXPEDITIONARY FORCES

Saturday, March 16th, 1918†
[Le Treport] France
My dear Mabel,

Not a very long letter tonight because I have so little to tell you. Of course, you know that the expected happened, and that Mrs. Harte died. It's very sad to think that this had to happen at this time, and that the doctor was denied the right you might say to go home to his wife and family. He has been in Paris all week, and has with him his brother-in-law – Mr. Ames – and his brother, Mr. Sproul, Mrs. H's brother-in-law – also Mrs. Ames. So he has as least some of his family with him. He was expected here today, but did not arrive. What he is going to do, I do not know but may yet get home.

We are busier than we were on account of two convoys this week. All gas cases, and we do not like gas cases. They are very distressing. Gas is bad stuff to get into. The Hun has been throwing a lot over lately, but I guess it is going the other way, too. It generally does. We are still much interested, and uncertain about the coming Boche offensive. He is getting quite active and may break forth in all his "glory" at any moment. If it comes, it will be big. If he puts it off, he will certainly have more Yanks against him, and by the way the Yanks are doing pretty well. Mr. B. won't like them any more than he likes the Ico's [Canadians] or the Anzacs [Australia-New Zealand]. The weather has been simply perfect for about a month – the best fighting weather you could get. When they start, I presume the weather will change. It always does, especially when the British start something.

Last Saturday night at 11 o'clock, we took over summer time. At 11 o'clock, it became 12 o'clock. So we are now 6 hrs. apart instead of 5. Our evenings are getting quite long, both naturally and on account of the hour we lost. In the Fall, we will go back to real time, so we have not shortened the war an hour.

Your letters this week came Monday. Also, ones from Dorothy, Ted, your brother Wallie [Walter], etc. I was very sorry to learn of Dr. Turnbull's death. I wrote to Mrs. Turnbull as soon as I could. But that kind of letter I do not enjoy. I have been intimately associated with Dr. T. for 27 yrs. Since 1891. Now that is a very long time. He was much interested in me and very kind to me, has written several times, and has sent me various things in spite of the fact that he was always very busy. Of course, you called and offered your services, but as always there's nothing to do. I am so sorry for Mrs. T. with her children all away and their husbands in this darn war. Ted tells me he is now a senior with various duties, and also V.P. of his class. He sure is some leader, and if he continues may have a brilliant future!! How near conscription age is he? If anything, will they allow boys like Warren & Ted to finish their education before pulling them into this show? I surely hope so. But you never can tell.

Now you called me "Goosy" because I smiled at your food situation. Of course, I know why it is, but do you think perhaps the "Professors" are, in their theories and enthusiasms, going a bit too far? As I understand it, things

---

* Enclosed with WTS letter of March 9, 1918.
† WTS to MWS, 12pp., on 6pp. note paper; opened by Censor 3575; passed by G.B. censor; postmarked APO; received April 9.

are piled mountains high awaiting for ships, the plans for which are being drawn. So I was just wondering if it would do much damage to eat a little of the waiting stuff in the meantime. At all events, the U.S. is feeding the Allies very well, and if it will be able also to feed itself, it will be fine. Most of our meat comes from South America, and it is very good. I am not always as Goosy as I sometimes ~~sound~~ seem. And Goosy or not goosy, as I told you, I cannot follow Mr. Prof. Garfield. And my *Literary Digest* tells me that few can. Of course, the Prof. may be right, but it is sometimes difficult to follow even right Profs. I don't read so much about him lately, is he still doing stunts?

By the way, what happened to the Income Tax? Medea paid Charlie's and sent him a copy. I hope you did not have to make out a return. It's a great nuisance, and in our case would not aid the Government much in financing the war. Your Liberty Bond will not be delivered until paid for; this we are doing monthly. But you have one alright. All of this is very interesting I know, but what can I write about? Nothing. So I will stop for the present. Tomorrow may bring a mail. XXXXXXX

Tuesday. 17th. Tomorrow did not bring a mail. Dr. Harte returned today. Poor man, they won't let him go home, so he just has to carry on. I have an idea that "Powers that be" think this is smart & efficient, but I can't see it. Believe me, if I were in his place, they would be glad to get rid of me "for the good of the Service." I don't think I would be much good to them. Charles has a grouch – sounding home sick & war weary. I took a nice walk with him along the Cliff this afternoon, and dined at the Cliff House. Had a nice time, but gee, we are fed up with this show. Ed returned from Isolation today. In about a week, he will return to work. He is very well, and I think enjoyed his rest or change.

I will now telephone:-

Miss Fuhrmann

Dr. Shoemaker, Le Treport, France

Please send me: -

About 12 dozen eye droppers from Baers. He will send them, same as last.

About 100 dark glasses from Short, Linden & Prospect.

Try to pay for them, but don't try too hard.

Toothpicks I think I ordered.

Thank you. Good bye –

That is all for now present.

Hope I hear from you tomorrow. Is this not Warren's birthday? Wish him for me many happy returns of the day. Is he 20 or 21? I rather think he is 20, but am not sure. His is the only birthday I know and that is because it is the ____'s day. I know one other – the day after or before St. Valentines' day. That I think belongs to Barbara. Bill was about a week before Christmas – and Dorothy toward the end of Feb. But I cannot tell you whether Bob and Ted were born in the winter or summer.

Good night my dear – Much love to you & the kids.

B.

Saturday, March 23rd, 1918*

[Le Treport, France][83]

Well my dear, as you know the German offensive has started and we are indeed busy. Our C.C.S. [Casualty Clearing Station] received hurry up orders and went straight away to the front. [Edward B.] Hodge & [Henry K.] Dillard and Miss [Isabella] Stambaugh. Thursday night we received our first convoy from the drive, and since then, they have been very coming in 3 or 4 a day – and will so continue. Everybody is in use, and everyone is working under pressure. It is a fearful show, but I really believe it is the start of the finish. What will be left at the finish, God only knows.

The casualties have been horrific on both sides, but much higher for the Huns. We were told today by a fellow that the German dead were a great obstruction to their progress; that they were in places piled 8 & 10 ft. high, and that masses of dead had to [be] literally blown away with explosives.

---

* WTS to MWS, 6pp., on 3pp. note paper; opened by Censor 3029; passed by G.B. censor; postmarked APO.

Rumor has it that one of our Casualty Clearing Stations was captured. In other words, the Western Front has gone now and is extremely busy. I am perfectly fine, though my guess is that this engagement will be the last great one, and while its loss will be unprecedented, the end has begun. I am tired and cannot concentrate on your usual Saturday night letter, so I must stop and go down to my ward.

By the way, before I forget it – I have not missed a week in writing, have something written in a week, but not lately. But we received notice that all mail for the U.S. sent between Jan 19th – 22nd or thereabouts has been sunk. Doubtless other mails have been lost. Some of our fellows had beautiful lace etc. on the destroyed boats. Then, I think perhaps the Base Censor has destroyed some of my letters; he sometimes does it when it is easier than reading. Well, good night. More tomorrow. XXXXXXXXXXXX Little loves.

Monday 25th. There was not time yesterday for letters. We never had so many patients nor did so much work. The Operating Room ran from 9 A.M. to midnight, and we had about 65 operations. Today we are pretty well caught up and are ready for the next convoy which is on the way. We hear the most exciting rumors of great things up the line. The Huns broke through all right, but they say few or none will ever get back. All those in a position to know seem to be entirely satisfied with the situation, and say that everything is happening just as expected. But you have better & fuller news than I have. Just keep your eye on this battle line as "the one."

Was very sorry to learn of Mr. Burnham's death. He was a real man, good & useful. Regarding note: See that interest is paid to date and then as you suggest, wait & see if you are called upon to do something, see [your brother] George about transferring the debt for the present. That is all that you can do just now. Don't sell your stock, I think. Another thing that you can remember is that Horace Smith was very anxious for you to know that he would be only too glad to do anything for you he could, "financial or otherwise." He meant what he said and if cornered, you need not hesitate to transfer your security to him. Of course, do it only if necessary, but do not feel that you can't be sold out. So don't worry. Personally, I don't think you will be disturbed because now all executors are afraid of the terrible Orphan's Court. I have known some that were.

My mail this week did not come. Last week, I received an interesting letter from Owen "from the firm" and also quite a newsy one from Burnham. If you will excuse me, I will stop for the present, and try to write again soon. It's hard to write during an exciting battle. Much love as usual. B.

30*

Mrs. Wm. T. Shoemaker
109 So. 20th St.
[Philadelphia]
Easter

"For remembrance sake"
Love and Easter Greetings
From yours
in France.
B.

---

* WTS to MWS, in envelope mailed April 28, 1918; received May 20; opened by Censor 6048; APO postmark 29 AP 18; 3x5 in. envelope with these lines written by WTS on the outside; inside, 5 lines on 3x5 in. once-folded note.

April 1918

Le Treport – The Big Evacuation

"Amiens is still in our possession" but "Rheims is said to be no more."

April 7, Sunday*
[Le Treport] France[84]
My dear Mabel,

I want first to congratulate you upon doing your Christmas shopping early this year. We received today many nice Christmas presents and while I know you would prefer that I do not open them until Christmas, my curiosity was too great and I sailed right in. Sorry. Next, I will have to enumerate so that you can check up.

400 cigarettes from Ted. A Victrola record. Cigarettes from Owen. Two silk handkerchiefs from Aunt Linie [Caroline] though Charlie, but not Christmas stockings. I guess they are lost. Two letters from you. One from Warren. One from Mrs. Mrs. Adie – Gen. Hospital, and a memorandum from ___. As it was my first mail for about 3 weeks, I was mighty glad to get in touch with you again. Everybody got Christmas presents so everybody is again happy. Then many English things came trailing through. "Old Virginny" is fine; I took it out in the Mess and played it. It's a good record, but is calculated to make one home sick. Charlie got about a ton of good things – a grocery store; a cigar store; a girl's finishing store; a 5 & 10 cent store; a paper & magazine shop – a candy store, and any other kind of store you have a mind for. We are supplied for [the] duration. There is nothing that you could ask for, that our room could not supply. All of which proves that it is a nice war in spite of what people say. Before I forget, I will answer some of your questions. First, however, many thanks for the record and cigarettes.

In regard to Warren & Ted – I am most anxious for both of them to continue their education just as long as they can. When the time comes, if it comes, they will be found on hand & ready as they should be, but I do not want them to force that time. Warren's university work along military lines, about which he wrote me, is excellent, and if he will continue that while he gets his engineering degree, I am sure he will be doing just what the Government wants him to do. And then, should the show continue, he will be of greater use as an officer in Engineering than he would be now. Furthermore, there is no scarcity of men. The same is true of Ted. I presume both are registered, and my advice is for them to stay put and saw wood at each school. Am very much proud that Warren made the Mask & Wig but cannot quite figure out how he expects to qualify as coxswain of a crew. He is too big and heavy, as small and light as he is. Anything you can do with the automobile is exactly right. I would however keep it running order and have Mr. Williams know that in emergency, or to ride the Sisters to the Hospital, etc., it is at their disposal. By the way, you spoke of Mrs. Page's car. Has she a car of her own now or did you mean the Page car? I fancy by the time the war is over, she will have a limousine, for I understand the stay-at-homes are doing a great business in our absence.

Summer arrangements: – By no means must you go to the office every day. Once or twice a week for as you think best will be alright. If you want to, get some young "needy one" for $2. a week, to answer questions, and take the mail to Owen. Hours 9-12. I have such on my books, but of course can never think of them when wanted. No indeed – after your winter, I want you to have a nice restful summer. And you will find that you will need a rest, too. Am glad you are going again with Owen & Greta. Perhaps you can store the car in the stable there and ride around Delaware Co. It would not be too expensive with your superior judgment & self-confidence – You would of course need Dorothy or Ted. But try tastefully to keep the car out of Cape May. Cape May did not seem to agree with it.

Now I suppose you want to know something about this end of the line. We have had a funny week. Shortly after we filled to overflowing and were going at top speed, we were ordered to evacuate. In about 3 days, we were empty. This great hospital without a patient in it was indeed forlorn. Then we were ordered to pack our X-ray, operating room, laboratory & quartermaster supplies. 45 of our nurses were sent away, and we were naturally

---

* WTS to MWS, 12pp., on 6pp. note paper; opened by Censor 3041; passed by G.B. censor; postmarked APO 8 AP 18.

looking for movement orders. This applied not only to us, but to the entire area. So we sat around for a number of days without a patient and nothing to do. Finally, they sent us a squeaky little convoy and we were told that we might unpack and stay a while. Well, that is where we are now. The reason if any for our little one act show would seem to be known by no one. Theories are of course freely indulged in at present; the situation is about like this:– If we move, we will not be here, but can be found at the place to which we move. If we do not move, we will remain here, but will not be at the place to which we were going to work. Now the place to which we were going to move is any other place but this; its real name is Somewhere in France, but you must be careful not to mention it, for the Boche might catch us –

The news today looks very good to me. If you read between the lines, I fancy you can see Germany breaking. Mr. Wilson's Baltimore speech was fine, and Gen Foch's interviews are very encouraging. I think I see Germany's finish. Her statesmen are certainly talking in a manner, formerly unheard of in Germany. They are getting from under, for they see the crash coming. Of course, everything I have said about this war has been wrong, but this time I think I will fool you. As I am going to stop on the other side of this sheet, I must commence my farewells. Oh, the Jan. & Feb. numbers of the *American Journal of Ophthalmology* came today. Thank you. Charlie's back also – he will tell you about it.

Took a fine walk this afternoon. Beautiful day – birds and flowers singing and flowers & birds in bloom. "The back bone of winter is broken"; we got off easy.

Good night my dear; take good care of my wife and children, and maybe next year or the year after, will relieve you and take care of them myself.

Much love, from

B.

April 13, Saturday night*
[Le Treport] France[85]
My dear Mabel,

We are still here and I might say doing very little; practically nothing. The war's history has not been very favorable, but we are daily expecting a come back in the way of something profound from up the sleeve, as it were. We are about 45 miles from the fighting line; the guns we hear and the flashes from the artillery we see at night in the sky. We did the same all last summer. It is some fight, but you need not worry about the outcome. The only thing is, that it would be nice if the outcome would come along sooner.

Your letter detailing the summer arrangements [was] received yesterday. It was all very interesting. Whatever you have done in regard to Warren is right, although in a letter which you probably now have, I advised a different course. You must however, not let that disturb you, for you know better than I what is best from your end of the line. And you are wonderful to let him go into service. If he wants to go, we should not prevent him, and I can even understand how he would naturally want to get into it anyhow. I think he will be back in time to finish his education, for this affair will not last many more years. I am glad he has the opportunity to sign up where he will be most happy. Ted should certainly graduate from the high school before doing anything, and if he had been more diligent and had not missed his class, he too would be able to get in. He may see also how important what seems at the time (to him) of no importance, might in time become. However, he may have his chance yet. Dorothy's proposals look so good and I think she will enjoy it.

Should ask Jack to write Barbara so she will be alright. As you say, Bob is your problem. I can't help you much in that. I suppose Uncle Sam will send John Deaver to Army Camp again. Bob, I am afraid would not get much out of Wallingford; I would not care much about having him at Becket all alone; he would enjoy Cape May; etc., etc., etc. I guess your faith is having something turn up will carry you through. Hope so anyway. I hope also that you will take a trip somewhere. Go to Boston and have a good time sometime during the summer. You will have lots of bids I am sure. Don't let the office keep you. Get Miss Miller, or Tisch-Marine to look it over several times a week. Put a notice on the door – patients see Dr. Creighton or whoever you like. Just remember that Dr. Geisinger, Dr. Buchanan and Dr. Creighton must have their vacations and ascertain when they will be vacating. Geisinger will rest for about 6

---

* WTS to MWS, 12pp., on 6pp. note paper; opened by Censor 6958; passed by G.B. censor; postmarked APO [date obscured].

weeks in Maine. Tom Holloway will also help you out if necessary; but don't let Shumway do anything for you. You see, if I were home, I would have to look after a great deal of Dr. Turnbull's practice, because he had trained most of his patients to see me in his absence. But as it is now, his practice will go to the four winds; the winds that stay home in war time!!!

I was amused at poor "Medea." She wrote to Chawles – "I have just been down to see the man about getting a new Steamer. I hope you will not think I am extravagant and will approve of what I have done." A Steamer is a good war car; it only costs $2,000. Its price is just the same as a 1st Lieutenant's salary. [drawing of a headless woman inserted here in WTS's letter]

Cartoon showing Charles Jack's wife, Mary, who WTS calls "Medea"

Somebody sent me a New Church *Messenger*, which I read and for which I enclose a clipping which disgusts me very much. It is just an account of this kind of stuff that I have no use for religious journals, etc. At a time like this to be unable to worship with a Unitarian or a Catholic, or ___ does not appeal to me. If the chap who wrote this would get over here on some of these battle fields, he would be darned glad to worship with the Devil, and he would soon think that the good Lord was not paying particular attention to his wrong little sectarian interests. Just read this over and see if it does not make you tired. I know a good many Unitarians that are better men than he is. Imagine old Dr. Furness or Dr. May reading this squid. Please send no more New Church *Messengers*. Was it the Pharisees who thanked the Lord that he was not like other men?

I got a nice letter from Mary Kille Warren – nice girl she. I must answer it tonight. She says that the family does not like [Herbert] Warren Arnold's marriage much. I don't care, do you? I am afraid that Miss Arnold will have rather a stupid time of it, but then perhaps she will like a stupid time. And as for being a Catholic, that is much better than a Disciple of Thought or some such thing, working under a patent granted to some poor inmate of an insane asylum.

<u>Sunday</u>. 14/18. Well my dear, as compared to yesterday, this is tomorrow, but not a thing has happened, and it is a very slow day. The weather is wretched; cold, windy – overcast & gloomy. The news from the front is punk. I cannot discuss that because I must not say anything disrespectful to certain parties concerned in this war. I will say, however, that I am getting fed up on "magnificent retreats." They may be wonderful, and they may show just how great & grand we are, but I don't like too many of them, don't you know? And I care less for dead Boches where we <u>were</u> than for live ones where we <u>are</u>. Of course, I love the <u>dead</u> Boches, you understand, but the live ones seem to be more important just now. They should be purified when they started, not where we used to be. Oh well, it will all come out right in the end.

Good by again, my dear. Much love as usual. Will get around to the children's letters before long.

Your <u>B</u>.

The war things are a little better. There would seem to be a general halt up along the line and the Boche shows evidence of playing a losing game. As you know it has been a hard work for the fighting man, and many a fighting man has gone West. Amiens is still in our possession, and I believe the <u>French</u> will see to it that it remains so. Of course, there will be nothing left of it on account of the daily shelling and the nightly bombing. Did you notice that the home of our old friend the Jack saw it entirely destroyed? Rheims is said to be no more. I am very sorry for it was an exceptional town for historic interest, etc. Nothing would stop the Boche quicker than by destroying some of his towns and I hope they do it. We are fairly busy again as convoys have started to come our way. Have been anxious to hear about Warren and his enlistment and when I do – possibly tomorrow, I will write him "Carry On."

That is the good word here for the fellow going up the line. I think I understood you to say that the Warren Brothers or rather Warren Nephews Company was scheduled to leave shortly for overseas duty. I wrote to Mary Kille last week and told her how glad I was that Harry, whom by the way, I have never seen, is a Sergeant Major. You know, that is a real job, and takes ever so much more brains and some then are require for Lieutenants, Captains & Majors. To be one, you must be some fellow.

Speaking of Lieuts., Capts., & Majors, the differences are all impressed thusly. A Lieutenant knows nothing but does something. A Captain knows something but does nothing. A Major knows nothing and does nothing. We have at last gotten Ralph [Straub, Pvt.] off the chimneys and into a good job in the Company office. He has a grand opportunity now, and if the war lasts a while, I feel sure he will be a Sergeant. He has a splendid record and is looked upon as being one of the most efficient men in the Unit. He has not a single mark against him, which was some stunt in the days of that [Major Matthew A.] De Laney.

When I get on my blue suit, boiler shirt, brown neck tie and black shoes, I will tell about Matthew [De Laney]. He was always very nice to me however, but I don't count. Our nurses are now back. You know 45 of them were hurriedly sent elsewhere. They are glad to get back too. We have the best hospital in France, and the way to be sure of that is to try some others. You know of course that the Presbyterian Hospital Unit under the famous Major Chesty Johnnie Jopson has been broken into pieces about the size of the Majors, and each piece is doing in France something in the way of work which he has never done before. And this after parking heels at home for almost a year in training camps. And poor Joppy, chest and all. Well, as my mother used to say, "Man's inhumanity to man makes countless millions endure." She would then say, "Thomas Moore," meaning Mr. Alexander Pope. Wasn't she just the bestest mother? There is only one as good – Dorothy's.

I see by the paper that "they" (the U.S.) are going to register every man up to 50 yrs. old. I hope I can get home in time to register. Now tomorrow, I hope to get some letters from you and others. If my hopes are realized, I will then be able to continue this, to quite a length. So as not to take a chance on boring you with too long a letter I will stop now. Good night. XXXXXXXXXX

Sunday night, 21st. Hopes not realized. Sorry, because I wanted to hear from home, you know, my dear, this game is getting deadly. Its failure to terminate, not show positive signs of terminating, gets on one's nerves. The possibility of another winter here is disheartening. And there are so many thousands of men [at] home who should take a go at it and do their share.

Then you know there are ever so many doctors over here than they know what to do with. It's a funny game. I positively can't think of another thing to say. Of course, I could tell you lots, but the censor;– he would object. Well, here's hoping for some mail tomorrow. Good night. Much love,

From

B.

Apr. 27, [19]18. Sat[urday]*
[Le Treport] France[86]
My dear Mabel,

After waiting for more than two weeks, mail arrived today, but my share was rather slim and ill nourished. Your letter of Easter Sunday, a note from Bob – funny one at that – a removal card from C.H.S. [Central High School] and a letter from Dr. Martha Tracy. But then I also got some 2nd class mail, so I guess it was not too bad. New Church *Messenger*, which I "discontinued," *Literary Digest*, *Am. Journal of Ophal.*, and 2 lbs. of lovely candy from Miss Jane. Nice of her to send me candy, isn't it?

Was so glad you received the daffodils and glad that you were surprised and pleased as I knew you would be. I take much pleasure in surprising you and anxiously awaited the results. I also sent some carnations to Lulu which I hope were received. Of course you had to get daffodils. As for sending me anything, you must not worry. I did not want anything and do not need anything.

By the way, the tooth picks came through. Thank you. Your letter however made me feel bad in spots. I am sorry you are so poor. Of course, we have run our establishment down to $7.- before, but that seems awfully little when you think of it. I am sorry my dear but the U.S.A. won't let me come home and make more for you, nor will it allow me to do better for you here by making me a General. The war I am afraid is going to clean us out, but we don't care, do we? If we live through it, we can settle down in some nice old country home and amuse ourselves talking about the past. And we have had a nice past, so what is the difference. Martha Tracy is Dean of the Woman's Medical College – Exit Clara Marshall. She notified me that I have been appointed Clinical Professor of Ophthalmology at a salary of $100. per year. So you see I am a "Perfessor" anyway. Before, I was only an Associate Clinical Pro., not nearly so good – and the Salary ~~though~~ thereof was nothing to compare. Well, that's very nice of the girls, and I appreciate it.

What I wanted though I won't get. I wanted the Drexel House – the Children's Hospital you know at the Lankenau. Turnbull had it, and I wanted to just naturally succeed him there. But, I guess I should have been home for that. However, this is confidential, so please do not mention it. I was much instrumental in getting Ralph Butler

---

* WTS to MWS, 12pp., on 6pp. note paper; opened by censor 6048; passed by G.B. censor; postmarked APO 29 AP 18.

there when Burton Potts died. But keep mum. I agree with you in regard to Warren, and have always tried to get him sorry when he lost time at the C.H.S. There are certain things that one must do now. One is to get an education. From what you say, I presume that Warren will be accepted. If it depends on Hollie Lester, I am sure he will be. I must take back all the weeps about our old friend S.S. *St. Paul.* I see that she nearly turned over in her bunk and was not submarined and sunk half as many times as reported. I am glad, for she has been a very useful little boat. Also read that the U.S.N. has taken over the Phila. Polio House. That is good, and I hope Mr. Daniels can do something with a good quarantine.

We have lost another of our Unit. Geo. Norris has been detached and sent to the A.E.F. That makes three: Gibbon, Cadwallader & Norris. They are gradually pulling us away. Sorry, but can't help it. Norris will be greatly missed. P.A. [Capt. Paul Austin] the fellow from Germantown, about whom I wrote recently, and of whom you spoke in your last letter, has not been heard from since the push started. We are afraid, but hope he is alright. Until then he wrote often and kept us advised. But for Heaven's sake, don't mention this and possibly start a rumor. Keep it to yourself, please.

Was very sorry to see the death of Geo. Woodward's son confirmed. He was in Aviation, and if you do it long enough, you get caught. It's too bad.

Now about this war. Today's communique was bum. We have not had a very good week on land, but some little Naval stunt, don't you think! The loss of Mount Kemmel was not good. Of course, you can put it down as a mathematical certainty that Germany will eventually be beaten, but suppose that happy time is 5-10, 15 or 20 years distant. It is not so encouraging, is it? If I were allowed, I would almost wish that we had started earlier, – at least to prepare. But, I am not allowed to wish that. Orders concerning hopes have not yet been issued, but until they are, I am going to hope that the Boche will be licked very soon. That can't be wrong until I learn better. Will tell you more about it tomorrow. Until then Good night XXXXXXX

Now this is tomorrow, Sunday, April 28th, 1918. Some place in France and nothing doing. The communique was better. We held very well and killed lots of Germans. But we did not get Kemmel Hill back. You know it took over a year for us to get it, and we lost it in a week.

I was invited by Mr. Treacher Collins, President of the Ophthalmological Society of the United Kingdom, to attend their annual meeting in London, this week – and be his guest at their banquet, but I couldn't go during the acute situation here abouts just now. Can't get away during a great big push, you know. I was not awfully sorry however, because it's no fun travelling here at present. You are handled like an express package, and can't do as you please. Then I am expecting a special mail which I will tell you about if it comes through.

Oh, funny: Owen wrote with sadness, about my comments on the new president of the Union League – Chester politician. He was to be sure a politician, and did live in Chester, but etc. Owen's choice, he said, was Charlemagne Tower. I replied that it was not a good time to elect to that office the Kaiser's old & fast friend. I understood that Mr. Tower's friendships made in Germany were so lasting that they continued to the present time. Well, Charlie received a letter yesterday from a member of the League, saying that there was talk at the League of expelling Mr. T. from membership. Poor Bunny, if he thinks a fellow is nice, it doesn't make any difference. However, my only point was that I did not see how Mr. T. could be elected president of the Union League just now, or for that matter for many years to come. He sure could not get my vote. That's all. Good bye. Much love from B.

April 27, 1918 [Saturday]*
[Le Treport] France
My dear Bobby,

You are some letter writer, and I am always glad to hear from you. You write something like your father in that the near writing is difficult for others to read. Your spelling is direct and to the point, simplified to a degree and is just as good as any other spelling. I am glad your mentioned Bo in your letter, because I was getting worried about him. No one has mentioned him for months, and I was thinking that he had perhaps been interned or interred. While I know he is American born, there is never the less a good deal of German in his family tree. His sympathies I am sure are Allied. Had he been in Germany, he would have been eaten long ago. All little tender dogs of his variety are eaten there to nourish the Army. Big dogs have to pull wagons and work.

You speak of a "thrift stamp" costing 25 cts. I never heard of one. What is a thrift stamp? Perhaps, if they had been around when I was your age, I would now be better off. If they are any good, you might buy me one. Your school term must be drawing to a close, and of course you have done well, and are prepared to pass all of your examinations. I hope you will have a fine summer where ever you spend it. You would have a dandy time here, but some trouble getting here. It would be as hard for you to get here as it is for me to get away from here. But no harder. If at any time you hear when the war is going to end, drop me a line. Now just as you do, I am going to stop short and say excuse shortness, etc., etc. Give my love to your mother, and your brothers and sisters and Bo – and if the war cat is nice, be a little to him.

Your

Dad

April 28, 1918 [Sunday]†
[Le Treport] France
My dear Ted,

It is time that I had written to you, for you have been very good writing to me. First, many thanks for the cigarettes which after many months arrived, but were just as good and just as much appreciated as they would have been Dec. 25th. Glad some made across, as I can't get them here. You see I can't smoke cigars which of course I prefer, because they are too expensive and I come into possession of very few. Accounts of you have been very satisfactory and I hope they may continue so. Please finish your course at the C.H.S. before you sign up for military work. If you are conscripted, that will be another mater. Hope that the war will not last, but if it does, you will have your chance to get in it. If it continues, there will be room for all, so don't worry. Then again, with Warren away, you will have to take care of your mother, which is a very important duty. Your class honors and basketball honors are very credible, but don't work the latter too hard.

If you have a German education, tell it to go to H——. The worst you can do in German, the better. It is no longer popular. The only time I use German now is when I am orderly officer and visit the German prisoners to see if they are alright or have any complaints. Most of them are as happy as clams, and could not be driven back to Germany. It a cushy job to be a German prisoner. Quite different from being a prisoner in Germany.

I am very tired of my job and want to get back into a real world. When I think I see the finish starting – Bang! It's just commencing. And so it goes. I would like to go to sleep and awaken to find it all over. And then shouting – there will be some.

Write again if you get a chance before your various duties.

Love to all, from

Dad.

---

* WTS to son Bob, 4pp., on 2pp. note paper; opened by Censor 3742; passed by G.B. censor; postmarked APO [date illegible]; received May 20, also with letter to son Ted.

† WTS to Ted, 4pp., on 2pp. note paper; opened by Censor 6105; passed by G.B. censor; postmarked APO 29 AP 18.

May 1918

Le Treport – Sudden Activity on the Western Front

"The push is on any day now."

May 4th, Saturday*
[Le Treport] France[87]
My dear Mabel,

The U.S. Government owes you $20- and you will get it, but it may require 50 yrs. I owe for no Liberty Bond installments not included in vouchers. I think the Q.M. has probably neglected to include foreign service pay, which would amount to about $20-, but how he did, I do not know. He should have paid the amount called for in the voucher or else held the voucher up. You are right to communicate with him at once, as you can attend to that from your end better than I can from this end. Let me know his explanations. Of course, you know what a Q.M. is, so do not be discouraged if you do not hear from him for a very long time. This is his first mistake with us however, so we must be lenient.

Saturday nights come along pretty fast, but they do not seem to do much good. There must be too many of them in this war, because I do not feel any nearer the end. You must be getting very tired of it, working along all alone toward an end that does not come. You ask when I got the flowers. You will get some beautiful roses Monday May 6th and I guess you will know why. June 6th, you will not get any of them, but I will think a whole lot of you on that day. Last year, I sent you a cable; this year I think I will omit that also. Well, the flowers are all paid for with my money just as if I was home. You know, I have "agents" to transmit that kind of business for me. The only mistake my agent made was to not get them from Pennock as I specifically directed. I have lots of ways of doing things no matter where I am. So now you know all about the flowers, don't you? And don't forget, they are paid for with my own money, which you never even knew I had. All of your guesses were laughably hopeless. Anna! Mollie C., etc. They could be real "agents." However, if you guess right, I will tell you.

I forgot last week to tell you about my Captain's commission. It is rather interesting. Many months ago, I received telegraphic notifications etc., and in a few days a paper in proper form which stated that it should serve in every way as a commission until such time as I received the formal document, with Eagles, Scrolls, etc., on it, you know. Well in a short time, [William J.] Taylor, [Francis R.] Packard, and the rest received theirs, but I never received mine, and did not bother about it, thinking that I would, someday after the war, would [go] to Washington for it just for fun. The other paper was official and for every purpose, the principal of which was to draw Captain's pay. Well, about two weeks ago, a letter came to the C.O. enclosing my commission, which was messed up and dirty, and resembled somewhat the Jackdaw of Rheims. The enclosed letter said that "these papers had been picked up in a street near the Front. Capt. W. T. Shoemaker did not belong to this Brigade." Very funny, wasn't it? I suppose, some fellow had been carrying it around – possibly a spy, and may be using it when necessary. I recall now that the mails were robbed in Paris some months ago. Anyhow, my commission has been at the Front, even if I have not. It is rather remarkable, that I should have gotten it at all under the circumstances.

While on commissions, Charlie & Ed's have not yet come through so they are still "as was." We have had several applications rejected, so they are never sure until you have them. The new rule is that a Lieutenant can apply for a Captaincy after 9 mos. active service. So, I am not eligible until May 15th, but even so, I will never apply and nothing will be done for me, so don't worry about that. Major is as high as we can go at present in the Reserve. In the wartime Army, you can be anything. A bill is before Congress to increase the ranks of the Medical Reserve Corps. It will surely pass, I hope, and then we can all be Generals & Colonels.

---

* WTS to MWS, 16pp., on 8pp. note paper; opened by Censor 6133; passed by G.B. censor; postmarked APO [date obscured].

Regarding the Dental Corps, that is new, and as yet but imperfectly organized. It is to be sure a part of the Medical Corps, but until one year ago, there was no Dental Reserve Corps. That is why Charlie is No. 1 and Ed is No. 2. – because they got off with us May 18th. Now there are 5000 men in the Dental R.C. – 1200 on active duty – mostly at home.

This afternoon we had the first public appearance of our Band. About 20 in it, and they did mighty well. Months ago, Dr. Harte bought the instruments in Paris, then after much time, the instruments arrived; then after more time, we got music from London. We had the material, so it started in about four weeks ago to practice and organize. Result – today a full-fledged brass band, with everything in it that makes a noise like a band. Great big toot toots, little tiny ones and all kinds of boom booms. When it comes to talent, we sure have it with us, for anything – Surgery, Medicine, (Ophthalmology)! Singing, Dancing, Acting, Piano playing, Violin, any old thing you want – Mandolin, Guitars, Football, Baseball, Tennis, Golf, etc., etc. It is a pity that we can't all be home doing something useful.

I am glad you got the note business straightened out with Mr. Burnham's executors. I remember some time ago, the note was reduced, just how much I do not know. Some executors would have been afraid "on advice of counsel" to go before the Orphan's Court with such an outstanding asset. I am rather disappointed that [our son] Warren shifted to the Medical Corps. I would have been better pleased to think if he had stuck to Engineering. But, I don't know anything about it, and also [your brother] Herbert does not know much about it. So don't bank too much on his military advice, as nice as he is. Once in, you have not hold or claim on any organization and are shot around anywhere that suits the petty officer on deck at the time. However, I guess Warren has done the right thing and I hope he will come through alright. He will at least be better for having been in it. Is Percy Whiting still persuading others to go in? As I remember, he was perfectly wonderful at that. Swell job.

Good night, my dear. Will write a little bit more tomorrow. XXXXXXXXXXXXXX

Sunday, May 5th. A very important event happened today; one which might have a considerable bearing upon the war. I took my first lesson in golf. Now think of that – me a playing at golf. Ed was my instructor and a very good one, and as for his pupil, he was *bon*. I never thought I would do it, but this war seems to undermine the moral of the most determined. Of course, even before the war and alcohol and golf had gotten a very firm hold on many of our good people, but since the war, conditions have gotten worse. Oh well, after the war, I will switch off and burn all the little golf balls & sticks. I would hate to become addicted to it.

You probably know, that you can no longer send me things from time to time. Such a ruling was sure to come, with the great abuse of the mail service that was carried on. One of the greatest offenders has been [Charlie's wife] Medea. She has certainly sent tons of useless stuff. The last convoy contained several pounds of ordinary cheese – just cheese. Well we have fine cheese every day in great quantity. Never missed a day. Then sometimes will come Tea. Tea can also be had here, if you care for it. Pounds of almonds, & other nuts. We are strong on them. Have them every day.

On the 18th of this month, we are to have a great party in celebration of our anniversary. One year from Philadelphia – fine eats, etc.

Now about yourself. I hear from time to time, that you have lost much weight. Is anything wrong, that I don't know about? Please do not get sick. You are doubtless worried and very tired and you should have a good Summer vacation, which I want you to arrange for. Now do not forget. And let me know exactly how you are. You see if I get it in my head that you are not well, I will begin to worry, and that might lose the war.

Good night, my dear. Be good as usual. Love to all the children & to you a plenty –

From B.

May 6th*

With love and many thoughts of you and ours.

W.T.S.

----

* WTS note to MWS, MN, on 5x2 in note paper, in an envelope, enclosed in the envelope with his letter of May 4-5, 1918.

Wednesday, Saturday, May 11th. 1918*
[Le Treport] Same place in France
My dear Mabel,

As the mail did not come through this week, and as last one brought me but one letter from you, I have few thoughts in my war weary head to put into this letter. Your letter was a nice one and told me lots, most of which I commented on last week. Forgot to mention the refund from the [Union] League, and to thank you for sending it overseas as it was under. It was of course very decent of the U.L. to do this, but I expect it sooner or later. You see the dues are paid January 1st in advance, and as I departed in May, the League was really about $60.- to the good. Was glad to get a portion of it back, and wrote to the Treasurer as you suggested and expressed my appreciation. Also am glad that you sent the proper acknowledgement.

The income tax is funny – most of the men here have heard nothing about it. Medea paid "theirs" and I guess a good one as they have some income between them. More than we will ever have. Our first year of war was not so bad, and I was more than pleased, but I have great fears concerning the subsequent years. You understand I say years. Why shouldn't I? I see no evidence of anyone thinking that perhaps someday, the war will be over. Several months ago, I was over at the Tank Camp. They had set out, boulevards of small cars – little fellows. "It will be very nice when we get fixed up" – said they. Read in the paper today that, Congressman so and so said that we would have to have an Army of 80 000 000 men in the next four years. Everything is progressing on a permanent basis. That is right of course, and need not interfere with licking the Hun, but I object to building forests out of tiny trees and thinking how nice it will be when the trees grow up and we are dead and gone. However, we are in the pink just now, and the Hun is being hurt. We are daily expecting his best effort, and hope to hurt him a whole lot when he makes it. The Americans are getting much praise for their fighting qualities. A good compliment came to us today from an Australian in our Hospital who came in the last convoy and who spoke about us. He said, "You fellows are a pretty rough lot – I would not care about running with them." Now you know the Australian – he is a born fighter – he fairly eats Huns, and enjoys it. I think our boys will eat Huns when we get around to it.

Now most import[ant] to the war is golf. I like it very much and am now taking lessons from the professional teacher. We have a rather nice course – right here at our back door, and I enjoy playing sometimes. We have also splendid tennis courts, but tennis is a bit too strenuous for me the way I would like to play it, so I guess I will shift to golf. It is very cheap, which is another advantage. I have been at it a week, so there are still a few here that I cannot beat. In another week, I suppose I will think I have the entire garrison beaten. It would be very funny if I should learn to like golf for I have always turned up my nose at it. But it is fascinating. If you get a chance this summer, just get the teacher and take a few lessons. It will be fine for you and you will enjoy it. And then, after the war, we can play together and have a fine time. It's not difficult you know; anybody can do it. The only thing is to start right and get a proper form first. Start wrong, and it is more difficult to break a habit than to make one. And knowing nothing about it, as we do, it is just as easy to learn right as wrong. Should you got to Cape May for 2 weeks – get a guest membership at the Club and go to it, with the teacher. Every morning, we get a guest membership at Spring Haven for the same time you can do it and have a fine time. Think it over.

The weather here is now beautiful. Little shoots – a shootin' on the trees, and flowers plentiful. It must be apple blossom time in Normandy, but I have not seen apple blossoms yet. I guess I don't walk far enough from home.

The Special Detail, that I spoke about and expected was turned down. I expected to go up the line to treat specially gassed eyes, as good as gassed. I had visions of getting up to the [Front the] first time, just a teeny bit this side of No Man's Land, but my reputations and propositions were canned. You see we had a new treatment, which I was anxious to try out. However, it is safer here, so I am not worrying too much. I know you will be disappointed, but never mind, we may have lots of opportunity yet. I hope tomorrow to hear definitely about Warren. If he gets in, I hope he will come over here some and take his chance at seeing something & learning something. Training Camp I fancy will not appeal to him any more than they do to me. I often think of Orlando, 7 mos. in actual training on the border, and then a year elsewhere, and still at it. 3,000 miles from the Enemy. I suppose he is being trained for the job of a retired General. He sure will be a good one when he gets his finishing touches. Howard also should be able to kill a few Boches by this time. They both must be very sick of the war game as played at home. By the

---

* WTS to MWS, 16pp., on 8pp. note paper; opened by Censor 6133; passed by G.B. censor; postmarked APO [date obscured].

way, we get service stripes – one for each 6 mos. Next week we are entitled to two. They are gold braid to be worn on the left sleeve.

Thus: - 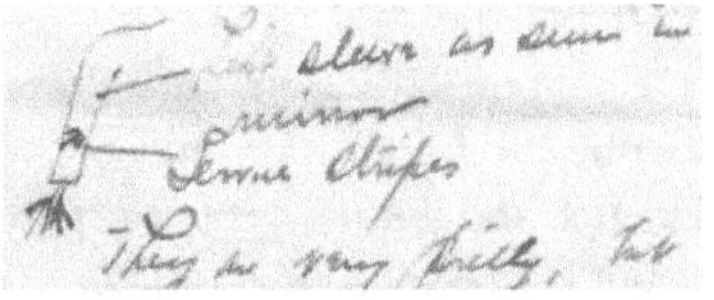

They are very pretty, but I don't want too many of them. Two will about do for me. Of course, you have seen our new dinky overseas cap. I have ordered one from Paris to play golf in. We do not have to wear them in this area. That is officers do not; the men are wearing them. Our Campaign hats are taboo. Never again for them. We wear our [visor] caps or can wear the overseas [cap]. The overseas has little to recommend it. It keeps neither sun nor rain off and looks like the Devil. But on a dark, dry day it is not uncomfortable.

Good night, my dear. More tomorrow – if I get a mail

XXXXXXXXXXXXXXXXXXX Special love this time.

Sunday, May 12th. Mother's Day.

General Pershing said I should write home today. He did not specify any particular mothers, so I will write the best one in the world. A nice mail came through this morning. A letter from Warren told me of his plans, and I was glad that after all it came out as I wanted it. I would much rather have him Engineering than have him in the Medical Corps. He has also had some experience in Washington with France's weather. We have had loads of weeks like that. He seems to quite enthusiastic and happy in his new enterprise. I wish him luck & give him my blessing. Whatever happens, see that Ted graduates in February. Do not let him miss. He tells me he is head of the house, and I bet he is a good one. Dorothy's experience at the Art School was fine. I congratulate her. She must be doing very good work.

The socks I wore for a few days, but they are too heavy, I think. You see, while we have inclement weather, we are not at the North Pole – and unless you have large shoes, which I have not, a heavy light sock is colder than a medium. That was my trouble, I had to take them off. I do not want anything much. I will now use for underdrawers summer only ones. One pair is going rapidly, but I think I still have a pair in my trunk. Have not had a neck tie on for a year. So I don't need neck ties. Some are pretty well gone, but I can get them here.

This letter is too long. I must save something for next week. Am awfully glad Owen has at last gotten a new house, and I guess it is a nice one. I do not think you will have as much from a week ending at Moylan as you would at Wallingford; out there is a nice treat. Creighton should be doing some work still; got a letter from him today. Hope he continues. Well my dear, I must stop and write other letters. I always fix you up first so that there can be no mistake. Good night; much love, and I wish I was home –

From B.

Morgan, Harjes & Cie*
31 Boulevard Haussmann / Paris, France
May 13th, 1918[88]
Dear Sir:

We receive to-day your letter of the – inst. and enclosed we beg to hand you, as requested: Fcs 403.70 notes & stamps place to your debit, with Fcs 3.20 insurance and postage. Please return receipt duly signed. Your account is thereby balanced. We remain, Yours faithfully,

Per pro MORGAN HARJES & Cie. [s] [illegible]

---

* Carbon copy of TN [Typed Note], with blanks filled in, in envelope addressed to: Captain Wm. T. Shoemaker / General Hospital No. 16 (Philadelphia) / A.P.O. 13 / Le Treport (Seine-Inf.); postmarked Paris May 19:30 18; enclosing similar notes for transactions earlier in 1918, on February 15, February 18, April 23, and May 5.

Wednesday, Saturday, May 18th. 1918*
[Le Treport] France[89]
My dear Mabel,

Just one year ago today, I saw for the last time my best girl and my children and started on this perfectly useless proposition of Army life. So many things have happened. I have seen so much, and have done so much "Carrying on" that I would like to think that I am better for it all, and that some good has been accomplished. But there is nothing good in war. It is all a sinful waste and time spent in its prosecution is time lost, but unfortunately time that has to be lost.

In celebration of our anniversary we had a dinner. I am almost ashamed to send you the menu, but it is herewith enclosed for the scrapbook. It was such a good dinner for a war. Dinner on the end of a great drive, and exceptionally good dinner. The distinguishing features were American cocktails & ice cream. The decorations of our Mess were very nice. A whale of an American flag across one end. British flags & other flags in perfect proportion. Tulips, Lilly of the Valley, etc., in profusion on the table, and in fact everything just as nice as could be. Champagne. Port wine. Cigarettes & cigars. We all enjoyed it and hope that we would not have to repeat it for a second anniversary.

The push is on any day now. They say Monday, but may be tomorrow or tonight. I believe we are ready. We now have for the first time troops in great numbers between us and the Enemy. Americans are here galore, and should any get hurt, we are to care for them. We have quite a few in the Hospital already. I have 4 in my ward. Of course, I know nothing, but if Warren comes over, he will probably come very near here for his landing. I hope he has studied his geography & should he enter this sector, let me know. I would only be able to spot him by accident. It is very nice to know that so many Americans are in front of us, and as I told you last week, they are making a good fighting name for themselves. I wish we had 5 000 000 of them right now.

Yesterday I went to Dieppe & saw a golf match. Ed & Mr. Reddie (the professional here) played two other fellows. It was fine. The country is beautiful and for 4 days we have had June weather. Glorious weather, warm, sunshine & clear. Everything in nature tumbling heels over head to get out and on the job. I passed a number of great fields of mustard in full bloom. Did you ever see it? The most brilliant yellow imaginable plunked right down in the landscape. Like a great golden carpet. Awfully pretty. The course at Dieppe is a very good one and in fine condition. The greens were like velvet. It is right on the Channel, as everything here seems to be. Normandy just now is alright, and by the way, I saw apple blossoms.

But then, Normandy has nothing on Delaware County or some of the places around home. Do you remember the Sunday morning at Belvidere, when I asked you if the flowers always behaved that way in the Spring? They were pretty, weren't they? The upper Delaware is alright and suits me.

The mail this week brought me one letter – a very nice one from you #81 - #80 has not arrived, and is I think the only one of yours not received. Surely, the only one for many months. I was sorry to hear about Nancy Page. I suppose it was more fibroid business, and the operation was probably a hysterectomy – which means the whole blamed business. Hope she is alright & feel sure that she is. Except for the Christmas present, I have not heard a word from Henry. I guess he knows that I think he is a slacker. Nor have I heard from Pres. I guess he knows the same thing. Two perfectly good stay-at-home doctors who will profit by the War. Mr. Langdon is behaving very well, and would seem to have the proper spirit. I am very glad he rang your door bell. You will find that the Land [Langdon] will send you through his various agents, the necessary where withal, even when you can't see the faintest hope. He has seldom failed me. Too bad about Uncle Jake, but I guess it is time; he must be over 80. You must be sure to have him comfortable and not in want. I can think of nothing more at present. So good night. More tomorrow.

XXXXXXXXXXXXXXXX

Sunday, May 19th – Well my dear, letter #80. came through today. It had gone astray although the address was perfectly good. What made you think I wanted bent eyedroppers? I never use anything but a straight eyedropper –

---

* WTS to MWS, ML, 12pp., on 6pp. note paper; not opened by censor; passed by G.B. censor; postmarked APO 20 May 18; received June 22, 1918.

Mrs. Baer knows that. They have not come, nor have the dark glasses. A new doctor arrived today for our Unit and I was glad to find him. Dr. Barnes, one of the German Hospital men, and a good one. He is a bully good man for us and I am awfully glad that we have him. [Henry K.] Dillard & [Isabella] Stambaugh are now in London, and their wounds are now healing nicely – hence the long delay in getting them back with us. We have no team at the C.C.S. [Casualty Clearing Station] at present. You see it has been a little difficult to maintain a C.C.S. in our particular locality. We may become one ourselves for that matter. It all depends on upon what happens when the hat drops. By the way that hat is going to drop soon. But you will know all about it just as soon as I will.

The weather is glorious again today. I played golf this afternoon and it is perhaps unnecessary to say that I played very badly. It's good exercise though & good fun, so what's the difference. Don't forget to send me Warren's picture. Did he finish his sophomore year so that when he returns, he can enter [as a] junior? I hope so. Can Ted get a scholarship and enter the U. of P. in the Fall? If not, don't you think we had better send him anyway. Put him in the Wharton School and give him a chance. He is too young to be a soldier. A year or two in College will mean a lot to him later on. Be sure and get him through the High School, and then be careful what you do next.

I think that I will get home about Christmas or not at all. If the Show goes over another winter – it's a life job, and I will abandon hope, and forget that I ever had a home. However, there is no use in bothering about that now.

Good night – Lots of love. You are doing wonders. Keep at it and maybe someday – – – – 1918, 1919, 1920, 1921, 1922, 1923, 1924, 1925, etc.

from B.

Wednesday, Saturday, May 25. 1918*
[Le Treport] France[90]
My dear Mabel,

Again, I must bore you with a letter and you surely must be bored, because I have so little to say. Our Area is now packed with American troops, and we have quite a number of them in the Hospital. Naturally, we have taken on more of a warlike appearance and are of greater interest to the Hun. He has been over us several times lately and as the moon is full & glorious, he will probably be around again tonight. Two nights ago, he did quite a lot of bombing in our neighborhood, but it was not for us. We are however sandbagging our huts, and for the most serious & helpless cases, are digging shelters into which others can scramble. And we are doing it rapidly too. Our patients fear the bombing very much and when the Hun comes over, they get the wind up. You cannot blame them, for this bombing business is a terrible thing. And when you have your arms & legs off and are absolutely helpless in bed, it is unpleasant to hear the bombs dropping around carelessly.

You have read of the Hospital not far from us that was bombed the other night with such dreadful casualties. That Hospital was on a rail-road or beside a rail-road, and near an important bridge. The Hun was after the railroad & bridge. If he disturbs us, which I do not expect, it will be pure cussedness, as we are only a Hospital. Our only danger I fancy is a poor shot at something else or and individual cuss who might take a shot at us just for fun or practice. However, we don't trust him and are not overjoyed when we hear him over head. Have you noticed how dreadful he thinks it is when we bomb Cologne & the Rhine towns? He thinks there should be some agreement among the belligerents not to do such wicked things. He invented the bombing of open towns, of Hospital Ships, of Clearing Stations and of General Hospitals. He also invented the bombing of helpless women & children. He has furthermore persistently practiced his inventions, but he never thought that the game could be played both ways, and when we play it in reprisal, he is shocked & horrified. Can you beat him for Hell up to date? He has not yet started his drive, and nobody knows why. Heard that he started this morning but have heard nothing since. I have in my mind's eye his Waterloo this time, but my mind's eye in this war is most trustworthy. We get a Convoy tonight, and then we may get some first hand information from the line. We are pretty full, but with mostly medicine cases. All sorts of things have come down, and many patients very sick.

---

* WTS to MWS, ML, 16pp., on 8pp. note paper; not opened by censor; passed by G.B. censor; postmarked APO 27 May 18; received June 22, 1918.

I have a new job. Our new labor battalions – British, Chinks, Italians & German prisoners are billeted in most all of the little towns in France. We are the medical men in charge of three or four of them in a radius of 10-15 miles. The M.O. [Medical Officer] in charge has to visit them daily or several times a week, or when sent for, and hold sick parade, inoculate for typhoid, etc. I am M.O. in charge, and travel to them in an ambulance and if necessary bring them in. It's a good summer job for a time and uses a couple of hours – always a good thing. John Flick was doing it, but is on the next C.C.S. team awaiting orders, which might leave any time. So Harte landed on me as being the most available M.O. with the least to do. Flick and & [Henry C.] Earnshaw are packed up and can move in 20 minutes. It's a funny C.C.P. team, but it will get away with it alright. You will be glad to know that Paul Austin is alive & well, and has been ordered back to us. That's fine. He certainly has had a great experience. He has seen everything from the front line trenches & No Man's Land to a Base Hospital. I am mighty glad he is coming back, because you cannot be up there too long without getting a bullet or a piece of shrapnel with your name on it. Received a letter from Mary Burnham yesterday. She writes a nice newsy letter. She thinks of me every night that the Seeline meets when I should be in the Chair. She tells me that Shumway has taken Turnbull's practice. The war is breaking nicely for him, isn't it!

More than once Turnbull told me that he wanted me to take his practice when he was done. But, you see I am not on hand to take it. Cruel war isn't it? I know one thing though, when I get home, he won't have <u>all</u> of it, see? But, I must not say "when I get home." I am not coming home. That is decided. As Burnham says, "They say the war will end in August, and next morning the paper says it will last 4 or 5 years." If anybody knew anything about it, he could make a fortune big enough to pay for it. B. says you are sylphlike and he recalls your school days in the 80's. I don't like you getting so thin, even if you do say you are perfectly well. You see, you are not the thin kind, and I am afraid – very much afraid, you are camouflaging. Are you really well, or are you only trying to make me think so? If you are sick, I will come home. I must say that very fast if I want to think so. General Pershing would never let me go, but General Gorgas <u>might</u> if you were sick. But I hope you are not sick, and then I can stay and waste more time here. I would love to be here when it finishes. It will be <u>great</u>. That would make it worthwhile.

Golf is progressing, but it's awfully hard to find balls in the daisies. A white golf ball in a field of daises is well camouflaged. Saw a V.A.D. [Voluntary Aid Detachment nurse] playing this afternoon with a <u>red</u> ball. That wouldn't help me much would it? I never did like red berries on green bushes, and I would not like red golf balls in green grass sprinkled liberally with white daises. Now a blue golf ball might be just the thing for me. I will think it over.

Most of us put on our service stripes last week. Order came out tonight to take them off as we had no authority for wearing them. Funny Army; off they came and probably next week will go on again. Witness the overseas cap. The style has been changed several times in as many weeks. Can't tell what to get now. I only want one to play golf in so that someone won't bang at my head for a golf ball by mistake. And it would be a great mistake. Goodnight. Want a letter tomorrow.

XXXXXXXXXX

XXXXXXXXXXXX.

Sunday night. No letter, so just for that I can't go far with this one. It is quite cold again, after a week of perfect June weather – temp about 80°, you know. No drive yet. Much guns all last night, but the Hun did not get over us. Perhaps he will come tonight. Had a fine golf match this afternoon. Ed & Bill Drayton against Norris, Vaux & Ich. Must not say "*Ich*," change it to <u>me.</u>

Received a copy of Col. McKendrick's doings with Kings & Presidents for G.C.W. Also a Convoy of N.C. [New Church] *Messengers*. One of them had a letter from Ed to his father. I showed it to Ed. He was embarrassed but pleased, and said it read pretty nice, "when it was all printed up." Another [from] the mother [Catherine Clark Warren] of your cousin, Lt. [James Reed] Warren. I think the son of Herbert Langford [Warren]. His picture was in the *Messenger*. Fine looking young officer. It's too bad. People that I know at home are dying almost as fast as they are here. Also saw that Ever Hooper died in church where had said he wanted to die. It is a queer place for that little function, but it evidently happened his way. I presume he is Walter's father-in-law. Paul Austin got back today without a scratch. He was reported to be dead, or a prisoner, but he missed being either. The University of Pa. Unit has at last arrived in France. Sandy Rundall is with that Unit. They did the talking but we beat them by a year. They are not near us.

The batteries from First, Linden & Prospect arrived yesterday. Probably on the way many months, but they are alright. Quantities of them. Very nice of F.L&P. to send them. Will write, but you also thank them. See?

Your story about Wimpy Stella is funny. Of course, you do not feel terrible, and of course you are proud to have your day in the line of duty. But those with small minds cannot understand that they have the "Medeas."

Well my dear, I have done pretty well, and you sure must be tired of this very long & uninteresting letter. I could tell you interesting things, but I am not supposed to, and have I stuck pretty close to the rules. You must hear a lot from Medea though, for Charles prattles along at a great rate, and says anything.

Good night. Be good. Love me, and miss me a lot. Do not get used to being without me, and rather like it, see?
Yours – with love

B.

On the day after WTS wrote finished this letter to his wife, the Germans launched a major offensive on the Western Front. Known as the Third Battle of the Aisne, it began on May 27 and it lasted until June 3. The Germans captured the Chemin des Dames and the bridges across the Aisne, and they crossed the Marne. They captured Soissons and Chateau Thierrey, and they were finally halted only 37 miles from Paris. For the first time in the war, the American Army took a major role in this battle.[91]

On the following day, May 28, the American Army began its first offensive operation in the war. It was known as the Battle of Cantigny. The town of Cantigny had been captured earlier in the spring by the Germans. The town was recaptured by the Americans, who successfully defended it over the course of two days. Although the action had no strategic importance, it showed that the Americans could fight well, and it provided an important morale boost to the Allies.[92]

June 1918

Le Treport – Father and Son in France

"The news today was pretty good. The Americans seem to be really in it and holding in fine form."

June 2nd, 1918 [Sunday]*
[Le Treport] France
My dear Mabel,

Last night was the first Saturday night I think since I have been away that I did not write. I was just starting when I was torpedoed by Barnes, who sat and chatted for 2 hrs. and then it was too <u>late</u>. It has been quite an interesting time but the Censor has gotten so fussy that I can't tell you much about it

It was an Air week. You remember that I told you last week that perhaps the Boche would come over that night. Well, he did, and has been around all week, while the moon lasted. And it was some moon. The weather has been simply perfect June weather. Just what June should be. We have been busily sand bagging and digging trenches and are now quite handsome. Of course, he [the German] does not want us, and we are not worrying about that, but it is not altogether satisfactory to have him around; - he might make a mistake, you know. When he gets to this Area the bugles sound an "alert." If he gets real close, the bugle goes off again, and we get busy. We have not had to "get busy" yet. A Boche plane can be recognized by its markings, which is very different from that of the Allied planes.

Last Sunday night he went over rather low and paid his respect to a town below us a few miles. Then he came back. Thursday was Decoration Day and we had five American graves in our cemetery to ponder. It was a very pretty and impressive cemetery. The entire garrison marched by the cemetery with the band. [Edward M.] Jeffreys gave a 5 minute talk and we saluted the grave & marched home. It was the first time that we had marched to music. It makes a great difference; the band did very well; it is improving every day. When we all get out, we have quite a parade. About 30 M.D's, 100 Nurses, and 200 Enlisted men. Then there are a few R.A.M.C. men with us. Quite a parade. But I wished so hard that we were marching down Broad St. instead of a Country cemetery off in France.

I now have another new job. I am to be Mess Secretary and Mess Captain. Funny job for me, isn't it? Boss of the Mess – it's Eats & it's soups. Ed[win Shoemaker] is my Assistant and is to have charge of the surrounding grounds. Guess who has the best cut flavor buds alluded to. So for a while the Mess will be under the care of the Shoemaker family. It's not a bad job, but I do not care much for accounts as you know. Then I am a very cautious C— [cook]. I spend too much money. We have a Cooking Mess Sergeant, who of course does the actual work of buying & purveying. Without a good Sergeant the job would be hopeless.

I am so glad you enjoyed the roses, Mabel. I had great pleasure in sending them, and of course wanted them to be superior. I will think of you a lot June 6th, not thinking, not L– [Love] dear for anything. Hope you will not be disappointed though Charles' Clubman that died. Too bad; what was the trouble? I think he has a boy & so over here, hasn't he?

Very interesting letter from Dorothy, telling me about Murphy Stella's Sailor's Party. It must have been very funny, but as D. says, also tragic. I think the entertaining business is carried a bit far. Honestly, I fancy but would spend time with you at Wallingford. It's a pity that she will not be working on the same R.R. Division. She would thru Wash less time in Cripon. However, I guess she does not care. I think it's been for you to go to Maryland the best part of the time. You would have more fun at Wallingford.

---

* WTS to MWS, ML, 8pp., on both sides of 4pp. plain 7x5in note paper, torn from a tablet; opened by Censor 6368; passed by G.B. censor; postmarked APO [date obscured by censor's label]. Dr. Barnes, of the Germantown Hospital, is not in *History*.

Now about this war. I do not like the doings of the past week or the Air days. It's about time that the Boche had a crimp put in him. It seems that every time he tries, he comes forward for more ground. And while retreats are very beautiful if perfectly made, and magnificent advances are interesting but are tiresome if repeated. A true magnificent advance would be a pleasant change for the sake of variety. The things "that are going to happen" never seem to come through. Nothing ever comes through but Huns. When enough of us get into France, we are going to dig a big ditch back of them so they can't get back home, then we will send them all to death. One thing seems pretty sure. The Hun will have to win this summer, or it's never for him. He promised the little Huns that he would, and if he doesn't, he will not, I think. That is why he is trying so hard, but he will never win out. Americans are pouring in here like mosquitos down in Jersey. And I hope they keep coming. They will surely put the end to Fritz.

So you see, if he does not pull his win this summer, we will have enough men here to wallop him good and proper. Good Bye my dear. I have love for you all I know, not much to be sure, never said above, the Censor is particularly fussy just now.

Charlie [Jack] put at the top of a letter – "France – same old place," and a D.F. [Damn Fool] Censor cut out "same old place." Now that fellow ought to be *pair extra,* don't you think?

Much love – to all

Your B

June 8th, 1918 [Saturday]*
[Le Treport] France
My dear Mabel,

Big surprise today – a letter from Warren from "Somewhere in France." I am very glad & proud to have him here. Your last letter spoke of him being still at home, and a recent letter from him came from Washington. He must have made a quick trip. He couldn't give the name of the vessel, but from what he did say, he was either on the *Leviathan,* the *Argentina* or the *Mauretania* – according to my figuring. He had some excitement also – potted one submarine and probably two. Wasn't that fine?

I am the first Officer of our Mess to have a son actually in France, so I am stuck up just about now. Padre Jeffery's has a son who may be here, but he does not know yet where he is. Dr. [William J.] Taylor has two sons in Camp at home. But our boy is actually here in the push. See? It will be a great experience for Warren and will do him a lot of good. He is about 200 miles from here, probably at Bordeaux or Brest, so of course I can't call on him yet. Perhaps later, I can arrange to visit him.

Was so sorry to hear about Elizabeth Ross. It was most unfortunate, but even so, they got off better than we did. From your description she is likely to have one or two inches shortening, in fact it would be very remarkable if she did not have at least an inch. But an inch is not too much. I wanted to write to Gev., but was afraid that perhaps he had not heard about it. We naturally knew all about it, but I was not sure, so decided to wait for your next letter. That has not come. Motor cycles are very dangerous things. They are wavier things for safety than are horses. It's all too bad, and I do hope Elizabeth will come through alright with the minimum after results.

Your last letter gave me many "strafings." First you think I am hard on the Army. Don't you believe it. When I speak of the Army, I do not mean any particular Army. I mean an Army. And any Army is you will admit a necessary evil. Quite necessary to be sure, but none the less an evil. Now when it comes to Armies, the one we are fighting is a very good one, you might say. One of the best, but do not think for one moment that the German Army ever was nice, or ever will be, until it is exterminated. So I am not hard on Armies, but I do not like them. That's all.

Strafe No. 2 was for calling Henry Page [Jr.] a slacker. He is certainly a very willing "stay at home." His natural caution has not allowed him to get into any excursions or discomforts in re War with Germany. Of course, he thinks he is too important to be spared, but believe me, more important men are in the service, and he would not be seriously missed. Same old thing. Some men are too big to stay out, and some are too big to get in. I do not mean that Henry would not get in if somebody told him to, but he will also stay out if somebody told him to. In other words, he would not be Henry if he had any initiative, or "get up & get." He would then be someone else, and I

* WTS to MWS, 14pp., on both sides of 7pp. plain 7x5in note paper, torn from a tablet; not opened by censor; passed by G.B. censor; postmarked APO 10 JU 1918.

wouldn't like him so much. Now, I have repeatedly heard bad reports about Pres. He is probably a Pacifist and according to accounts has verged on the unpatriotic. I am sorry if that is so. The difference between Pres & Henry is that Pres doesn't want anybody to do it and Henry is satisfied to let the other fellows do it. I thought that Nan & Henry were mad at me, so will write – and apologize. I only hoped last week that they would have an "active & interesting Social Season, with lots of nine dances, bridge games & dinners." As we understood it, that is what the folks at home were doing last winter. The papers gave the social events just as usual and we had our amusement over their doings, and were not always complimentary in our comments. You see, the nights were long & dark, & wet, & cold & windy & lonely.[93]

Of course, I know as you say that everybody is now working and the ones at home are winning the war as much as those others over here, but you know the difference between the ones who come across and the ones who play safe, [Edward A.] Shumway, I have no use for, and _____, little use for; Bill MacKinney, no use for and so on. Creighton will do what is right when the time comes. He is very conscientious and loyal. He only lacks initiative and push. In that way he is like Henry. Dinner time XX.

After dinner: – I must tell you what happened. Ned Hodge set them up to cordial at our end of the table, [J. E.] Sweet, [J. Howard] Cloud, Charlie, [John R.] Cruice & [Frank C.] Knowles. Then they stood up and he proposed a toast to "Father & Son in France." Well of course, the others soon caught on and I had lots of congratulations – "*prosits*" etc. Quite exciting. We have grown about 2 inches I think. I feel very proud and very happy. Will write to Warren tonight and congratulate him.

Had a very hard luck case this week. A dandy British officer 20 yr. old – recovering from scarlet fever, was playing golf last week with a Canadian nurse. The golf course adjoins Isolation. He drove off first and went for his ball. He found it and turned just in time to catch his partner's drive fair in his eye. The eye ball was ruptured and I had to enucleate [it] yesterday. But he is so full of cheerfulness that it makes you feel bad – he is sure he won't miss his eye, and is perfectly satisfied and getting all the fun out of it that he can. He is wonderful. But it's awfully hard luck isn't it? To come to war and lose your eye playing golf is scarcely a fair deal.

Now a line about the War. You notice that the Boche is not doing much just now. He is not going to do much more is my guess. Everybody is looking to Foch, and I fancy he is the one to look to. He is a great general and simply must have something up his sleeve for Fritz.

Perhaps tomorrow, your letter will come, and if so, I will write you more. For the present, however, you must be satisfied with English kisses XXXXXXXXXXXXXXXXXXXXXXXXXX
Congratulations on your 24th wedding anniversary [June 6]. Wait for the 25th. Good night.

Sunday night. No letters, so I can't say much. The news today was pretty good. The Americans seem to be really in it and holding in fine form. Oh, for another million and 50 000 air planes might now. The sun would then set on the Hun.

We have had a quiet week so far as air raids are concerned. I do not mean our place in particular, but other places. Of course, as I told you, we have never considered ourselves in much danger. Our preparations are merely routine, and necessary because the Enemy is without soul or honor. He cleaned out Etaples [sur Mer] completely, but this location and surroundings are different. The submarine seems to be cabooted. At all events it has not interfered very with the transfer of troops as the Boche thought it would. It has a very hard job and I would not care to be a member of a submarine crew. It used to be 50 x 50 – Now it must be a much greater hazard. If I were on one, I think I would stay submerged. As soon as they come up, they get winged with a depth bomb, and down they go for keeps.

Well, good night.

Much, very much love

from B.

O C Q M., A E F
June 4, 1918 [endorsement on letter dated May 10, 1918]
Finance Division / Office of the Depot Quartermaster / Washington, D.C. / 245.01
From: Depot Quartermaster
To: Captain W. T. Shoemaker, M.O.R.C. / Base Hospital No. 10 / Thru Chief Q.M., Amex Force
Subject: Pay account

1. Your pay accounts for the months of May, June, and July 1918, forwarded under the provisions of Army Regulations 1259, as amended, have been received for payment at maturity.
2. Your vouchers for May and June will be paid in the sum of $204.20, each, $5.00 being deducted for allotment to the Liberty Loan and $10.80 for War Risk Insurance.
3. Voucher for July will be paid in the sum of $204.45, $4.75 being deducted for allotment to the Liberty Loan and $10.80 for War Risk Insurance.

By direction of the Depot Quartermaster:
[stamped] P. L. Smith / Lt. Col. Q. M. Corps, U.S.A.

The Battle of Belleau Wood[94]

On June 6, the first large-scale battle fought by American soldiers in World War I began in Belleau Wood, northwest of the Paris-to-Metz Road.

In late May, the third German offensive of the year penetrated the Western front to within 45 miles of Paris. U.S. forces under General John J. Pershing help halt the German advance, and on June 6, Pershing ordered a counteroffensive to drive the Germans out of Belleau Wood. U.S. Marines led the attack against the four German divisions positioned in the woods, and by the end of the first day they suffered more than 1,000 casualties.

For the next three weeks, the Marines, backed by U.S. Army artillery, launched many attacks into the forested area, but German General Erich Ludendorff was determined to deny the Americans a victory. Ludendorff continually brought up reinforcements from the rear, and the Germans attacked the U.S. forces with machine guns, artillery, and gas. Finally, on June 26, the Americans prevailed, but at the cost of nearly 10,000 dead, wounded, or missing in action.

June 15th, 1918 [Saturday]*
[Le Treport] France[95]
My dear Mabel,

They tell me that this is Saturday, but it does not seem very different from any other day. However, I will take them at their word and write to you. First, I congratulate you upon having two boys in the Service. Of course, it was alright for Ted to enlist if he wanted to, but I am very much afraid his education is at an end. You had better not count upon the furlough in Sept. for him to return to school. The government does not care a tinker's d--- about his education and the word of the one who told you that he could return to school is probably not worth a plugged coin. Nothing short of a written contract on the stationery of the War Department would satisfy me. And even that would be a one way contest to be broken by them if they see fit, but not by you. I doubt very much if the fellow who told Ted that story of returning to school had any more authority than I have. He was after recruits. He was not so interested in how he got them. That is the system in war times. If Ted has enlisted – he is in. He no longer is an individual, and his education and his personal desires are eliminated for [the] duration. I don't want to discourage you, because you did the right thing, and he is in the right place just now. But, I don't want you to be disappointed

---

* WTS to MWS, 14pp., on both sides of 7pp. plain 7x5in note paper, torn from a tablet; not opened by censor; passed by G.B. censor; postmarked APO 17 JU 1918.

when school opens. He will enjoy it and have a fine summer. Being at Cape May [for basic training], he can at times get a good dinner at Uncle Josh's and at times you can see him there, or he can come to you.

You certainly have done your part with your family, and I am proud of you. I guess the "Example" that I was anxious to set for our children was followed alright. Won't it be fine when your chickens all come home to roost. They can talk war for a generation. You can sit by the fire and knit and I can help you. You know, I will have nothing else to do. Have only received the one letter from Warren. If he landed at Brest, I hope he thought to call on Gev. Ross, but he probably never thought of it. Your girls are also distinguishing themselves. Dorothy's honors are fine – and I am just as proud of her as can be. Of course, I knew she would get a scholarship, but I had not thought of the prize. I hope she will keep some of these things for me to see. Barbara, I am very proud of because she is alright, and because everyone says she is so pretty. This you must not tell her.

My pride in Bob is due largely to the fact that he writes somewhat after the fashion of his paternal Dad. A letter from him today was read in spots. I will tackle it again tomorrow and perhaps get a few more words. All that I gather was very interesting. And also, Ted's pole vaulting stunts are very gratifying – Good work. I bet though he will get sea sick when the time comes, and then wish he had taken the Army. Altogether my dear you have five fine children and I don't begrudge you the small part that I contributed in their training.

Many thanks for your cable which I received June 13th. Good to hear from you in that way. I did not send one because I had no money. But I did not forget you, that's sure. Speaking of money, reminds me that I am penniless in a strange land. Can't say I am hungry, nor thirsty, nor without clothes. Just without money, but as usual with plenty of friends. You ask how much I want or need. I have been trying to take as little as possible from you for you have far greater needs. I would say that $50- a month would be sufficient, $25- of which I get here as a Field allowance. In other words, you might send me $25-, $30- a month (about $1- per day.) My mess bill averages 200 F., then I have wash and cigarettes and other things of various kinds to buy. Last month cost a little more than usual because I bought golf clubs, etc., and shoes, and had trousers repaired and so on. Then every once in a while someone wants money for war prisoners, or I want to go away for a day – or some down town, or buy a nurse some food or something else. Etc., etc. etc. My principal difficulty is not having any tidy sum salted away for an emergency. My $100- I used as you suggested. Col. Lister has been trying to get me to go to Boulogne for a visit. Well, I tell him I can't get off, but really, I can't afford it. But I don't mind that, for I don't care whether I go to Boulogne or not. But when I buy real things like shoes, they cost real dollars, and a fabulous number of francs. But if you will send me $25-$30- a month, I feel sure I can be perfectly rich and comfortable.

I am really not spending much money but I would like to spend nothing so that you would have more. Pity, we can't be a Major. That would help a lot. Of course, when I have to buy something very big like a suit of clothes, I will let you know in advance. But I am alright in clothes for months to come. I think I told you that my summer underclothes are going fast. One reason that I will now confess to is that I wear them all winter. You remember that I thought I wanted winter underwear and you like a dear good wife sent me some nice woolens. Well, I still have them, but have never worn them. I will someday though, so don't worry. I have at least 4 pairs of drawers & undershirts. But I think probably I have still a new pair of each on my bunk. So, 3 pair would probably do. Socks, I bought from our Q.M., 25cts. a pair – perfectly good. Handkerchiefs I am long on, but one's nose runs all the time in France. I must have 50 of them. Some are too fine to use. Martha (French girl) for instance gave me half a dozen with my initials embroidered on them, because I attended to her brother. Then I baksheed 6 or 8 from the Red Cross – the kind we give our patients. They are cotton khaki and very slippery after use, but they show up well, and are quite as good as a coat sleeve. Then I have silk handkerchiefs also. So you see, I have no excuse for a dirty nose.

Tomorrow, I take over the Mess – as President, Secretary, Captain or Boss, which ever you like. That will be interesting but for me something new. They wanted me to take it last winter when Cadwallader left, but I declined. I enjoy my morning ride to my labor camps. The Chinks are quite interesting. They are funny people. Awfully glad to hear that Elizabeth Ross is doing well, and that they have the cuss who injured her. I hope he stays in jail for [the] duration and then some. You did not tell me what Nancy Page's operation was. I guess it was an altogether of fibroids. What you say about Ralph Straub may be true, but I do not think that is Ed's trouble. I think Ed is probably just snobby, and thinks that an officer like himself is far superior to a private. But that may not be it either, because his best friend _____, from Haverford, is a private – and his orderly in the Dental Shop. I don't know what it is. Ralph is alright and a splendid worker.

Now my dear, I must stop and write to someone else. Sorry, I have talked so much about clothes and nothing, but perhaps tomorrow I will add a little more about something more interesting. Good night. Keep a "missin'" me.

XXXXXXXXXXXXXXXXX

Sunday, June 16th. I said perhaps I would write more, but I have changed my mind. Nothing has happened but a beautiful day. I was busy and am tired, so can't write – just now. I'll write sometime this week to ___, to Dorothy, Ted and the rest. Good night – much love as usual – of the best kind – [the addition on June 16 is nearly illegible]

from B.

"The Austrian offensive of June 15 failed to achieve a breakthrough … Among the Allied advantages on the Italian Front was superiority in the air. More than six hundred Allied aircraft, Italian and British, caused havoc to the Austrian forces. … On June 16 the British and Italian forces continued their counter-attacks, the British taking 728 Austrian prisoners, and the Italians releasing two hundred of their own men who had been taken prisoner on the first day of battle."[96]

June 22nd, 1918 [Saturday]*

[Le Treport] France[97]

My dear Mabel,

Once again, I must write you about nothing, for nothing has happened since last time. Have not heard again from Warren, but received a letter from Herbert in which he says he is going to try after he gets settled to come here for a visit and bring Warren with him. You know I suppose that upon their arrival the Unit was broken up and sent every which way. Warren is no longer in Herbert's bunch. I do not know where any of them are, as it seems to be against the rules for them to tell. Will however write to him and ask where Warren is. Am hoping that his most although will tell me in some way.

It would seem to be all darn nonsense, but the Censor department must keep busy and earn its salary. In its effort it does some funny things. For instance, if the Huns should bomb a place known as Ca. – it would seem reasonable to suppose that inasmuch as they did it with nature of fore thought, they knew that they did it. Many other special pets must get theirs before the common herd can be civilized for just a weary little advancement. [J. Howard] Cloud, for instance is 46 yrs. old and a Lieutenant. He can't even get a Captaincy, and I presume won't get it until everybody of importance is a General or Colonel or some mighty thing. I need the money probably more than any man in this Unit, but the Government doesn't mind that. And so long as we can get along, I don't mind, for I care nothing about the rank, beyond the great pay for less work. If then, I should write and tell you that they did it, the Censor would object, for fear that the Huns would learn from my letter that they did it. At the same time, the papers might be full of it. Funny game but with all its seriousness, War is a funny business.

You ask in your last letter, if I had not better try for advancement. No, I will never ask the powers that be for a thing. I do not know how it is done, except that to frankly ask, would mean a turn down with flourishes. Harte, Gibbons, Norris, Hutchinseme [Edward Hodge] and others are made Lt. Colonels, R.A. So don't worry about it. I will get no promotion from this side, and have little chance to get one from the other side. It's alright, I will be glad enough to get home a private. As I told you before, when I get out of the Army, I will gladly present to the War Department all my insignia, stripes, clothes, etc., and be a plain M.D. as of old. The M.D. is the only thing they can't have. That is more to me than a dozen Generals, Colonels, Majors and everything else put together. I am afraid you will think a bit pessimistic and grouchy. I'm not, I am perfectly happy and satisfied, and would not have anything else than just what I have at the present time. But that need not mean that I like the business. You know, it's a part of character building to do things that we don't enjoy or like. The slacker won't do it, you know, hence he does not build so much character, and is consequently of less use in the world. That's all, I am just commenting as it were.

Now, what shall I tell you of more interest? Have been pretty busy this week with my new jobs. It's great fun to be Mess President. I enjoy it so far. Handle about $1,000 a month. Keep about 30 men happy with their Eats; supervise 8 or 10 boys to do the work – "Keep house," etc., etc. Now, you think I could not do all this, but I can. You

---

* WTS to MWS, 14pp., on both sides of 7pp. plain 7x5in note paper, torn from a tablet; not opened by censor; passed by G.B. censor; postmarked APO 24 JU 1918.

don't know how well I can do things when I have not you to do them for me. Then every morning at 8:30, I get in an Ambulance, and visit my Chinks, about 4 miles from here, and 2 or 3 times a week visit a battalion of Engineers about 6-7 miles from here. Then of course, I take care of all the eyes. My ward is at present full. Not all eyes – but other things too. Then every day, I must play some golf so as to learn how. I am getting along pretty well. Will soon be some golfer.

The news just now looks pretty good. The Italians are doing well, and for some reason the Germans have delayed their attack up our way. They are ready, and so are we, and you might add from us and then some. The Americans are at the present time a big factor, and I think Wilhelm II is beginning to notice them and wonder what he is up against. If he has been watching carefully, he has noticed a very rapid growth, and has not begun to see yet the full growth. His plan to win the war before American could get into it has failed completely, and he must realize it. And it must be a bitter pill for him to swallow. The same old question however is with us. When? That is different. There is no use in giving a true prognosis. It can't be done. It always fails. The prognostication that I hate most however, is the fool who says 10 yrs. – 8 yrs. – yrs., etc. Of course he is a fool because such a length of time includes so many factors as to make the proposition incomprehensible. Just put in for example – crops and inventions. Both may anytime win or lose the war for either side. And who is the guy who can tell what the crops will be for the summer of 1923? So I put the pessimistic chap who decides on 8-10 yrs. as the duration, down for a D.F. [damn fool]. It might last that long you know, but he does not deserve any credit for thinking so.

Now I must write Warren & Herbert. I want Warren to take full insurance, and will so advise him. In case Miss Jane does not get the letter I wrote her last week, tell her that I was advising her to consult Creighton or [William] Zentmeyer if necessary, but not to bother about [Edward A.] Shumway. When one buys a practice, he does not buy the patients. He only buys the records and good will. By the way, I notice new laws in certain states, compelling all able-bodied individuals to work. Does such a law perhaps affect brother Alfred? It would be hard to disturb the well-earned rest and quietude of such a self-made man, even in war times. I have not heard a word about said brother since I left home. Perhaps he is in the Resting Corps or the Keeping Corps. Well anyhow, I hope he has purchased plenty of Liberty Bonds. That would help. Also you did not tell me about Elizabeth Ross last week. I fancy she is doing well. Also, I need money very badly as I also did last week.

Good night, my dear; dreamed about you last night – I arrived home unexpectedly and came in on you and Dorothy. You were down in the cellar working. Upstairs was a little curly-headed kid as Barbara used to be. I hope that part of the dream will not come true. But she was about 4 yrs. old, and seemed to belong to me alright enough

Good night my dear –

XXXXXXXXXXXXXXXXXXXXXXX

Sunday night. Have just written to Warren telling him to take $10 000 War Risk insurance, naming you as beneficiary. It will cost him almost nothing and will come in very handy later. Played 18 holes of golf [in] the afternoon not too badly. But it is not so easy as it looks. Don't forget to learn how, so that you can play with me after the War. When we get on our feet, we will join the nice long club house at Chestnut Hill, where we took a Sunday night dinner with Miss Stovell [a] couple of years ago. The Club is less expensive & sporty than the others and is very nice and accessible. All of this is a long time after the War in case we don't go to the poor house instead. Of course, I still have my eye on the poor house, but it may be full. You never can tell.

Tomorrow, I am going to Dieppe to try to buy some water for our Mess. Syphons, you know. We have run out, and can't get it here now, because the necessary chemicals to charge it can't be had. Must stop now, and write to someone else. Have not decided yet whom. Good night, my dear. Much love from B.

June 29, 1918 [Saturday]*
[Le Treport] France
My dear Mabel,

We have a very buy week ahead of us – i.e., I have. Monday is Dominion Day, which as you remember is a great day of field sports – celebrated by Canadians the world over. Last year, I wrote you about it and sent you a program, but little did we think then that we would be here to celebrate another Dominion Day here. But, so it is; I wonder if we will have to take in a third one next year. Then, comes July 4th, which we celebrate and which has me on the jump. We are giving a base ball game, a Tea from 4 to 6, and have a play "Officers 666" by our own dramatic company in the evening. But the Tea hits me as being Mess President; it is up to me. Now you never saw me arrange a Tea for the Entire Area in France at which about 400 or more people will come. But that's what I am doing just now. Our menu will be 5 kinds of sandwiches; Tea – Iced Tea, Chocolate – fancy cakes, strawberry ice cream – cigarettes, and a champagne punch. It will be outside, we will put up a large marque tent – and camouflage our lawn & surroundings with flowers, tables, chairs, etc., etc., - and two bands – regular band and jazz band will be working and everybody in the Tea class will have to be in attendance.

Have sent invitations to all the Hospitals – the Chaufferines in the V.A.D. Nurses – Y.M.C.A., Salvation Army – Tank Camp – Hostel – Mayor & Council of Treport, and Everybody in sight. I have 15 nurses for a Reception Committee – and will have 20 more to help collect & plan the "scenery." Thursday morning I am going to take a bunch of Nurses off in the country per Ambulance to gather us a ton of flowers, which by the way are plentiful and beautiful just now. Poppies you know and of course other things. I wish you could be here to help us arrange them. It will be fine when we get fixed up, and I know just how to arrange it, but it if it should rain, it won't be so nice. Of course, the party will have to go on anyway, as everything does in the Army, but it will have to be mostly in the Marque and & Mess Hall – not so nice, but it won't rain. It's lots of fun to run it, and I enjoy it, but I will have to get away with it and give everybody a good time, or I will get in Dutch. Then you know we have to give a party up to the dignity of Base 10 U.S.A. Things of course are very hard to get and are expensive but my Mess Sergeant is cracker jack and I only have to tell him what I want and put of the money and he is there with the goods. For instance, there is no ice here, so we have to send a truck to Dieppe and get it. Strawberries are fine, but in such quantities, we have to order in advance and send a motor [truck] after them. Tables, I wangled from the wards. Chairs, I borrowed from the Canadian Red Cross. In addition to the cakes that our own cook will construct, the Sisters are also going to make some. Well, that's enough about our party. Next week, I will tell you how it came through. Here's hoping.

A very sad thing happened Thursday morning. One of the Chaufferines became mentally off and had been placed in [Ward] No. 3. She watched her chance, and at 6 o'clock in the morning jumped out of the window, but did not hurt herself, so she got up and before she could be caught, ran to the cliff (very close) and jumped over. Down she went for the 350 ft. to the edge of the Channel. The cliffs are perpendicular; if you go over, you only hesitate once and that is a free fall stop at the bottom. She had a full Garrison funeral this morning – the biggest we have had. It was too bad; she was a nice young girl (30) – her parents are living in England. But she could not stand the strain and became melancholy, introspective, & self-centered. The Chauffereines were to give a tent party last night and we were to have a party & dance (Our Sisters) tonight, but of course, both were postponed.

Received a letter from Warren today. He is at Subligny-Villeroy – some distance S.E. of Paris. He is in Forestry at present – I tell you this in case you do not get his letters. He seems very happy & contented. If I had any money and my bills paid, I might run down to see him, but there is no hurry. I will see him in the course of time. I might take Charlie and Ed down some time during the summer. When I say take Charlie, I mean get Charlie to go along. Not really take you understand. We are getting more Majors. Mitchell's commission has come through, and Newlin, Packard, Cruise & Charlie are up for it, & waiting. They will come through presently. Perhaps when everybody else is a Major, I may be allowed to be one, but surely not until everybody else is fixed. And there are a great many others to be looked after, so I need not worry for a year or two.

---

* WTS to MWS, 12pp., on both sides of 6pp. plain 7x5in note paper, torn from a tablet; not opened by censor; passed by G.B. censor; postmarked APO [date obscured] 1918; marked "Officer's Mail."

The war news is pretty good. Austria I think is disintegrating. She will end up a Republic. She is no good and never was any good. Germany, I think is getting scared. She is laughing at our Army to keep her courage up. But, it is very unwise to laugh at a million men, and have a laugh at 5 or 6 000 000 before you get through. They are nothing to laugh at. I think she is scared, and is apt to go to pieces before long.

Well, I have given you lots of "local color" as Dorothy would say. Must stop for the present. Lots, lot of love from XXXXXXXXXXXXX

Sunday night. Nothing ever happens between Saturday & Sunday, so I can't write much. The weather has been wonderful for about a month, which make me think it will rain July 4th. We need rain very much, but it is not so nice to have it. Played golf this afternoon, very badly – am getting worse and worse. You certainly did forget to send me any money lately.

Have not a cent or a centime – can't buy a cigarette or a cough drop – can't buy a bottle; can't buy the *Daily Mail*; can't buy anything. Just have to live on rations and not change my clothes because I can't pay the laundry lady. It's a terrible war. What you say about pay vouchers is correct, but can't do anything in that way until September, because my vouchers from May, June and July are at home. July's will be paid to for August 1st & August's can be paid any way I say in Sept. – so then I can do as you suggest, but by that time, I will be in a terrible condition, not having bought any of the aforesaid things. Well, here's hoping –

Good night, my dear –

Will tell you all about the party next week – Just a little more love than last night from B.

Funny for Funk & Wagnalls to send the bill here. – Pay it please. I enjoy the *Digest*. It is very optimistic.

[enclosed]*

NOTICE

We have been honored with your subscription for "The Literary Digest" in the past, and trust you have found it of such value that you will wish its visits continued another year. Subscribers wishing their files to remain unbroken should remit promptly, as foreign subscriptions are discontinued at the expiration of the time paid for.

FUNK & WAGNALLS COMPANY.

* TN on 5 x 4 in. note paper.

July 1918

Le Treport and Paris

"The drive has started as you are at this moment reading in the evening paper."

July 6, 1918 [Saturday]*
[Le Treport] France[98]
My dear Mabel,

Last week I said I would tell you about our party this week. Well, it was a great success and everyone for miles around is talking about it. It is said to have been the best, & best managed function ever held on the top. All of which is due to the way everyone worked with the proper spirits and an unbounded willingness, and also to a little common sense on my part in using the ways & means at my disposal. You know how I love to give and arrange a party. Well, I had my chance without interference or restrictions of any kind, so I had a bully time doing it. Our surroundings were fine, a big Marque tent, with the flags of the Allies on it, and a huge garrison flag of our own, the prettiest I have ever seen. Then, as I told you, I sent a number of nurses off in the morning in our ambulance to gather flowers and boughs for camouflage, and they did it alright – beautiful flowers in quantities. Then the night before, nurses and men went off and gathered flowers. It is not dark now until about 11 o'clock. Then we placed them and fixed the tables, etc., and made the place look beautiful.

Wednesday morning, I went to Dieppe in the truck which we now have and bought 400 lbs. of ice and other necessaries, also some red, white & blue ribbon and 15 little medallions with which we made badges for the Committee of Reception – 15 nurses that I appointed to help & flit around and serve the guests, etc. I guess they were Aids, but they looked mighty nice. Then we had our dining hall nicely decorated & converted into a place for "gentlemens." The gentlemens seemed to enjoy it. – Cigars – cigarettes and real American cocktails <u>with ice</u> and anything else that one might wish for on the 4th of July. Our two bands, and a very fine band – British, supplied lots of music. And to make a long story short, from 4 to 6 we had some party and everyone enjoyed it. But that is only a part of the day's program. At 2 o'clock, there was a base ball game, which of course I did not see, as I was busy with the afore mentioned party. The score of the base ball game was 18-8. Against us. So we did not play ball so well. Then, every patient in the Hospital had a <u>*bon*</u> dinner or spread, gotten up by the nurses. The Enlisted men had a grand banquet, about 250 all at once. The Sergeants had their party also. It was some 4th.

Our Mess boys worked so hard and so well, that tomorrow I am sending them all on a picnic in the truck – from 3 o'clock to bed time. They are going to Dieppe where they will get dinner and have a good time. Everyone (workers) in the Mess is going. The Cook, Assistant Cook, Sergeant Branden and the Mess boys and the little French boy that helps in the kitchen. With the assistance of two patients, I am going to run the Mess myself. The boys are tickled to death, and they sure deserve an outing. Just one thing of many. They made 155 qts. of strawberry ice cream in two 4 qt. freezers – gallons of punch, and 1,500 little cakes; quite as many or more sandwiches, and innumerable other things, and served as regularly with our meals just as if there was nothing else to do. We had between 300-400 guests so it did not take long to eat us out of house and home. That was because the Canadians were very hungry & seemed to like the ice cream, so there were many repeaters, up to 3 & 4. But we held out strong until about 20 minutes to 6. If we had had more than 20 minutes to go, I confess I would have been worried.

Oh, I forgot to tell you, you may some day see this party, because the British official photographer was on hand taking <u>movies</u>. They are going to the States, so don't be frightened if you should see me walking around running a party. He also took the ball game, which had an audience of several thousand. It will be interesting, won't it! Excuse me telling you so much about our 4th, but it went so well that I became enthusiastic. But it was very trying, watching the weather and planning for "if it should, which it was apt to." But the weather was fine and thank Heaven, I watched it very carefully and had a penny up my sleeve in case of rain, but it would not have been so nice.

---

* WTS to MWS, ML, 12pp., on both sides of 6pp. plain 7x5in note paper, torn from a tablet; not opened by censor; passed by G.B. censor; postmarked APO 8 JY 18; received July 30.

I enjoy being Mess President, but believe me, it's running a hotel and being proprietor, day clerk, night clerk, bookkeeper, cashier, etc., etc., all at once. We have 30 officers here and sometimes more. They come and go, guests come and go, and today in particular, we had an unusual number of "arrivals and departures." I tell you my dear, when I get home, I will know how to do so many things, that you won't know me.

Now, of more interest to you perhaps, I have arranged to visit Warren, sometime this month – if you have sent me any money. I am wangling a duty leave – not a regular leave, and will only be away 5-6 days – perhaps a week. Charlie is going with me. He has to see some jaw surgery in Paris, and I have business also in Paris. Warren is about 75 miles S.E. of Paris, so I hope to see him & Herbert and Harry [? Shoemaker]. I will have to go after the middle of the month – after I get our Mess bills paid. The first of the month is a busy time for the Hotel business. Do you know that butter costs us 70.cts a pound? We had a great deal of margarine, not real butter also; - enough to keep our boarders happy. It's a critical bunch of boarders, too. If anything is not quite right, they promptly tell me, as if it was my fault. I smile, and say sure, I will fix that – tell Branden & forget it. He fixes it. Here I am talking about food again just because butter is expensive. That won't do –

Well, I am going to see Warren. Oh, yes, I forgot to tell you that in the evening of the 4th we had a play – very good and well done. Program enclosed [not preserved]. Now I must stop until tomorrow. No letters from you this week. Perhaps tomorrow. Good night my dear.

XXXXXXXXXXXXXXXXXXXXXXXXXX

Sunday – A beautiful warm day. The boys got off at 2:30 for their picnic. I am so glad they have a nice day. I am doing a little of everything down to keeping bar. Just made Col. Harte a weak punch which he has to have every afternoon because he gets so tired with all his worries and responsibilities, and he is getting in much trouble I hope.

Every one was disappointed today because we got no American mail. It is much overdue but perhaps tomorrow. Here's hoping. War news continues good and the Allies are acquiring a feeling of great confidence. I do not think it is misplaced.

Am looking forward with much pleasure to next week or the end of this week, when I will see Warren. It will be a nice little trip. My dear, I am darned tired of this war, and unless it stops soon, "We are broke." Good bye, much love to you and the kids.

Your

B.

July 10th, 1918*

Le Treport, France

From: Captain W. T. Shoemaker, M.C., U.S.R. / To: Depot Quartermaster, Washington, D.C.

Subject: Allotment.

1. In accordance with G.O. 53, 1917, G.H.Q., Amer. E.F., I hereby allot $180.00 of my pay per month for three months, commencing August 1st, 1918, to Mabel W. Shoemaker, who is my wife and whose address is 109 - South 20th. St., Philadelphia, Pa.

W. T. SHOEMAKER / Captain, M.C. U.S.R.

July 13th, 1918 [Saturday]†

[Le Treport] France

My dear Mabel,

A letter from you early in the week which was quite blue. Sorry, and today a nicer one from you and also very interesting letters from Dorothy, Barbara & Mary Buchanan. Dorothy & Barbara described their work and Dorothy seems to have quite a lot of it. Barbara is a splendid letter writer. She told me about her school. Your letter satisfied me a lot. First, I can't spell "socks" properly. Well, that's alright.

---

* CC of TL from WTS to Depot Quartermaster.

† WTS to MWS, ML, 12pp., on both sides of 6pp. plain 8x7in. note paper (a new form of paper, first used in this letter); not opened by censor; passed by G.B. censor; postmarked APO 15 JY 18; received August 2.

Then you objected to my calling the war useless. My dear, you don't understand me. It is a most useless and harsher War. Of course, I do not mean that it is useless for us as it exterminates the Boche. Far from it, it is a great necessity. But why should he have caused this terrible sacrifice? Every other nation was carrying on in a perfectly orderly and peaceful manner. No one was harming Germany. He was, before the war, able to go & come where & when he pleased – but never again. He had what he called "freedom of the seas," but not what he meant by freedom of the seas. He means by that, nobody on the seas but Germany. Her gain when all is over, will be, I hope, complete ruination. The Allies have nothing to gain, because everything was right & fair before. Therefore, I say it is useless, however necessary it might be.

Then you seem to object strongly to my calling Page & Whiting slackers. I know of ~~too~~ no better examples. It is all very nice to decide that you are needed at home, but there are thousands of us over here who also have homes and families and are needed at home. As nice as Henry [Page] is – he is a weakling and lacks the moral courage to respond to his Country's call. His "duties" at the Lankenau Hospital are small indeed, and you can take it from me, the Hospital would run just as well without him as with him. That is camouflage to disguise lack of force, individuality and guts. Excuse the word, it's expressive. Pres as I once told you was different. He has the courage to do even wrong. I admire that in a way more than I do Henry's conscientious devotion to himself and his poor helpless family that really needs his personal attention and fat salaries. – My dear, I am afraid you will think I am disagreeable tonight. Well, I am disagreeable and just in the mood to write a very poisonous letter. I am reaching the breaking point, and am apt to bust at any time. There has been very little to do in this hospital for several months. I am doing as much as anybody, but there is nothing doing. We are being broken up in a way. Orders have come for Earnshaw, Packard, Gerhard & Knowles. Norris, Gibbon & Cadwallader have long since gone. Drayton is also going. All are going A.E.F. By and by, I suppose they will pinch me, but I hope not. A.E.F. has no charms for me. I could if I went probably get a promotion, which I will never get here under the present C.O.

If I have to put another winter in over here, I will go crazy. It is the damndest, most monotonous job you can think of and to think that you were of some use at home and could be of some use at home, and that you are of no use here, does not relieve the monotony. If things continue this way, I will be patient until about October. Then I will ask to be relieved and will probably be court marshalled [sic] for being so independent and impatient. Court marshall [martial] would be a change. However, I am here with all the ability that I have at the disposal of the Government, but nobody wants it. I have been paid 14 mos. salary for doing what any recent graduate could do. And what is true of me is true of every man in the Unit. Base 10 is entirely too high grade for the work it is doing. Of course, that is the meaning of detaching men for the A.E.F. as [the] organization progresses. But the immense organization necessary, naturally is slow. So you see, I can't blame anybody. I just have to pick at things as they are and must be.

So far as my profession and life work are concerned, I am ruined, but I don't mind that so much. I figure that our family is almost raised. Dorothy, Warren & Ted are self-supporting or can be. Barbara will get married to some very rich man. Bob is easy, and I can certainly make enough out of the work to take care of you and myself. So I have stopped worrying about the loss of a practice that took me 20 yrs. to build. What's the use. After this war, life will be too short. Those who "were needed at home," will invite us occasionally to dinner, and perhaps sometimes give us a few of their discarded clothes. You can make them over for us, and we will be just as happy as can be. I am still running poisonous. I did not know I had so much pessimism stored up during the week. Well, it's nearly out now. Charlie is much worse than I am. He is serious, and his peculiar mentality makes him almost dangerous. I feel sorry for him, because he is not used to adversity, and always has the feeling that he can force things his way by pure brute strength. When his brute strength fails, as it always does, scratch the surface – he looks licked. Funny but pathetic.[99]

Now I will tell you a pretty little story before I say good night. It will make you feel that I am not so bad after all. The Chinks at the Camp where I go every morning, working along the roads, got into the habit of catching little birds in the wheat fields and keeping them in a crude cage or bag until they died. But we stopped it, or thought we did. But this morning, as I approached one of the huts, I heard the "pips" of very little birds, so I said to Dr. Lawrence, the C.O. of Chinks' Camp, "They have some more birds." Sure enough, a Chink produced a bag with half a dozen little birds huddled together – only a day or so old, one of them dead. I ordered him to put them carefully back in the wheat field across the road. As I came out of the hut, I heard a bird in a tree across the road chirping away and said to Lawrence, "That's the mother, watch her come down." The Chink's back was no sooner turned after ~~the~~ he put the little fellows in the wheat, when down flew the mother to them, mighty glad to get her children back. Awfully mean to take them away from her when they were absolutely helpless, wasn't it? The next Chink that catches a bird gets fined 10 Francs.

I paid our mess bills – about $1000 – and am now ready to go see Warren; but alas, your money did not come, and it's not enough anyhow, as I have to pay my own Mess bill, and as Frank Knowles is going away, must pay him 200 F. that he loaned me. However, if I get your 50-, I will go anyhow, and borrow from some one else. Have now made you an allotment of $180.- for August, Sept & ~~October~~. I kept $29.- and some $10- as an out for insurance. The money will come to you instead of to the Girard Trust Co. as heretofore. The Liberty Bond is paid for, and should be delivered to you. It is yours, but do not lose it, as it can't be registered and any body could sell it. I can't get my $29.- until September, as it is August pay. I may however get "commutation of quarters light & heat," which has never been allowed by Congress, in which case I will get 200- or more dollars, as it dates from April. So in time, I will have lots of money and can send you more. For quarters, etc., I will draw about $50- per month, maybe more. But until it comes through, I am down & out. Good night my dear. Excuse the mean disposition which I have tonight. I feel better now.

XXXXXXXXXXXXXXXXXXXXXXX

All for you. –

Sunday – another deadly day. Rain in showers but I got in a golf game partly between showers & partly in the rain. No money yet from Morgan Harjes, so I can't take my little trip. Received a letter from you this morning. No. 89. No. 90 came yesterday so they got twisted. You want me to write you twice a week. My dear, I could not think of a thing to say, but will try if I get a chance. I doubt if you have gotten all of my letters, for you make no mention of things which I remember having written about. Barbara tells me that Ted might go to California – That would be fine. I am so glad that he was not "needed at home." He will have the opportunity for a broader usefulness.

I can't write any more tonight. We just received a convoy. I will go to my ward and see if perhaps I drew any patients. Most of the patients I think were medicine [influenza]. That is all we are getting. We get lots of Americans with nothing the matter with them like this – a patient (American) comes into the Hospital, "What's the trouble?" "Well, my eyes kind of smart a little sometimes and the light hurts them." "How long has this been?" "Well me eyes were always weak." "You see alright?" "Oh, yes, I can see good." Then the doctor says to himself, or half to himself, with great disgust, "There's a war on, get to H--- out of here and fight." Another favorite complaint is, "I can't see at night." Answer: "Neither can I." Still another yammer is – "I can't see with my right eye." "How long has that been?" "Since I was a child – I was hit in it with a stone." "How long have you been in the Service?" "Three years." "Well, the eye is blind and always has been, and I could not make it see in peace time, let alone in war time."

Of course, they don't like to go back to the front, and you can't blame them. Good night.

Your B. XXXXXXXXXXX

The Second Battle of the Marne lasted from July 15 to August 6, 1918. Conceived as an attempt to draw Allied troops south from Flanders to facilitate an attack in that region, the offensive along the Marne proved to be the last the German Army would mount in the conflict.[100]

July 16th, 1918 [Tuesday]*
[Le Treport] France
My dear Mabel,

You wanted a second letter this week, so I will write – just a teeny one, principally to tell you that Miss Furhmann wanted to send $25.- home and I was so much in need of money that I seized the opportunity. So kindly send to Mr. Rutan your check for $25.- upon receipt of this. Sorry to have to annoy you in this way, but your money for Drexel has not arrived and I have nothing, out of which I must pay my mess bill, pay Frank Knowles 200 F, and go see Warren. Just how this is to be accomplished, I do not know, but I am figuring I can't go to Warren until I get

---

* WTS to MWS, ML, 2pp., on both sides of 1p. plain 8x7in. note paper; not opened by censor; passed by G.B. censor; postmarked APO 17 JY 18; received August 2.

some money, so probably won't get off for a month or two. The drive has started as you are at this moment reading in the evening paper.

The first day of it was a good one for us. I hope we will continue to be good. But I am so fed up on "magnificent retreats" that I am going to wait and see this time without ventilating any opinion.

Well, I have nothing more to say, so I will ring off.

Much love – from B.

July 20th, 1918 [Saturday]*
[Le Treport] France[101]
My dear Mabel,

Notice came today from Morgan Harjes that they had $50 for me, so I will now tender some more, pay my bills and try in the next few days to visit Warren. You see, I can't go near the end or beginning of the month on account of Mess business, so this is just my time, if the C.O. does not forget that he told me I might go. He is away now, but I expect him home tomorrow or Monday.

Tomorrow, if clear, I am going to Dieppe to see "Marché Eglies" and "Chateau d'Orge" (not responsible for the spelling). Will send you pictures of them. They are said to be beautiful places, and well worth a visit. Knowles, Cloud, Whitaker and myself have hunted around for them for some time, and weather permitting, we will have a nice little trip. These places are 4 or 5 miles out of Dieppe. Therefore, I have never had the time when in Dieppe to take them in. That is where my Mess boys went two weeks ago on their party as I think I told you. They had a wonderful time. They gave me their pictures taken in [a] group, which is a little too big to send you, but which is in my trunk and which will eventually adorn our scrap book.

This has been rather a bad week for me in the tooth line. Two worthless molars, a first and a wisdom tooth, were declared unfit for service and retired, the one Thursday and the other today. Both have been of no use what so ever for more than a year, but only a menace and handicap. They got acutely bad, and when I was tired of the pain and annoyance, I discharged them. Just at the same time, or a couple of days ago – a first tooth threw a pericentral abscess. So with one on each side and one in front playing in concert, the first tooth will get well for a time and will last for [the] duration or longer, but his days as the maverick, which his God gavest him, are numbered. Anyway, my mouth feels like a shell hole at present, but will be alright tomorrow.

The news yesterday and today is the best for a year or more. Foch seems to be the man for the job and a continuation or repetition of that kind of "Franco-American Soup" and the Boche will be fed up and fit ~~for~~ to kill. We have waited a long time for such a counter attack, and everybody is much pleased. Of course, it won't have anything to do with the end of the War, because the War will never end, but it helps some, believe me. [pause]

Was interrupted, so have lost my train of thought. A mail arrived late – this afternoon, and will be distributed tomorrow morning early. I just know that there is a nice letter in it from you, so I can write more tomorrow night. I see by my *Journal of Ophthalmology* that my friend & classmate Wallis Parker is a full Colonel – in Washington, and ___ is a Lt. Colonel. Washington has certainly looked after the stay-at-homes.

Censor or no Censor, the members of the Medical Reserve Corps who volunteered early, gave up all and came to France when Balfour & Joffre etc., ~~who~~ were begging for doctors, have received dirty mean treatment by the powers that be. We have to stay here, and see our former students and assistants come over and out rank us. You will admit the injustice and you can imagine the friendly feeling which we have for the administration. However, there is no use in kicking. Goodnight. XXXXXXXXXXXXXXXXXXXXXXX

Monday, July 22nd. No letter from you yesterday or today, so I got left. Well anyway, yesterday was a beautiful day and we had a fine trip to Dieppe. Chateau d'Orge is a fine old ruin. It was built by the uncle of your old friend and relative Wm. the Conqueror. It is in a beautiful forrest [sic], Touche d'Argues. Marché Eglies would seem to be a very attractive place to eat. Out of doors on the back of a trout stream is a grove, etc., etc. We had a wonderful

---

* WTS to MWS, ML, 8pp., on both sides of 4pp. plain 8x7in. note paper; not opened by censor; passed by G.B. censor; postmarked APO 23 JY 18.

dinner there and a beautiful ride home in a mixture of sunset & moonlight. The pictures which I enclose are not good, but the best to be had.

I am still in trouble with my mouth. An abscess has developed in my cheek, which keeps up a constant shelling of my disposition. It seems that every time I come to envelop for a prolonged stay, I have a run in with my teeth. In the last 2 weeks they have been giving me a great deal of trouble. This blooming abscess seems to be a secondary infection; it is not now connected with any teeth, because the offenders have been extracted. Would you love me just a little with false teeth? Of course, I would not expect you to love me so much, but just a little, yes? I am still on my own you understand, so you must not stop loving me yet.

My trip south is beginning to look doubtful. The C.O. has not returned and now I can't go before Wednesday at the earliest. After Wednesday, it is too late, because I must be here before the end of the month. Just because Morgan Harjes was so slow in getting my money, so if I don't get off by Wednesday, I will have to wait until after the middle of next month, and then I suppose the fighting will be up in our sector, in which case "*pas du trip*." If we go, I will write you from Paris. I had to borrow more money, but that's alright. You see before going, I had to pay my mess bill, 184.F and Knowles 200.F because he is going away. So I have enough to get to Paris and there I will collect your $50.- so hope to bring some back. Then I want to give Warren some, and take him something besides.

Well my dear, no more just now. The news continues *bon*. May it continue to continue. You are at Wallingford and I hope will rest up and enjoy yourself. Please let up on the work or you will be no good next Fall.

Good night. Much love to you and the kids.

Your

B.

July 24th, 1918 [Wednesday]*

Paris [France]

Well my dear, here we are in the Great City – on the Big Vacation. For the first time since June 12th, 1917, I sleep away from the top at Treport and beneath sheets. The sensation of independence is wonderful. I will get up in the morning when I get d--- glad and ready.

Only one disturbing feature; I am not well because of an abscess in my cheek, secondary to a bad tooth which I told you about. It hurts me all the time, and threatens me in an unkind way. Yesterday, Charlie opened it, but did not get it. It hurts like Sam Hill, and you know, I can't stand being hurt. Tomorrow, I must go out to the American Ambulance at Neuilly sur Seine, and in all probability take a general anesthetic and be fixed up right. I hate to do it, but I cannot stand this constant nagging pain. If I have that, I can proceed on my journey & see Warren Friday or Saturday. I have a week at my disposal, so if I do not have to lay up too long in the Hospital, will be alright. Ran into Dr. Baer of the University Unit in the Hotel Continental. He has been detached and left for Boulogne tonight. He told me that I was listed by the A.E.F. for Paris – I know nothing about it but Allen Griswold in Charge of the Opht. Dept. A.E.F. told him that he was going to bring me to Paris. We will wait and see.

You will be pleased to know that my recommendation for a Majority has been sent in and my hopes and prospects are good. Will tell you more about it later.

So if in the next few months, you find me a Major & stationed in Paris, you need not be surprised. But if you don't, you would not be disappointed, because I am in the Army. Enough said. Will write again soon. This is just a note you understand.

But I am on my big vacation, with a very sore face, so I don't know whether to laugh or cry. However, I am Carrying on, and will so continue until I see your son Warren & report.

Much love from

B.

----

* WTS to MWS, ML in pencil, 5pp., on both sides of 2pp. of folded watermarked stationery note paper; not opened by censor; passed by G.B. censor; postmarked APO 25 JY 18.

July 26th, 1918 [Friday]*
Troyes [France] Grand Hotel Terminus[102]
My dear Mabel,

Left Paris this morning at 8 o'clock and arrived here 11:30. Next train out of here for Sens leaves 1 o'clock tonight, so we have hung up for the day. Travelling here is not travelling that we have been used to all our lives. It takes ages to go a few miles. Subligny-Villeroy is 5 miles from Sens according to Warren, so sometime tomorrow I hope to drop in on him. Paris is in full bloom and very beautiful despite the war. Of course, it has gotten more or less shabby, because the inhabitants have not had time in the last 4 yrs. to keep things up to standard. But still, the parks etc. seem to be nicely kept. The entire country around is crowded with Americans. Paris is full of Americans. Also, yesterday we went out to the American Ambulance at Neuilly and looked over the whole show. They are at present very busy. But I am very glad not to be there, and hope when I get picked I will not be detailed there. I don't want to go there. In fact, I much prefer being B.E.F.

Will not see much of Warren, because it takes too long to go & come. We only have 6 days and a day to go & one to come. If we over say, it kills regular leave. That would not bother me, but Charlie has put in for leave and does not want it stopped. Charlie has changed very much in the last five months until he is getting well nigh impossible. He should go home, and I hope he gets sent home. He started this trip with one of his "spells" and I am afraid is not enjoying himself too much. However, it's all in the war game, and I can't help it. Herbert, I think is not with Warren, so I probably won't see him. If I had much time and more money, I would look him up, but unless he is close at hand, I will not be able to connect.

For the first time in about 10 days, I am completely free from pain. My abscess, I think, broke last night, and I trust will get well. But for the last 10 days, I have had it good and proper. I am tired of the game, and am ready to come home and let some of my "good friends" in Phila. come out and enjoy it. Send out Shumway, Langdon, Carl Williams, or Gresham or Head, or a score of others "needed at home," such as for instance Page & Whiting, if they could be spared. Good bye, my dear. Will write soon again. Enclose pencil; pen & ink not at hand.

Much love to all
from
B.

---

* WTS to MWS, ML in pencil, 2pp., on both sides of 1p., hotel letterhead paper; stamped: "Passed as Censored A.E.F. A.579."

August 1918

Le Treport – The Battle of Amiens

"The news is fine and I think the tide has turned."

The Battle of Amiens (Third Battle of Picardy) began on August 8, 1918. It marked the beginning of what came to be known as the "hundred days," a string of Allied offensive successes on the Western Front that led to the collapse of the German army and the end of the war.[103]

August 9, 1918 [Friday][*]
[Le Treport, France]
MINSTÊRE DE L'AGRICULTURE ET DU RAVITAILLEMENT – 1918
Chaque ticket quotidiene cette feuille correspond
à 100 grammes de PAIN   Les tickets ne peuvent etre utilizes qu'au jour indiqué.
31 100 grammes   30 100 grammes   29 100 grammes

Obituary of "Dr. William H. Greene, Head of the Stephen Green Printing Company, and former professor of chemistry at the Central High School, who died yesterday … Dr. Greene's death is attributed to the excessive heat."

August 12, 1918 [Monday][†]
[Le Treport] France[104]
My dear Mabel,

I am so sorry my dear about the money situation which you mentioned in your last letter received this morning. You were counting on commendation of quarter's money for taxes and I did not know anything about it. The tax proposition is a hard one, and many a time I have struggled to raise the 400 & some dollars before August 31st. They have not been raised for two years. I paid the same amount the last time I handled it. I will send you some more money just as soon as I can, but it won't be $400. As I told you last night, this action came in after I had sent my pay vouchers home, deducting $29. for myself. Then Miss Fuhrmann's $25., I had to take because I had nothing; then your $50 dividends came, and so on, all of which I explained to you.

You see it all happened because I ran too close and because you passed me by, which put me in debt for a time. I have been greatly handicapped by not having money aside from my actual expenses. For instance, I am in trouble with Col. Lister because I could not go to Boulogne to see his work. He came for me several times, and I make just as many excuses, but I could not go because I had no money. Then he got sore, and in Paris, Geo. Derby, who had been with him, asked "Why wouldn't you visit Col. Lister?" More excuses as usual. Could not get off, busy, etc. – None of it [was] true.

---

[*] An envelope from WTS to MWS, containing a slip of tissue paper and two tickets, cut from a larger page; and a clipping from the *Philadelphia Evening Star* (9 August 1918), which must have been inserted in the envelope at a later date.
[†] WTS to MWS, ML, 6pp., on 3pp. 8x7in. note paper; not opened by censor; passed by G.B. censor; stamped APO 13 AU 18.

Then I might be detached at any time, and certainly would not like to borrow the money to go with. So must have to keep some on hand. I feel awfully sorry for you my dear, but you realize now one reason why I hate the Army. And that reason is only one of many. Then you asked me some time ago if I was not a little hard on the Army. Do you think I am?

It's too bad about Ted. And there is a funny side to it also, which you would appreciate if you saw some of the "fit" American soldiers who have been sent over here. I rather wish you had taken Ted home if possible, and had him fixed up where the fixing is good. I fancy your doctor at Cape May is alright. He certainly was decent to you. If he is not a regular, he is probably a very good man. And from what you say, he is I guess a civilian doctor. Murphy, whom I know, was the foremost surgeon in America, and if this man was with him for 10 yrs., he must know his business. Ted will however get very little training before he "returns to school" in September. But why should we care; the Boche is on the run and liked to stand still, and the war will soon be over.

The news is fine and I think the tide has turned. We have waited a long time for this; but it seems to be here now. I am afraid they won't but I hope the Allies will push him all the way to the Rhine before they discontinue the present operation. His losses in dead, wounded, and prisoners are terrific. Ours have been moderate. He can't stand many crashes like this one.

Good night, my dear. Truly sorry, because I felt so bad about your poverty. You have done so well, that I wanted you [to] have all you needed, without the necessary worries.

Burton Chance!! Enough said – miserable, grafter, and around white-livered dud. I hope they send him to Salonika. They might send Shumway & Langdon to keep him company.

Now I am getting disagreeable again, so I had better stop.

Good night again. Much love

from

B.

August 17, 1918, Saturday night*
[Le Treport] France[105]
My dear Mabel,

A real old fashioned mail came through yesterday containing your letter about Ted. He is entirely alright and I am very glad of it. I trust he has had a good operation and have every reason to believe he has had. Now if I were you, I would get him back to school on a furlough if possible. You see, he does not know much and it would be a shame to get so near graduation and then miss it. Tell him the war will wait for him; that there need be no hurry, and that the later years of it will be much more interesting. Of course, he will not be fit for hard duty for about 6 weeks, and by that time school will have started. I am very anxious for him to graduate from the C.H.P., because I may not be able to send him to college.

An interesting letter from Dorothy told me of her work & plans. She has done well and I guess the hard work has helped her. The pictures were very good; I have them together up on the wall beside me. Quite a fine collection I am getting. I hope you went to Boston with her. It's a shame for you to have to stay so close to home and business "year in & year out."

Mr. Harvey wrote me a funny letter. I really do not know what he was driving at. There was some vague suggestion that I go into "Church work," although he admitted that I was never a good Church man at home. However, my policy is this: There is no scarcity of religious work out here. Padres are thick and are doing good work. They need no help from the M.D.'s in this line. I find that the men die just as happily irrespective of their special doctrinal beliefs. It makes no different what religion a man might have, the war situation seems to have raised him above such small and non-essential things as sectarianisms. The bullet hurts the New Churchman just as much as it does the Presbyterian. But I must say, if I were doing it, I would have it hurt the Presbyterian very much indeed – much harder than anyone else. But that's just an aside. Well, to make a story stay short, the Padres are doing great work and no one is neglected. And they are doing the right kind of work for a war. There is no

* WTS to MWS, ML, 12pp., on 6pp. 8x7in. note paper; not opened by censor; passed by G.B. censor; stamped APO, date illeg.

"salvation" business & repentance stuff and pleasant theories of the hereafter, etc. The dying soldier does not want that. He has his communion if he wants it; he has his last message cared for; and anything done for him that can be done.

Now that in very brief way & of course incomplete way, is the practical religion's handling of war vicissitudes. I am afraid that what Mr. Harvey has in mind is the religion of peace times, far from the casualty zone. But there is a war on and it's a big one. But as I said above, I don't know just what he was talking about. But I was very glad to get his letter, and will answer it, but not so specifically as I have answered it to you.

Before I forget it – ask Langdon the name of the patient of mine who consulted him. He is very good to return you the fee, but he also should give you the name. He has, I am afraid, a reason for it. At all events, force him to bite. You have a right to know. You need not expect anything from _____ – I know Harold. I am grateful to the few good fellows you mention who play the game on the square, but believe me, my dear, they are few indeed in comparison with those of the other type. It is very discouraging to find so few men who ring absolutely true. The war will never regulate the bad ones. War never correct meanness.

Speaking of <u>meanness</u>, did you know that the Huns sank a Hospital ship in the Channel about 2 weeks ago. Some 200 were killed. The ship carried many patients from our hospital. Some of whom were drowned and some saved. The picture from a London patient is that of one of my patients. Fry was blind. They had taken a trench and he was digging in it with a pick. He struck a hand grenade, and the thing exploded in his face, and destroyed both eyes. He can see a little light with one eye, but mighty little. He was a wonderfully cheerful patient, as you can see by the expression on his face. The whole ward was interested in him and gave him a fine "send off." I saw him off, and shook hands with him, just as he was being put into the ambulance and wished him "good luck." He could not see and did not know who I was, so he said, "Good bye, old man." Everybody laughed, and he went away in fine spirits. Well, he had his good luck, because he was saved from drowning.

You know, of course, how we work our patients for evacuation. We have A. B. C. & D. A. is absolutely helpless and needs constant attention. He generally has a long splint on his arm or leg. Compound fracture for instance. B. is helpless but has not a long splint and could "beat it" if necessary, or help himself a little. C. is a cot case. Cleaned out completely. Some cases are pathetic. I sent one to a hospital in England for reconstruction. After operating on him & removing bullets, shrapnel, pulverized teeth, bone, etc.

But then the morale of the Allies has been bucked up 200% since we started the last Show. The Americans made it possible and have been as good as old wheat. I bet the Highest of Himland [i.e., Kaiser Wilhelm] is doing some hard thinking. It must be so; the hand-writing on the wall, but it is so ominous & distasteful that he is struggling not to see it. Do you remember Macbeth? Well the Kaiser is enacting little sketches from that interesting drama. I can just see the Kaiser & the Mrs. Kaiser, and the Crazy Prince and the whole Court having pleasant little scenes at home – each blaming the other, and all wondering what they will do next. Then they decide to go out and sink a hospital ship or destroy a defenseless hospital full of sick and wounded patients.

Have not heard from Warren since I saw him, although I wrote to him immediately after my return. He is not good at keeping me advised. I hope he treats you better. Mrs. Melhuish wrote me another nice letter in answer to my letter to her. She is very cordial. She will be delighted to greet us as Warren's nearest dear relative, should he get to England. She says if he is Fanny's nephew, of course he is hers, why not.

Now this is supposed to be a very nice long letter, in which I have not socked anybody very hard. I must stop though or I might start on the Army. Will hold it however as usual until tomorrow, so that I can tell you how I sleep tonight. The weather is at last turning bad. We have had ages of perfect weather, so we should not complain.

Am still busy, but not so busy, as the push is slowing up. The Huns are still hesitating, and beating it for the old Hindenberg [sic] line. There they have concrete fortifications that are not so easily disposed of. Well anyway, good night – XXXXXXXXXXXXXXXXXXXXXXXXXXXXXXXXXX Keep a thinkin', a lovin', and a missin' <u>me</u>.
Sunday – Just over the edge and down the other side a little. Ted's latest letter arrived this morning. He told me of his prospective operation. Also a letter from Anne with shots of Lucretia. My but she is the image of her mother at that age. Can't you see it? I have been wondering where in the world you will get the money for taxes. That's an awful job. It generally takes a cataract of two to pay the taxes, and you can't do cataract operations. I wish I could do a couple here for you.

If you see Furbush, you might ask him if it is possible for me to be transferred to the U.S. I can exchange places with some of the very important Majors, Lt. Cols., etc. I should after 18 mos. be nearer home to salvage my practice and personal affairs, all of which are being destroyed without an adequate return. My usefulness here does not compare with my usefulness at home, and the losses incurred by my prolonged absence, in fact my work for 20 yrs., has been practically thrown away.

Well, goodbye. Much, very much love. Your B.

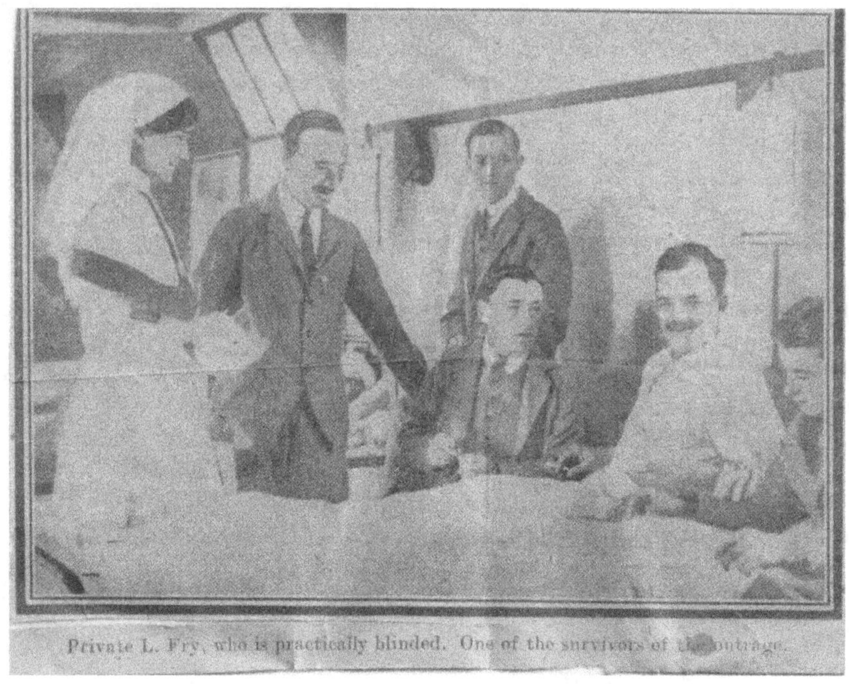

Private L. Fry, who is practically blinded. One of the survivors of [illegible] outrage.

British newspaper clipping, inserted into WTS's letter of August 17, 1918, showing Pvt. Fry. WTS tells the story of his being blinding in combat, his treatment, departure for England on an Australian hospital ship, torpedoed by a German U-Boat, and his rescue.

August 25, 1918 [Sunday night]*
[Le Treport] France[106]
My dear Mabel,

This is Sunday night. You did not get a letter last night, because I was busy. Just as I was about to write, I had to go operate – with Dr. Taylor on a brain case from my ward. Convoys are still coming and we are full up. The news continues good and is getting better. Your cable came this morning and I was glad to learn that Ted was convalescing. Of course, I knew he would be alright, barring some unforeseen occurrence, but was glad to have such recent information. I am very glad he had been fixed up, because with his trouble it was soon to come sooner or later, and there was no better time.

You flatter yourself or me when you think that your kind & courteous treatment from the doctors was due to the fact that you were my Mabel. They do not know me from Adam, and are probably just naturally decent and kind "docs" who would be nice to any one as nice as you. Now, if they had not been nice to you, then they would have learned me from Adam. But of course, that was unnecessary. Ted's picture is good and satisfactory. Was and is. Have one of him in uniform. Just to show off you know. Charlie gets back from leave tonight – or rather is due, so I

* WTS to MWS, ML, 8pp., on 4pp. 8x7in. note paper; not opened by censor; passed by G.B. censor; stamped APO 26 AU 18; a slip of paper inserted, with one crossed-out sentence on it.

suppose he will be here. His promotion has come through, so he is now a Major dating from last February. I am glad. It will probably improve his morale, which has not been good, as I think I told you. It may also help Medea's morale. Gosh, Mabel, if you hadn't more brains than Medea, I would go crazy.

A letter received this week from _____ tells me that he is here and where he is, but I have not been able to find the place yet. I can't tell you, in case his letters home do not get through. He has been here since the latter part of June and was first stationed at Montmorillon. Then from his letter, which he did not date, he "moved yesterday" to La Courtine. He expects to stay there two or three months, and then go to the front. With 9 months on the border, more than a year in camps at home, and 3 or 4 months training over here, perhaps someday they will let him go out with a gun and hit a Boche. They must be training him for the next war. Now my family over here consists of Herbert, Lewis, Harry, Orlando, Ed, Charlie, Warren, Jesse, Arthur Jack (kind of), and Jack's boy (also kind of) and other too numerous to mention. We are well represented don't you think. Warren, you know has moved. He is having a great time and seeing some real country for the first time. His little Captain (very conscientious) carefully cut out the location, so I don't know just <u>where</u> he is. But I have a pretty good idea. He is down near Switzerland or Italy, in the Vosges, or near the Alps.
He said after describing his altitude and the mountains scenery that he was in the foot hills of the

That's what the little Captain did but anybody I think could supply the word Alps. After all, it would not help the Boche much to know that some of his enemy were in the foot hills of the Alps. The Alps have awfully big feet, all covered with hills just like corns, you know. But the little Captain wants to do the thing right. He is perfectly right, you understand, but when he gets tired of this darn Censor job as I am, it will be easier for Warren to tell me where he is. Anyhow, he is down that way – 150 miles from where he was. He said it made the Upper Delaware look like 15 cts. I know the feeling when one first sees real mountains. Connie Melhuish wrote about him and wants him to stay at their house should he get English leave. But I told you that. Excuse me. I am tried. Tired of the war, and of the Army, and of France. Tired of everything. Will never go to war again, will never go in the Army again, will never go to France again, will never do anything again. That's the way I feel about it now. Perhaps when the Boche surrenders, I will feel differently.

Am sending a little lace thing [not in envelope]. I have more of them so don't laugh at the littleness of your gift. I have a whole dozen, all different. They are very old, some of the pieces 80 years. It is called bonnet lace. Worn years ago in the top or crown of the bonnet that the French ladies wore. Martha [his French teacher] knows all about it, but I don't know much about it. She told me it was nice and rare and that you would like it. If you don't, send it back before I send the rest of the "set." By the way, do you know that you can as my wife purchase anything you like from the Quarter Master in Phila., Gray's Ferry Road, or the Frankford Arsenal, or any U.S. Q.M. They have everything from coal to canned fruit & vegetables, clothes, under clothes, blankets, etc., etc. – Everything. Of course, they have no delivery, so you must take the car when you shop. You will find – household goods, etc., ever so much cheaper than you can get them elsewhere. I enclose your "Introduction." You must only buy for yourself or immediate family. I think it would be well for you to look into it.

Now if you will excuse me, I will ring off.

Very much love of course

from <u>B.</u>

~~So many of the three wounds tear through the orbits into the brain though that we can't get them all well~~[*]

----

[*] Enclosed in the envelope with the letter of August 25, this sentence was written on 7x4 in. page, and it was then crossed out. The page is labelled in the upper left corner: <u>2</u>.

August 31st, 1918, Saturday*
[Le Treport] France
My dear Mabel,

I made out a voucher today for my commutation of quarters and tried to send it to you but find that money cannot be paid in the U.S. but must be paid here. However, I will when I get it in a few days, send the check to Morgan Harjes and them send you $54.44 through Drexel & Co. Phila. The reverse of what you have been doing for me. I expect to be able to do this every month or nearly so. This will come to you in [the] course of time, and with your allotment will help.

We have been very busy in the hosp., having for the month of August received more patients than ever before. Now have about 6,000. That is heavy work. Today is the first without a convoy for several weeks. Then a good many have been away on leave and other duty. Withal, it is very much honorous [sic] and tiresome. If we could only make the very good news 10 times better and finish this think up, it would be worthwhile. Your American papers must be very exciting with the news of the last month. You know our papers at home always head-line so extraorgantly [sic] and color up so elaborately. The news has been good and no mistake, but still we only have about our half of March losses back, and we are not in Germany yet. The best part about it is that the Hun's year has been spent in vain and at a tremendous sacrifice. When this year is up, he will find himself where he was, or worse, and will find he has a greatly depleted man power. Then he will observe that we are not where we were but better and that we have a greatly increased man power. Then I hope he will put 2 and 2 together and collapse.

There is very little news that I can tell you. Ed is a Captain, dating from last February, and Charlie is a Major, also dating from February. DeLaney recommended them at that time, and is much better than R.H.H. [Harte] could do, or rather would do. Ed has been running the x-ray in addition to his dental parlors for several weeks, and has been very busy for him. And as Charlie has been on leave, he is a little more busy for him. He has handled the x-ray very well, much better than did Knowles, who really knew nothing about it. Or perhaps, to be more accurate, not much about it. Of course, Knowles was a skin man and not an x-ray man, but Ed put it all over Frank on the unknown rays.

Received a letter from Warren this morning. I still do not know where he is but he seems very well satisfied, and contented. He is still sawing wood. I think he is over on the Vosges. Our nice warm summer weather has departed, and it is now blowing a gale and is quite cold. It looks very much as if we would soon have to settle down for another winter. That is just what I don't want to do.

Have not heard from you for a very long time; two weeks, I guess. And you will be home when you get this. Just starting the winter campaign. Wish I could start it with you. Well, perhaps the next one a year hence, if you have luck. I can't seem to write tonight; can't think of anything of interest to you. Guess I will have to stop. Oh no, I just thought of something that happened as long ago as this afternoon. I was coming up from Treport on the funic[ular] and sat beside an American nurse who is attached to [B.E.F. General Hospital] #47 adjoining us. I did not know her, but she had her arms full of peaches, grapes, etc., for her patients, so I walked along with her and carried some of her packages.

Chapter II. "Where did I come from in the States?" "Philadelphia, of course." She had only passed through Phila. from New York to Illinois where she lived, but she spent much time in Boston – had I ever been there? Well, I had. And I got my wife from West Newton, etc., etc. Oh yes – lovely place, she knew West Newton, and the Newtons of all other directions.

Chapter III. But she used to go to Allentown. Oh, so, I had to go there a good deal, in fact, had a cottage or part of one at Waveland – you know just next door. Yes, she knew Waveland and even Nantucket. Well my big brother-in-law, Mr. Albert Warren, had a nice cottage or house at Allentown. Albert Warren! Oh yes, he was a great friend of the "So and So's", and the gentleman thought that Albert Warren was just about the best man ever. Agreed. But Mr. Warren is dead. So is Mrs. "So and So." I can't think of the gentleman's name, but the nurse's name is Miss Wallace. All bromides about the size of the world strictly forbidden. Miss Wallace won't think I am very nice, because I was not feeling any too good and was at first very uncommunicative. I made her speak first and as my dear old mother would say – "did not show an interest."

---

* WTS to MWS, ML,10pp., on 5pp. 8x7in. note paper; not opened by censor; passed by G.B. censor; stamped APO 2 SE 18.

I took her to her Mess and when I handed in her stuff, she asked my name. I told her, and she said – "My name is Miss Wallace." I said "Thank you," and turned on my heel and left. I bet she thought I was a grouch and a queer one. She started out by saying, "I am attached to 47." I answered, "So, you are very close to my place." Of course, the 20 American nurses attached to 47 all want to be with us at 16. It's a darn shame to be attached to a British Hospital a hundred yards from an American Hospital – all of our nurses and men have such a good time and know one another so well. 47's American girls naturally feel out of it, and are out of it, for we are one crowd; for it's an entirely separate, distinct and different place. All of 47's people are British – Matrons, nurses & men. British. It would make you homesick, wouldn't it? Well, if I see Miss Wallace again, I will give her the best smile I can crack after 15 mos here. Then she won't think I am so queer.

Good night, my dear. Sleep light and dream of me. Keep a lovin and a missin XXXXXXXXXXXXXXXX

Sunday night. Not much my dear, because it is late and I am tired, and no mail came, and the war is still on.

A convoy late this afternoon, made it necessary for me to go to my ward after supper, dress my cases, and go to the operating room from which I have just come. The cases were pretty bad this time. They are having an awful [time] now up front.

Got my check for quarters this afternoon. Will send it to Morgan Harjes tomorrow and you may look for $54- in about 2 weeks from now. Let me know how it comes through. Nothing more now. Same as last night, are the final instructions. Much love from B.

September 1918

Le Treport – First American Push

"It's great news and is just beginning."

"On September 3, Foch gave the order for continual attacks along the whole length of the Western Front."[107]

Sept. 7, 1918, Sat.[*]
[Le Treport] France[108]
My dear Mabel,

You will remember in my last letter I complained because I had received no letters from home. Well another full week has passed, and still no letters. It is now more than 3 weeks since I have heard from home. Of course, I know you have written, but something is holding the mails up. It is not nice to go so long without a word. I get quite out of touch with things and feel entirely separated. I sometimes feel that I am entirely separated and will never again have a home and family. It surely does not look as if I would have for some time to come anyhow.

The *Daily Mail* published a map today which I have been looking for for some time, and I can see why it is not published more often. It was a map of the frontier. Well Germany in the entire northern section is anywhere from 80 to 130 miles away from the present fighting line. The ground gained in the remarkable success of the last 6 weeks looks like a small spot or streak on the moon. If we continue to gait at the same rate, 18 mos. or 2 yrs. might see us up to the Hun's frontier. But of course, that is not a fair way to calculate it for when he begins to beat it, he will beat it much faster than he can be driven. At all events, our "magnificent retreats," on which we were fed for a year, are being balanced by "magnificent advances." That sounds better, to be true. And the crop of prisoners has been very Good, as well as the crop of dead. And then again, I think the Hun has at last admitted that he cannot win. He likewise claims that he cannot be crushed. That is where he is wrong. If he keeps on, he will be, but when will the darn fool get that through his head. You see, if he can't grasp that fact, you cannot look for me for a few years yet.

Tell Mr. Wilson to keep sending men over here. He has some yet, before he has to ask those who are "needed at home" to take care of their wives, practices and hospitals, and themselves. As a country, we are very lucky to have so many men who are not "needed at home." I am sometimes thankful that I was of such little use at home. And then sometimes I wish that I had had a hospital position or two, a practice that could not even in war times get along without me, a wife or even as many as 2 or 3 children. But alas! Just because I was independently wealthy, and was not a great enough doctor to be necessary to my patients, and had a wife and a few children who were just at well off without me as with me – just because I was not "needed at home," I have been canned for several years sojourn in France, working some of the time, and wasting most of the time.

You see my dear, when I get no letters for 3 weeks, I have nothing to write about and much grouse, as the Englishman says. Have been quite busy for a few weeks, but have not been very happy because the thing seems so hopeless. Much is due to a [scratched out word, probably "damnable"] ~~punk C.O.~~, but I can't help that. I mustn't say that about the C.O., so I will cross it out. I am afraid I am not writing you a very nice letter, but I don't feel particularly nice tonight.

Morgan Harjes have sent you $54- less a few small charges. Let me know when you receive it. You must now be at 109 [So. 20th St., Philadelphia]. Hope you received my letters about Ted in time to be of use to you, and that he is returning to school. That is where he belongs. Next summer he can enter the war; it will then be at its height, and

---

[*] WTS to MWS, ML,10pp., on 5pp. 8x7in. note paper; Registered Letter w 2d British stamp; not opened by censor; passed by G.B. censor; stamped APO 9 SE 18; stamped "Registered Philadelphia SEP 23 1918." The return address is carefully written: "Wm. T. Shoemaker / Capt. M.R.C. U.S.A.," probably as a counterpoint to the address given by Bishop Harvey in the letter.

he can have a fine time for the next 3 or 4 years. Tell him, if he ever gets over here to look me up. My permanent address is France. Anybody can tell him where I am. Am one of the few Captains left in the American Army. All of the others have become Majors, Lt. Cols., Cols. & Generals. And I understand that no more are coming over, as they are no longer Commissioned at home, the lowest rank there being a Major.

Your friend Mr. Harvey or Bishop Harvey addressed me Capt. Wm. T. Shoemaker, M.D. R.A.M.C. He is a great Bishop. I don't belong to the Royal Army Medical Corps. Then again, we don't in polite society say Capt. – M.D. – anymore than we say Mr. – M.D. I guess he was thinking of the spiritual world where all the good and all the bad in a man shines out in full view of the surrounding angels rubber [sic] meeting around. The Capt. would be the bad in me, and the M.D. would be the good in me. That's what the good Bishop had in mind. By the way, I have not answered his letter yet, because I know not yet what he was driving at.

Major [Charles] Jack has returned from leave much improved. He had his kind of a good time, and is in a much better frame of mind. Ed [Shoemaker] wants me to take a leave with him, but I can't afford it. Would like to go in a way, but am not much interested in anything. The leave I want is home. Don't forget to ask Furbush or Gen. Gorgas if it would be possible someday to be "needed at home." I know they have no use for Captains there, but they might make an exception in my case, as I am still within conscription age, and could not possibly get away from them.

Jim Talley paid us a visit this week. He looks quite old and serious, but seems well. Was glad to see him. You know he is at Brest in Navy [Hospital] #5.

No more tonight, and if no mail tomorrow, still no more. Good night.

Much love XXXXXXXXXXXXXXXXXXXXXX

Sunday – One letter from Dorothy written at Chestnut Hill, Mass. But none from you. Now that is too bad. Was sorry to hear about Pet's Mr. Cornell. Never saw him, but heard a lot about him at one time as you remember.

Did I tell you that the eye droppers arrived last week? Many thanks; also, dark glasses about two weeks ago. Funny thing about the eye droppers. I never used anything but a straight eye dropper; and do not intend to. Have no use at all for curved or bent ones. But every time Mr. Baer send me droppers, he tries so hard to get the bent ones, and then apologizes because he can't get them. I hope he never succeeds.

The news today is still good as you have noticed & forgotten by the time [you receive the letter]. Now I have nothing more to say. Perhaps next time – next Saturday, I will be able to write you a more interesting letter. Until then Good Bye. Much love and then some from

B.

Sept. 14, 1918, Saturday*
[Le Treport] France[109]
My dear Mabel,

The other day while soaking in the bath tub, I composed a little verse about the Hun. Them I came home and wrote it. Here it is:

The Hun

They call me Hun, the wicked Hun;
I would the name be blest;
For to rival old Attila, and be his tribe,
I've done my best.

I love to murder women,
And children, when I can;
And if he's quite defenseless,
I love to murder man.

They talk of being human,
Of love for fellow men;
In a world of Civilization,
With God's Commandments Ten.

Of what they speak, I know not;
It may appeal to some,
But while I've never tried it,
I'd rather be a Hun.

---

* WTS to MWS, ML,12pp., on 6pp. 8x7in. note paper; Registered Letter w 2d British stamp; not opened by censor; passed by G.B. censor; stamped APO 16 SE 18; stamped "Registered Philadelphia SEP 30 1918." Received Oct. 1, 1918

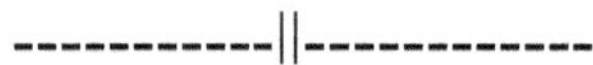

Several of your letters have drifted in, but that's all. The last one was a real nice long one. In it you express pleasure because as you say, I have been recommended for promotion. Possibly I told you, but forget it. The situation is as follows. The C.O. never told me that he had recommended me, and I doubt if he has. A communication came here from H.Q. saying that for certain reasons, there had been no promotions or for promotions in our Unit. H.Q. realized that impression had been done and that the general high commissioning of our juniors – assistants, students, etc., who stayed at home, etc., made our positions intolerable. Or, to that effect –

They were now ready to make amends, send on the recommendations. The C.O. forthwith went to G.H.Q. around with a bunch of names of men here who deserve more than they have. I was later told indirectly that my name was among them. That is all I know about that. But here is the funny as well as the characteristic part of it. We were told later on pretty good authority, that the bearer of these important papers, when he arrived at H.Q., forgot all about them, and never mentioned them. That seems quite likely and is what I would expect. So that is all there is to that "rumor," and if the war ends someday, I think your Captain will return as he left. I surely will so far as Base 10 is concerned.

But after a reasonable time, I am going to take the matter up personally with Gen. Ireland and tell Base 10 to go to H---. You can draw your own conclusions when I tell you that. The Harvard Unit came over at the same time we came. Every man in it has had his promotion for 3 or 4 months and there is not a Lt. in the Unit. Mike Nolan was detached for a time and went up the line. His C.O. recommended him for a Captaincy and it came through in just 3 weeks. He returned to us promoted. Dillard went to England, was promptly made a Captain, and has since been recommended for a Majority. He will probably return to us a Major. It would seem therefore that it is not altogether an unwillingness on the part of the Staff at G.H.Q., but an unwillingness, or an indifference, or an inefficiency elsewhere. At all events, I can't afford longer to stay here if I can make more money elsewhere. Gen. Gorgas & Col. Furbush are in France. I think probably Furbush will come here to see us. If he does, I will have a chat with him regarding my future plans.

I don't want to go A.E.F., as you know, but I don't propose to stay here as a blind pawn, and pull chestnuts out of the fire. The work I have is not too bad, as not a soul in this Hospital knows an eye from an ear, and I am at least nominally responsible for a great many eyes. And of course, you can't run an establishment of 2000 – 3000 people without an eye man – And then I have had a great deal of important operating. If I get out, another eye man will have to replace me. Now if Creighton could only come over and relieve me, so that I could return to my home tomorrow, it would be *bon*. It would also be a fine job for him. Just what he should have. I am so glad that you thought to offer your service to him, in helping conserve his practice. Am also glad that he has signed up. He will be in a much better position afterward. Do what you can for him. Let him use my office all he wants and in any way he wants. He has been very loyal and faithful to me. Interesting that Carl Williams has at last come in. This is also good. I hope Lauther gets exempted – and Grisham also.

Are you aware of the first American push? It's great news and is just beginning. I know they would make good when they started that push, and it has been in the air for some time. I wonder if the Boche still thinks there are only a handful of Americans here; and "very poor and untrained troops." He has changed his mind a number of times in the last four years. England's "contemptible little army" has given him a great deal of trouble. The British Empire has put up 7,500,000 men, 69% of whom, or 4,300,000 came from England. It's a dangerous thing not to take 7 ½ million men seriously.

Your letters seem to want to take Ted out of school. I am sorry because I wanted him to return to school. All of that "doctor talk" that they have given you is junk. Ted is alright and quite healthy enough to spend 6 mos. in the High School. However, it's up to you.

Sunday night. – A more beautiful day after a week of rain & coldness. Played 24 holes of golf this afternoon, and feel now quite stiff and tired. 24 holes for a Shoemaker are a good many. Played 6 with Barnes & 18 with Eddie. Beat them both, but of course Eddie give me a big handicap. We now have a new X-ray man so that Eddie can spend his time on teeth, and work less hard. The two jobs were impossible for one man. Ed wanted now that he is a Captain, to pay more rent for his office and apartment at 202 [So. 20th St., Philadelphia], but I advised him to wait until after the war. Of course, he pays a ridiculously small amount for about 7 rooms – an entire floor, and a part of another floor, but it was very nice of him to want to hike his own rent. Then again, he and Martha are making more together than ever before, so that the war to them is not a financial hardship – but a money maker. You speak of renting your room – please don't do it. It will not be necessary, I am sure. Also be careful how you rent my office. I might come home and be much embarrassed by an occupant of my office. Of course, you can't give a lease. Any friend of course might have them; but beware of a stranger. But I don't want you put out of your quarters anyhow.

In the *Messenger* of July 31st is the list of "Our men in Service – ", etc. Mr. Harvey is probably responsible for the mistakes in the Pennsylvania delegation. First, he included everybody he knows – men not members of the Society at All. For instance, Ralph Straub, Arthur Jack, and others. Then he puts Dentists into the Medical Corps instead of the Dental Corps – and finally he puts us in R.A.M.C. As I told you before, we are not members of the British Army but belong to the American Army. Being attached to the British has not changed our citizenship. So tell the Bishop that we do not belong to the Royal Army Medical Corps. We have no royalty in our country. If he wants to list Ed properly he should say D.R.C. U.S.A. And if he wants to list me properly, he should say M.R.C. U.S.A.

But I believe now we have all lost the R. [Reserve] There is to be but one Army and we are all to be Regulars. We are all sorry because we value the R. very highly. In the Medical Corps it is an indication that we know something about medicine. However that may be, if true, the Bishop in his revised list need only drop the R. In our Unit, the men still have the R on the Collar (U.S.R.) because the new order has not yet reached us. Elsewhere, I believe they are wearing U.S. only. Of course, it's better to have one Army instead of a Regular Army, a National Army, a National Guard and a Reserve Corps. They have overcome the difficulty in rank which was a serious thing, by carrying certain ranks for the duration of the war only – after which the regulars will return to their original rank. For instance, our friend Moncrief is a full Colonel now but after the war he will be what he naturally should be, according to seniority, service, etc. And of course all good Reserve men like myself promptly resign and have no rank. It's a good scheme, but we will miss our R.

Good night my dear. Miss me a lot as usual, and look for the end of the war any time. Send Creighton out here to relieve me. That will be nice. Much love your B.

Sept. 21st, 1918 [Saturday]*
[Le Treport] France[110]
My dear Mabel,

It is Saturday night again, and while I will start a letter to you, I will not be able to tell you much because I have nothing to tell. Now nobody writes to me anymore and I am doing well if I get one letter every 8 or 10 days from you. This is sad because everybody else gets lots of letters all the time. About once in 10 days, as I say, my box will draw one lonely little letter. But I don't mind because I am so fed up in this war, that I have ceased to think of anything. Have been mechanical as it were, and have lost interest in the present, past & future.

Another good man left us this week. Ed Hodge is head of a Mobile Unit which is a sort of moving C.C.P. From our place it takes 5 doctors, 22 nurses and 30 Enlisted men. There were all detached and go A.E.F. I was very sorry to see Hodge go. He is a lovely fellow and was one of our really good ones, you know. Up to date, Gibbon, Norris, Cadwallader, Earnshaw, Knowles, Packard and Hodge have gone. Wilmer, Outerbridge, Nolan & Hetherington are to go in a day or two on the Mobile Unit. So you see Base 10 is no longer the Base 10 that it was. As I told you last week, I would like well enough to go, but not A.E.F. I wish I could get a good job with B.E.F. Drayton is another one who has gone.

I have heard several times unofficially, and it is probably not true, that after 18 mos. foreign service, a man can be transferred to home duty. If it is true, me for the U.S. I guess I would not be any happier there so far as the work is concerned, but I'll take a chance. And I could probably hold together a few remnants of a trusted practice. "After the war" bothers me, because in spite of what a few encouraging kind friends tell you, I know that I will never again have a practice. And you know we cannot afford yet to retire. So as usual the reconstruction will be as bad as the war. XXXXXXXXXXXXXXX

---

* WTS to MWS, ML, 6pp., on 3pp. 8x7in. note paper; Registered Letter w 2d British stamp; not opened by censor; passed by G.B. censor; stamped APO 24 SE 18; stamped "Registered Philadelphia OCT 11 1918." Received Oct. 14, 1918.

Sunday night. This has been the dullest, gloomiest day we have yet had in my remembrance. Rain all day, cold – not a thing to do; few patients and none very sick. In fact, it is just a bore. Even the paper contained no news of importance. Things along the Western Front have quieted down. Of course, the East was pretty good, and was talked up to fill space.

Everybody expected American mail today. Nothing doing. I tell you this place is getting so dull that it is dying of dry rot. Haven't heard from Warren for a long time. Guess he is alright or I would hear. Charlie had to return to England as a witness in a Court Martial; he left last night and will probably return sometime during the week. One of our officers got into trouble while on leave at the same time Charlie was there. He was sore because he had to go back, but that did not make much difference, as you might imagine.

My dear, I just can't write tonight, so I will hold over until tomorrow when perhaps I will get a letter from you.
XXXXXXXXXXXXXXXXXXXXXXXXXX

Monday night. No use my dear, that "perhaps" above did not make good. Still raining. One piece of news came through today. Dr. Harte has been ordered home to take charge of the Walter Reed Hospital, Washington. He went to England on leave this morning, so will not go home for several weeks. Mitchell will be C.O. When he becomes C.O., I will get him to recommend my transfer to Shoemaker Barracks, Phila. I think they need a good M.D. in charge there, don't you?

I am sorry that I have written such a blue and disagreeable letter, but I will try to do better next time. You should by this time have 12 little "dodads" the last 3 supposed being in one container. If any have been lost, I will make them good with others. Am Orderly Officer today, so may have a night of it. No convoy arrived as yet however. But one is apt to come along about 3 or 4 o'clock.

Good night my dear.

Perhaps will soon! My last word from home is dated August 20th. One month & 3 days (letter 100). So you can understand how I feel – quite "out of touch."

Good night again. Much Love your B.

"On September 22, [Allenby] made a second aerial attack on the Turks descending to the Jordan … The Turks had been overwhelmed by British air power and by the speed of the Allied cavalry. … On September 25, Australian and New Zealand cavalrymen crossed the river Jordan and entered Amman. … On September 25 British forces entered Bulgaria. Two days later a Bulgarian republic was declared."[111]

Sept. 28th, 1918 [Saturday]*
[Le Treport] France
My dear Mabel,

It's better tonight and I hope I can write you a more cheerful letter than the last one was. Two nice letters came from you day before yesterday, Nos. 101 & 102, both quite interesting. Before I forget it, however, I must correct some of your spelling, "Ordnance" is the word I refer to you. You must not call it "Ordinance" – that does not sound educated. My roommate uses three words that particular that amuse me: "Admirality," "Ordinance," and "strategetic," and also a fourth, not a war term, "Muriel" for "myself." It sounds great to have "the British Admirality" handed to you in a very positive & measured tone – *a'la* Uncle Louis. And one thinks of the Village School Master, when he listens in wonder to the: "stragetic advantages" of such & such a military position. And when one hears of the "Ordinance," one suddenly remembers he must not spit on the side walk. Of course, the

---

* WTS to MWS, ML, 10pp., on 5pp. 8x7in. note paper; opened by Censor P2722; passed by G.B. censor; stamped APO 30 SE 18; received Oct. 21, 1918.

"muriel" decorations are very like what we used to know as mural decorations. Dorothy will know about them; they have them at Schools of Industrial Art.

Am sorry Ted had a set back, but presume he is alright by this time. Hope so any way, for I want him to go back to school. What you say about _____ is interesting. I do not know what could have happened. I guess she has just naturally come with her own. She surely was never given half a chance by her father, who deliberately neglected her *de facto*, at raising children. I would not call _____ a great success. _____, I am afraid will not be so fortunate. I have several letters from her that are indeed nice letters, but indicate certain weaknesses or a definite "backwardness." This, not in a spirit of unkindness, and not to be repeated or read to the family or to the Bishop or not even to be published in the *Messenger*. You understand. Funny! You remember the slip about "No. 494" that Owen asked you to send me. The day after I got your letter, I was Censoring my pile and ran across the same thing, which a Yank was sending to his family in the U.S. because he thought it was pretty good. Now which side of the Atlantic it came from originally, I do not know, probably this side.

Mighty glad you got the taxes paid, but am sorry we had to increase our indebtedness to do it. If Liberty Bonds will help you any, you can sell some. They have you know performed their war use. By the way, did you ever get yours? It is paid for, and I specified definitely that it should be in your name and delivered to you at 109. Investigate that little thing, and ascertain where my $50. went. Your comments on your friends who suddenly became so rich are interesting. While they may be alright now, I have never forgotten that England had them on the black list, and that it is said that great quantities of their valuable products reached Germany in ordinary letters – camouflaged as it were. This was probably or surely before we entered the war, but I have never a great amount of confidence in the professed sympathies of the whole d.--- family. I have learned one thing, and I think the world has learned the same thing, and that is, that a thing is not true because a German says so, but on the contrary is apt to be untrue if a German says so. As a people, high & low they are <u>liars</u> in every application of the word.

Mrs. Straub sent over a funny one to Ted. Said she: It is quietly gossiped and rumored around the City of Brotherly Love, that the Government will not allow Charlie Frazier to come abroad with his Unit, and that Mrs. F. has to report weekly to the Alien Enemy Board, or some such board. That's fine; I hope it's true because as you remember Mrs. F. was a very pronounced and noisy pro German, and Charlie had a similar reputation. The gentleman who brought you from the Lankenau Hospital to Broad St. Station – some time ago, his automobile is another well known P.G. [Pro-German]. I'll bet they would not allow him to come over her. We don't need him.

Everybody over here is bucked up with the happenings of the last 2 mos. Things are getting better every day. The East seems to be a washout and everybody here is looking for Turkey for Thanksgiving. The Boche must be worried sick just about now. Of course, he is very mad and will fight very hard on the Western front, but that makes no difference. The Yanks thoroughly enjoy it and in the course of time "with the aid of the French and British" will have him backed off the map. I think with Bulgaria and Turkey out, others will quickly get from under and finally Germany will stand alone, deserted by all her allies. But she will not have to stand long. She is going to get the biggest licking that any nation ever got, and when it's over, she will hurt for a century. It may take one year, two years, or three years, but it's coming. Gen. Foch is the man predestined for the occasion, and will live forever. He is 67 yrs. old and a contradiction of Osler's former theory. He is a religious man – an ardent Catholic. His chief characteristic in this show, which would seem to bring him a good return, is his silence and ability to keep his plans strictly to himself. That worries the Huns, because he has heretofore generally known long in advance just what the Allies intended doing. Not so now, and Mr. Hun is very nervous.[112]

Well my dear, I have done pretty well tonight. Haven't strafed anybody; have been real good natured, and in fact am quite proud of myself. I had better stop before my thoughts turn to the Army. Good night. Much love
XXXXXXXXXXXXX XXXXXXXXXXXXXXXXXXXXXXXXXXXXXXXXX

Sunday night. Have just returned from the operating room. We received quite a bad convoy this afternoon, and others are on the way. Of course, a good many fellows got hurt in the last few days, more are getting hurt just now, and so it will be tomorrow & the next day, etc.

Did you happen to see in a Phila. paper, probably the *Bulletin*, an interview with a boy named Hamilton whom we sent home because he had nephritis. It's a scream and contains not true word. He was my orderly for a time, so I know him well. He has never been off the base, is a hopeless liar, and I am afraid also a hopeless annexationist. He used to get me souvenirs until I discovered that the souvenirs never had a clear title. So you need waste no worship on that hero.

I am about to send your allowance for November, December & January. I wanted to allot the full pay of $220.- but am not allowed to allot more than my base pay of 200. This I will do. Will pay my insurance out of the $20 remaining, and keep the balance. Will then send you some or all of the commutation of quarters money as I am now doing. Will probably not be able to send you all, but will do the best I can. Will try to give you $25-30 per month. If I was a Major, I would have $900.00 more per year. That would help a lot. Well, there is no use worrying about this.

Ralph Straub started on leave for England & Scotland today. He will have a good time.

Good night. Be good and keep a "missin" me. Much love and then some

Your

B.

October 1918

Le Treport – A Trip to Boulogne

"The war is rapidly coming to an end. The news from day to day is electrifying."

"On October 4, having informed the Reichstag of the need for peace, and having obtained Austrian support for what he now realized could not be delayed, Prince Max telegraphed to Washington requesting an armistice."[113]

Oct. 5th, 1918 [Saturday][*]
[Le Treport] France
My dear Mabel,

You will be glad to know that my promotion came through Oct. 1st, so I am now a Major. Four of them arrived by telegraph before breakfast; Newlin, Cruice, Cloud & myself connected. The best of all was Cloud's. You know, he has been a Lt. since we left home and has never raised a rank. He is 46 yrs. old and one of the hardest and most conscientious workers here. We were all glad that they bounced him right up to Major when the time came. I presume others will come through in about a month, as they let them out, I understand, periodically. I hope so, for every man here deserves a promotion. You can never have more money which then some will be of use.

My base pay is 3000. Foreign service 10% = 300. Commutation of quarters, about $66 per month = 792. Field allowance (British) about 30. = 360. Total about 4432.- I will send you the full pay voucher and as much of the balance as I can. You should get at least 3600.- So that will not be so bad.

If I get picked A.E.F. – I will lose the Field Allowance and will have to pay more to live, but even so you will be better off. And then again, the war is rapidly coming to an end. The news from day to day is electrifying. Bulgaria out, Turkey out, Austro-Hungary next (I hope) – and then finishing touches on Germany, who must be about tired and ready to quit. If not, she soon will be. The combination against her now is able to push her anywhere. It's a great time to be living and the end of the war will be worth seeing. It can't be far off. Watch for it. The fighting just now is terrible. Our casualties are necessarily heavy, and we are getting our share in [B.E.F. No.] 16 – We are very short handed and everybody but the dentists are busy. As luck would have it, Captain Viner of No. 3 went on leave Tuesday for 14 days – and as a consequence, I have the eyes of the entire top. As then a great many men had injuries, I have had a great deal of operating and the work – I don't get though now until night – sometimes late. But as you know, that part I like. I much prefer being busy. Then of course, the 1st of the month had to come at the same time, which gave me the usual amount of book keeping etc. for the Mess.

Also, my Chinks got sick today (not all of them) and I had to take 3 or 4 of them up to Noyelles [Noyelles-sur-Mer] where sick Chinks go – That is a very pretty and interesting ride – about 25 miles from here – 5 miles beyond St. Vallery [Saint-Valery-sur-Somme] – the other side of the Somme – Have been up twice in the last week. Tomorrow I am going to take an Officer from No. 3 with a piece of steel back of his eye to Boulogne to see Col. Lister. As I have never been to Boulogne, I will enjoy the ride but what is to become of my ward, full of heavy cases? Miss Fuhrmann is my sheet anchor – she will dress them all. I could not get along without Miss F. She has now worked with me 16 mos. so she knows her business. [pause] Operating on the bad cases that came in today. I will postpone until Monday morning. Best I can do. – But Viner certainly took a fine time for his leave – Every bed on the top full – and bad cases. Boulogne is 70 miles from here, so it will take me all day to make the trip in an ambulance, which is not quite so class as our little Studebaker.

---

[*] WTS to MWS, ML, 9pp., on 5pp. 8x7in. note paper; opened by Censor 6906; passed by G.B. censor; stamped APO [date illegible]; received Oct. 30th.

Have resigned as Mess Captain and will turn over to my successor who will be Eddie – just as soon as I get my bills paid and my money in. That will be some time next week. – You see they took my Sergeant away and put him on the Mobile Unit which left Thursday. – Also, two of our Mess boys – It was done in such a mean way, that I decided they could get a new Mess Captain while they were about it. I did not care to be bothered with the Mess after the way they shot it up. Four months of it were about enough for me anyway. Eddie will enjoy it for a while, and then the war will be over. See?!

Johnny Cruice – left yesterday – A.E.F., but the day before he left, he very quietly married one of our nurses – Miss [Sara S.] Hopkins – a very new girl from Johns Hopkins – she wangled her leave accordingly, so off they went together to Paris. Now John is 44 – an only son – tremendously set in his way – quite selfish and all that sort of thing – *a'la* John Phillips only more so – and a Roman Catholic. The lady is 15 yrs. younger – very nice, but very light in the head and not an R.C. – She has a lot to learn about Johnny, believe me.

Three of our nurses have married – one [Elizabeth Voltz] fell for DeLaney – one married a private – and one married Johnny Cruice. I think the middle one probably has the best of it. The other two are about 12 or 15 yrs. younger than their husbands. That is often alright but it gets to be quite a difference later on. However, war brides are very nice, and I wish them all Joy & Happiness. –

No letter this week. Will tell you tomorrow about Boulogne – and perhaps again – answer a few letters from you. Good night. – Same instructions.

XXXXXXXXXXXXXXXXXXXXXXXXXXXXXXXXX

Monday night. Couldn't write last night – Got very late after a ride of about 160 miles. Was naturally tired & of course cold. Very interesting trip to Boulogne. My, what a busy place it is. All kinds of troops, officers, nurses, V.A.D., WAAC's, Y.M.C.A.'s, Red Crosses, etc., etc. Hustle & noise enough to set you crazy. Went through Etaples, and Abbeville, the places that were bombed so much & so persistently by the Hun when he was in that business. You will remember that many nurses, patients & doctors were killed at Etaples. Abbeville was also much shot up. But the Hun's day for that is past. We are now bombing his houses and doing it properly. He now thinks that such practices are against the Hague Convention, and is horrified. Funny, he also thinks that shot guns are very barbarous, and quite beyond the pale of "civilized warfare." His "peace offer" of yesterday is his last spasm. Mr. Wilson will hand him a few well chosen words, and then and there, you know the rest. And won't it be fine?

Today, I had to go to Dieppe to see a lot of patients in No. 5 Stationary Hospital. Started the day on a funeral at 7:30. Saw my Chinks as usual, went to Incheville, operated, dressed my war & did other things – so I am tired – if you will excuse me I will stop for the present. No letters from home as yet, so I have no comments to make on these contents. Must now go to my ward – after which I will turn in & get ready for tomorrow. Good night. Much love and all that. – from

B.

XXXXXXXXXXX
XXXXXXXXXX

Oct. 13th, 1918 [Sunday]*
[Le Treport] France
My dear Mabel,

La guerre est finie! Perfectly wonderful, isn't it? Of course, it isn't "definitely" finished yet, but that is a matter of just a few days. And it is not ending by negotiation but by complete military victories. I fancy the fighting will stop in the next 48 hrs. or less. My, how glad everybody will be. Today's papers brought confirmation of the Hun's acceptance of Mr. Wilson's terms. When they are accepted, the rest is easy and I think will be promptly fulfilled. The whole business spells unconditional surrender. You can't do better than that. Well, you know more about this than I do, so I won't bother you about it. The American papers must be a blaze of glory. –

As usual, after two weeks or more, I got your single letter, not a very nice letter it was – Also a lot of 2nd class stuff about religious material – *Messenger – Helpers*, etc. The eye journal also. About the *A.M.A. Journal* – never

* WTS to MWS, ML, 10pp., on 5pp. 8x7in. note paper; opened by Censor 6742; passed by G.B. censor; stamped APO 14 OC 18; received Nov. 7th.

received a single copy here, nor do I care for it at present. We get it of course, and can read it any time I want. So you now need not continue it unless it looks as if I might be getting home. Most of the croakers here too here seem to think that it will take a very long time to get back home after the war is over. Some of the pessimistic ones have us here one & two years. I personally think that after Peace is declared, we should be home in 3 or 4 mos. But that is only a think. Of course, to get our whole Army home will require perhaps 2 yrs. – but we doctors of Base 10 – having been here 18 mos. – are not the whole Army, and I do not feel as some that we will be the last ones home. Never can tell though in the Army.

Yesterday was a very funny day for me. I had a "Jinx" and it stuck to the last. Took 3 patients to Boulogne. It was a very rainy day, but I made a good start at 8:30. About 5 miles beyond Noyelles, we were rounding a right angled curve, when right in the middle of the road was a Frenchman in a horse & carriage. Frenchmen like he would not give an inch. The driver had to ram slap bang into the horse & carriage, or turn out, naturally. He turned out more & more until finally we plunged head long into a ditch about 5 ft. deep. – We had 3 patients, a driver, myself, my orderly and a Red Cross man. – We were all more or less "spilled" but fortunately no one was hurt. The Ambulance was practically on its side, hopelessly ditched. Well, I started back toward Noyelles to telephone for more cars and assistance. After walking about a mile, a French lorry came along and I hopped it.

Back in Noyelles, after an hour's efforts, I raised the Camp Commandant of the Treport Area, and got the message through for two Ambulances, one to take me on to Boulogne, the other with mechanisms to take back the wreck to Treport. They came out in quick time, and picked me up at Noyelles, where in the mean time I got something to eat, and sandwiches & cheese to take to the party sitting along the road in the rain where the accident happened. Well, we gathered up our party and continued to Boulogne, and the workers spent 7 hrs. getting Car No. 1 out of the ditch, and turned it back uninjured, getting here 10:30 last night, I am told. So I did not get to Boulogne until 4 o'clock, - wet, cold & tired of it. However, I deposited my patients at 83 General [Hospital], then took my driver & orderly down to the hotel where we had some eats – *omlette* & steak – potatoes & coffee – and wine!

At 6:30 p.m. we started back – in the rain and darkness. Five punctures did we have and I reached our Mess at 1:30 this morning – The girl tried to start her car, – nothing doing – Just out of gasoline – So she filled up & went to the garage – about 2 squares [kilometers] from here. – I found your letter, so after taking a rest – settled myself in bed and proceeded to read it. – All nice & comfortable like. At 2:30 I pulled the string, out went the light, and I thought I would go to sleep. 3 minutes later, a ring – and an orderly – You are wanted in Ward 11 – patient has a hemorrhage. So I got up & dressed and fixed up the hemorrhage – and returned to my bed at 3:40 – So you see, my dear, I can sometimes have a fine [time] even over here – Do you remember the one that went with us to Boston on our last motor trip? He was a good one.

Speaking about trips, Ed does not want now to go to London, but to the South of France. He thinks of going in December. With the war over, I don't know how it will work out, but we will "see about it" as Father used to say. Doules is a province in France. Warren is in that province I believe, although I have not heard from him for a month or six weeks. Have been anxious about Orlando and some of the others who must have been in the recent shows. We had a great many casualties – 2nd Lts. are always in the thick of it. Hope he is alright.

You speak also of Commissions dated back. That does not help any, because an officer draws pay for a particular grade from the date of acceptance. In other words, I was not a Major until Oct. 1st, when I received and accepted the Commission. Charles & Ed's were dated back 6 mos. but they get no extra money. The only way it helps apparently is in the way of seniority about which we are not worrying.

I do not think that Ted's doctor Franklin H. Martin is the one interested in the American College of Surgeons. In fact, the one who organized it. That Franklin H. is about 60 yrs. old – is Editor of the *Journal of Gynecology & Obstetrics*, is a general all around promoter & mixer. He is a great politician and was even seriously mentioned for Lieutenant General to General Gorgas. He is much too much and much to be operating on just Ted. Furthermore, he would not be under anybody, because he always makes himself "it." Possibly the Cape May fellow is a son.

Regarding the College of Surgeons, I told Dr. Martin that I could not afford to contract unnecessary indebtedness at present, and while I filled out the enclosed application [not retained], I did it as the easiest way to dispose of it. So if you get some bill for [a] large amount from the College, don't worry. I just know what they will do, but I have it fixed so that I cannot [be] obligated.

Wish you luck with your criticism for the Shakespeare Club, but don't work too hard. Hope you get the other 6 doodles, and think you will like them. I just sent them to you for fun, not because I thought you need them, but just because I wanted to send you something. See? Now that is about all I have to tell you this time. Except that I will mail tomorrow pay vouchers for November, December & January – each for $275. They will be paid as they become due directly to you. Will also send you later some other moneys.

The way I am itching, I fear that I have at least gotten "Couties." (Lice, you know). There are plenty of them here, in the Wards, and in dressing the recent arrivals from the front, one is apt to get them. Have never had them much as yet. But I saw or itch coutie-like.

Good night, my dear. Things are looking brighter and may be, it will not be so long now.

Much love from

B.

Oct. 20th, 1918, Sunday*
[Le Treport] France[114]
My dear Mabel,

Could not start your letter last night because I was too busy, and when through was too tired. The most important thing to tell you about is Jesse, and unfortunately I as yet know nothing about him. A cable was received from Margaret on Thursday, telling me that he had been seriously injured, and asking me to investigate and report. Col. Harte wired for me to G.H.Q. for Jesse's location & injuries & condition; two days later, a reply stated that up until then, they had been unable to locate him, but would continue their investigation and would try again as soon as possible. Yesterday, I sent a wire to the Red Cross Bureau of Registration through Padre Jefferys. Have received no reply as yet. It is part of the function of the Red Cross to locate men, etc., and they have special methods and facilities for doing it, so that I will probably learn something through the R.C. before I do through G.H.Q.

In the meantime, I suppose poor Margaret is worried sick, and wondering why she has not heard from me. Naturally, I do not want to cable until I have some information. Another thing is that he was apparently wounded Sept. 17th, and my first information was just one month later. Margaret's cable was undated, so I don't know whether the delay is between the War Dept. & Margaret, or between Margaret & me. It's all too bad, and I am sorry. I have tried to have my various nephews, etc., over here, to remember my name and address, so that should something happen, I could be notified at once, when it might be possible for me to be some service to them. Jesse, I presume, did not know that I was here, or even forgot that I enlisted. On the other hand, his injuries might have been such as to make it impossible for him to know anything. I hope he is not dead, but I have my doubts. And if he has been injured like some of the boys coming in here, I hope he is dead.

Our recent casualties have been very severe. While I told you last week that the war was over, there were yet a few details to be attended to, and in the meantime, the hardest kind of fighting continues. But it will not continue for long. The Boche is coming to terms with an unconditional surrender, not because he wants to, but because he is beaten to a standstill and has to. If he would only stop at once, he could save a half million lives, and his results would be the same. Last week was some week at this end of the line. Ostende, Zeebrugge, Bruges, Deai [in Dunkirk] and lots of other places liberated including Lille. The entire Belgian coast up to the Holland frontier is now free. That is very wonderful work. And of course, all along the line there has been a steady advance. Your prophies in a recent letter were about correct. He has no time to dig in until he gets to Germany, and there his grave is already dug.

Received a letter from Ted last week, and today a nice letter from Hoffer. Very sorry to learn of the death of Mr. Duer and Mer a'Becket. Did not know Mer a'Becket has just died. Mr. Duer, I was quite fond of. He was an awfully nice man. I was amused at the "difficulty" which Ted had in returning to school. You will remember when he enlisted, the glib way that some ~~enlisting~~ recruiting officer told him that he could return to school in Sept. Verbal promises in the name of governmental authorities are not to be taken seriously; in fact, they would be worth nothing. You will find that the chap who recruited Ted, as I told you, had no more knowledge about that than yourself had. But it was alright, you understand, and I am glad Ted got in, but also am glad that he is returning to school. Warren, I have not heard from. It must be 6 weeks or more now. I suppose he is alright, but he should let some one know, once in a while.

You must not worry about Col. Lister. He is a very nice man, and one that I would naturally like, but from the time we first met, we have never gotten along well together. I do not treat eyes the way he thinks they should be

---

* WTS to MWS, ML, 10pp., on 5pp. 8x7in. note paper; opened by Censor 3919; passed by G.B. censor; stamped APO [date illegible]; received Nov. 7th.

treated, as many as he has, I just don't worry much about it. But then he is a big Colonel and I am a little Major, so according to Army rules he has it on me. However, we are friends you understand, but he probably does not think much of me. I should worry. –

Monday evening. Was badly torpedoed last night, hence this addition. Ted told me in his letter that one of my patients landed in the Hospital at Cape May. Did you know that? Who was he, and what was his complaint? I repatriated several who should have never been in the Army. One in particular I remember was blind in the Infantry; had carried a gun which he couldn't see, let alone shoot. Had always been so, but that did not make any difference. Have a pretty cute story today. Guess you have heard it, as it came from Amiens. One of our soldier boys going home said when he saw the Statue of Liberty in New York Harbor, "Old Girl, you certainly do look good to me, but if you ever want to see me again, you will have to turn around."

Got a letter from [brother] Charlie today in answer to one which I must have written him a year ago. Very interesting letter; was glad to hear from him. No news from Jesse yet. Mrs. Krumbhaar's brother was here a couple of weeks ago on a visit – he went back to his work up the line and was killed a few days ago. He was a young fellow named Dickson with a wife and two children. – Too bad isn't it? Why should it be necessary to kill so many men & boys now, when Germany is obviously & admittedly beaten? If she had a chance, I could not blame her for going on. But she admits that the war is lost for her. Charlie says Buck Haines has entered the service. I suppose he was drafted. Now they must get Nellie Griscom (a grifter like Haines) and especially must they get Bill McKenney, who has been very fervent about coming forward. And Bernard would make an excellent soldier. He told me, he guessed he was a slacker, but he did want to go. And he looked it when he said it.

We received 5 new M.D.'s in the last 3 days, so are not quite so shorthanded as we were. A number of our men are sick though. Grip or Influenza is very prevalent. Barnes, Beebe, Mitchell & Sweet have been down with it. Sweet is now around, but is not well. Taylor is not at all well, and should go home. A great many of our Enlisted men are in Hospital with "flu." Altogether we have been hit pretty hard. They took Hodge, Wilmer, Hetherinton & Nolan for the Mobile Unit – Flick & Roberts for the C.C.S. Brown for A.E.F. Outerbridge also for the Mobile. And no reinforcements until two days ago. In the mean time we had the deepest drive of the war. But we are still carrying on. Good night, my dear. R. a L.M. and a'M.M. Love,

say the Statue of Liberty will have to turn around.

By By, Much love your B.

Friday, Oct. 25, 1918* No. 8.
109 So. 20th St., Philadelphia, Pa., U.S.A.[115]
My dearest Billy,

Just think of it! This is the third letter this week. It will be a short one, though, and it has one special purpose, which is to ask you to have a little friendly chat with one of your patients, if he is still there, because of a mutual friend. The patient is Corporal James Young of the 315th Infantry. His parents have just received a cable that he was wounded in four places and has lost an eye, and is in Base Hospital Ten. The mutual friend, or rather friends, are the Smith twins, of the Shakespeare Club. Amy Smith called me up this evening to ask me if I would write to you to look him up, and I told her that if he has lost an eye and is in that hospital, he is your patient, and I would surely ask you to have a special little talk with him. The boy is only twenty-two, I believe, and isn't it sad for him to lose an eye?

Martha came in this evening for a few minutes. She says that Ed said that maybe you and he would go to Biarritz and play golf instead of going to London. I don't care where you go, but I wish you would go somewhere, for you need a change if anybody ever did.

Just after I finished my letter to you the other day, Helen Hoffer came in for half an hour. She is blooming as usual. Cousin Emma is visiting Persis in Washington, and Helen had Mrs. Burnside in the car with her, but she wouldn't let her come in, although she would have liked to – Helen said, "No, I have only a few minutes and I want

---

* MWS to WTS, ML in pencil, 5pp. on 2pp. of folded note paper; postmarked Philadelphia, Oct 20, 1918.

to talk to them alone." Isn't that truly Hofferesque and honest? She says she and Mrs. Burnside are absolutely honest with each other. Mrs. Burnside wanted to be remembered to you.

Grace Hoffman Barnett's second son, Sellers, was married a few days ago to Esther Hill, to whom he was devoted when they went to F.S.S. together, Dorothy says. Eunice Schoff was married on Wednesday. She was to have had a church wedding, and we were invited, but the invitation had to be recalled because of the epidemic. I am enclosing a very nice letter from Eunice thanking us for a sandwich plate with silver handle that we sent her.

Martha says that though Ed's commission was dated back, he does not get back pay, and supposing he would, she bought $500 worth of Liberty Bonds, and now she has to hustle to pay for them. Tough, isn't it?

I forgot to tell you that I saw a picture of Dr. Franklin H. Martin, and it is not at all a picture of Ted's surgeon at Cape May. The picture is of a Major with white hair, and quite slender. Ted's doctor is a Naval Lieutenant with black hair, and quite short. Either there is a different middle initial, or else the Cape May man must be a Jr. or a relative, for both are from Chicago, and both are named Franklin. Of course, the older man must be Secretary of the Academy of Surgery. Well, my dear, nothing has happened since I wrote two days ago. Oh! yes, there has, too. The Board of Health has decided to allow churches to open next Sunday and schools on Monday. That of course means that the epidemic is at last nearly over. It has been a terrible scourge, the worst the country has ever known.

Good-night, dear. Consider yourself very much kissed, so much that you would cry, "Now, don't hurt me!"

Good-night – again, with much love From

Mabel

Oct. 24, 1918 [Thursday]*
[Le Treport] France[116]
My dear Miss Hoffer,

Was very much pleased to receive a letter from you. ▬▬▬▬▬▬▬▬ [censored] Have thought of you many times, especially when I felt like teasing someone. Was sorry to hear that Mr. Duer & Mrs. a'Becket have passed on. As I remember it, Mr. McCollins is the survivor of a coterie that has been known to me and advised by me for many years. Mr. Burnham, Mr. _____, Mr. Van Dam, Mrs. a'Becket and Mr. McCollins. Mr. Duer, while perhaps not exactly of it was close. I am one of a similar coterie – Who would be the last man?

If this war continues, count me out. I think your sister is wonderful to learn to do what she is doing at her age, and it must be a pleasure and satisfaction to all of you to have her with you performing useful things in spite of her affliction. You may remember my friend and patient Cyrus Chambers. He took the keenest interest and pleasure in fighting his handicap, and was much pleased and indeed quite proud of the things he could do without his sight. But he was far along in years. The sudden cares are the boys in the twenties that have been totally blinded in this horrid war. I have passed many of them to Blighty, where they will enter St. Dunstan's and start a new life. A different life to be sure, but let us hope a life not without some pleasure.

I hope you have a good map and are keeping your eye on the rapid movements of the Hun just now. He is at last moving in the right direction and I hope he will not stop until he reaches Berlin, by which time he will come to rest. Our boys are doing wonderful things but are not making the gains for nothing. The work done, however, is worth the price, even though the price is very high. We have with us at the present time influenza and pneumonia as double troubles. We are as a consequence shorthanded, and very busy.

We all see the End of the war, but sometimes I think it must be somewhat like one of our big mountains out West. If you don't know, you think it is 2 miles away; if you know about it, it's 40 miles away. Whether it's 2 or 40 though, the Allies are going to see it through.

I am afraid this is not a very cheerful letter, but I don't feel cheerful tonight. Have lost 3 patients today – more will go West tonight. However, I wanted to answer your letter, and felt that there was no time like the present.

Best wishes – Cheerio! –

Sincerely,

Wm. T. Shoemaker.

---

* WTS to Helen Hoffer, 4301 Spruce St., Phila., ML, 4pp., on 2pp. 8x7in. note paper; opened by Censor 6671; passed by G.B. censor; stamped APO 25 OC 18. How this letter got into the file of letters exchanged by WTS and MWS is not explained.

Oct. 25th, 1918, Friday*
[Le Treport] France[117]
My dear Mabel,

This is not Saturday night, but I have nothing to do, so might put in some time starting your weekly letter. Besides, I feel like writing to you tonight, so that is another good reason. Your letter No. 2 came this morning. Quite interesting to read about your parade. My but times are changed. I would as soon think of parading myself as to think of you parading. Guess you looked alright anyway. Was Dorothy's place on wheels or on a float? She is sure coming along to drive horses down Broad St. Hope you have pictures of the parade – including your section and Dorothy's.

Do I know where Schmidt-Kemper hangs? I know where he ought to hang. – Down over the end of the month – near the tent cards – under Mr. Lanenau [Lankenau Hospital], etc. Now of course if house has been cleaned – I have no idea when he hangs. You understand now that is not sarcastic. Do you remember the girl who used to change my rugs every morning to suit herself. I would change them all back again, the moment I entered the room.

Well my dear, we are having a rather strenuous time in some ways. Influenza is rampant and often fatal. Our men have it and some are dying – or have died. Some of our M.D.'s have it – several nurses have it. We are short-handed and of course from the nature of the fighting very busy. One of our saddest cases is Frank Doughney [Frank X. Dochney] – a boy from the City – who has worked with me practically all the time he has been here, or rather that we have been here. He was my orderly in [Ward] 12 for months and was a good one. He was fond of me and I of him. In fact, he has been a general favorite among the boys & nurses. He is engaged to a girl at home (his home) named Cara Franklin, and I think has written her every day or most every day. And at the same time, he would write to his mother or father. I censored most of his letters for him. Well, he started four days ago with "flue"- and pneumonia, and has been dying all day. Will probably pass on sometime during the night. I have been to see him half a dozen times today and will see him tonight, but that does no good. This morning, one of our Mess boys came to me and said Doughney was dead – as usual I went over to see about it. He was not dead, but knew me. This afternoon he does not know me or care for me. My two nurses, Miss Fuhrmann and the "Little One" (Miss Metz), both very fond of him, have been sitting beside him most of the day, keeping watch. Everybody is awfully sorry and broken up. Everybody likes him. His poor mother will be heartbroken; she lost her husband about 2 mos. ago. Doc [Dochney] got a letter telling him that his father was dead. He was operated on and died in the Lankenau Hospital. Deaver doubtless operated.

So you see my dear, I feel very bad. Yesterday, I lost 3 patients. As a result of battle casualties and pneumonia, we are having very many funerals. Have not yet learned about Jesse. Have sent another telegram to the Chief Surgeon. But no answer come. Yesterday I felt that I should cable Margaret even though I know nothing. So I cabled: "Cable received Oct. 17th – am investigating. No information yet, will try again." That is the best I can do, but I am afraid Margaret will be disappointed. But I guess she knows by this time from Washington.

President Wilson's reply to the Hun came out this afternoon. It is a corker, and in my useless opinion will shorten the war, and bring sooner the inevitable unconditional surrender. I love the way he speaks of the "King of Prussia" – intimating that the King of Prussia is no longer to be considered the Emperor of Germany. It's a wonderful document, and leave nothing to be said on the other side. The German people are invited to get rid of their Emperor, and failing in this they are invited to fight until they are satisfied. I think I know what they will do. If they don't, I think I know what will happen to them.

Your little story about Lauther is amusing. Like others of his type, I fancy he is drafted, or in great danger of being drafted. The draft is beginning to catch some of our slacker friends. Bunk Haines is another who "decided" to offer his services. By and by – our friend Bernard will probably arrive at a like decision. Of course, those over 46 escape for the present, at least.

Well my dear, having made a good start, I will stop for tonight. Good night. XXXXXXXXXX

Saturday night [October 26] – Well dearest – I left you last night and went to my ward – finished up there and went to #35 to see Doughney. Miss Metz was beside him. We stood there together for a few minutes and poor Doc was

---

* WTS to MWS, ML, 10pp., on 5pp. 8x7in. note paper; opened by Censor 3919; passed by G.B. censor; stamped APO [date illegible]; received Nov. 7th.

dead. He did not die. He simply ceased to live. I was glad I was with him at the end. "The Little One" collapsed, so I gathered her together and took her home. It's all very sad, and I feel very bad. He will have a Garrison funeral tomorrow at 8 o'clock. There is little can be done in war times as special attention to the dead. A shot, a short *bon,* and a hole in the ground: – a five minute service at the grave, and back again to work. That is the routine. Of course, flowers are in evidence when there are friends or relatives to supply them. Doughney will have many. In fact, I sent Miss F. down to get some for me, but the "boys" had previously brought up every flower in Treport. So Miss F., Miss M., & I could not get any. Well as a special mark of respect, the C.O. ordered Doughney to be buried in uniform, with his Overseas Cap. The boys, or his friends, also lined his coffin with some suitable material, which they bought. These little things do not amount to much except as they reach those who think of and do them. Then Miss F. & Miss M. slipped over to the mortuary against rules & regulations. You understand, and saw him. They said "he looked lovely." Miss F. had placed a flower up near his neck. You know nothing can be put on the uniform. Do you know, also, that every soldier is buried with his arms straight down at the side. The idea is that he must be in the position of "Attention."

Well, I guess you have had enough of this solemn and sad topic. Took a Chink to Mozelles this afternoon. He was a sick Chink. Had a nice ride, but it was a bit cold coming back. The days are so short now, that it is hard to get back from anywhere before dark. Charlie has had more trouble with his teeth. Lost another one today – making 7 buried in France. With the exception of the front ones, which of course he can't eat without, he has practically nothing left. [pause]

Sunday night: – Was torpedoed about two inches above, so will finish up this very long letter tonight. Have not been able to locate Jesse. A telegram yesterday said that he had been removed from #309 Field Ambulance to Base Hospital 14. We could not locate Base Hospital 14. Telegraphed them however to information. Answer this afternoon says – "Private Jesse Warren 310 Infantry not on our records as patient." Our next move was to telegraph again to the Chief Surgeon. But it would seem to be impossible for me to get in touch with him. Will keep trying however until I learn something. Which will probably come from home. Have not heard from Warren yet. I wonder if he is sick. My last word from him was in August. I wrote to him a few days ago and asked about it.

Good night my dear, I can think of nothing more to say. In your cheerful letter, you also seem to see the End of the war. I hope your opinions are in this case, as usual, more reliable than mine. I am dead fed up. – But we will keep on going and maybe someday the end will come.

Good night. Much love. Same instructions

from B.

Liberty Bond Parade, Philadelphia, Penn., September 28, 1918. View to south on Broad St.

200,000 people were exposed to influenza. Within days, all hospital beds in Philadelphia were occupied by victims, and 12,000 died. Both Mabel Warren Shoemaker and her daughter Dorothy were in the parade.

The Union League Club, of which WTS was a member, is two blocks further south, on the right.

Wednesday, Oct. 30, 1918* No. 9.
109 So. 20th St., Philadelphia, Pa., U.S.A.[118]
My dearest Major,

Congratulations! Your letter telling of your well merited and much belated promotion arrived this morning. I am so glad that at last you have somewhere near your proper rank. As you say, the extra money will not come amiss, but you must be sure to keep plenty at your end and not be too generous to us here.

Give my congratulations to Dr. Cloud. I am so glad that he is promoted too. I just called up Mrs. Cloud on the telephone, but she was out, so I gave my congratulations to her mother, Mrs. Perot, to transmit.

It is time to get dinner ready, so I will close and continue later. [pause]

Same evening.

Mrs. Cloud called me up on the phone. She is much perplexed, as she received her husband's captain's commission Sept. 28th to date from August something. She wonders if the majority mentioned in the cable with yours was a mistake. Probably by the time you receive this, it will all be straightened out.

Aunt Linie telephoned her congratulations on your promotion. She says she is just as much pleased as if you were her son, and I am sure she is, for she is very fond of her youngest nephew.

I had a nice letter from Warren this morning, too, describing the nice camp they have made for the winter. It must look very pretty, like a group of log cabins right in the mountains. He says the buildings are covered with "slabs" made from the first sawings of the logs, with the bark outside. Warren's letter last week told about a wonderful hike he and another boy took right into the Alps. Isn't it great that he is seeing such beautiful places, for he is one who appreciates scenery?

Dorothy is much excited over Martha Delafield's approaching wedding. Dorothy and Olive are to be bridesmaids. The dresses will be lovely creations of turquoise blue, and we didn't see how Dot could pay for it out of her very small allowance of $20 a month, so I was trying to plan to give it to her for a Christmas present, and now that you are a major with more money, I will be able to do it. About that, fine? I will give Barbara her Emergency Aid Aide uniform too as a Christmas present, for I did hate the idea of having her sell her only Liberty Bond.

Emergency Aid Aide uniform - Example in Smithsonian Museum

Speaking of Liberty Bonds, you must write and thank your brother Charlie for another one. He has been sick – grip, I believe, but is alright now, only a little weak. He came around here day before yesterday, and handled me a fifty dollar bond to put in the lock box for you. He said he wanted you to be in the Union League honor roll. I told him that Owen had put you there, and so had I, but he said that was alright, the more the better. Now you have four bonds, and I have two, while each of the children has one.

Edmund Boericke is upstairs playing the mandolin to Dorothy's piano accompaniment. He is stationed at the works at the Primor now, so he comes around occasionally. Last year it looked as if it might be serious, but this year I think not, for he is not so constantly on hand. He is an awfully nice fellow, but – ! Dot says he says, he is not so German as she is herself, that his ancestors were Holland Dutch, and only lived in Saxony for one generation, and now they have been Americans for two generations on his father's side, and his mother, he says, was a pure blooded Russian. That is the first time I ever heard anything about his mother.

Jimmie Fellows has not been around this fall, though he called Dorothy up on the telephone several times when she was out.

You know Barbara has callers now, too. She is very much of a young lady, and a mighty pretty one, too, the awful portrait picture she sent you notwithstanding.

---

* MWS to WTS, ML in pencil, 9pp. on 3pp. of folded note paper; postmarked Philadelphia, Oct. 31, 1918.
To: Major Wm. T. Shoemaker / General Hospital 16 (Phila., U.S.A.) / British Expeditionary Force / France.

Did I tell you that Walter Hitchcock died last week at Norristown? He was the only influenza victim in our church, and he had been at Norristown for years. It is said that more people died of influenza in Philadelphia alone than have died "over there" among our entire Army since the war began.

Yesterday, the Shakespeare Club met at Emily Claflin's apartments at 1519 Pine St., so afterwards I stopped in to see Nancy Page. I was much surprised to find that Henry is a patient in the Lankenau Hospital. Nan say he was absolutely worn out working with the influenza cases, and so when his finger got infected in some unknown way, he became very ill. Dr. Deaver and Harvey are treating him. He was in bed at home for three days, but then Harvey & Dr. D. insisted on his going to the hospital, for at home he would get out of bed upon occasion. He has been in the hospital three or four days now, and Nan says he is much better. Yesterday, she was going up to eat supper with him, he in bed, and she on the other side of a little table beside the bed. She says that when she was up there Monday, Henry growled and grumbled so that she knew he was getting well. She says he is an awful patient, who won't behave.

Tomorrow I resume Red Cross factory work, and shall be glad to get at it. Every outside thing being closed has given me a little extra time for home sewing and I have made Dorothy and Barbara each a dress. Dorothy's is a new blue serge, and Barbara's made over from one she has worn two winters already. They are both pretty and the girls are much pleased. I sent the coat of my brown suit to Mr. Kee, and he made it over to fit me beautifully. It did hang on me the way scare-crow's clothes hang, for it was made too large for me at my stoutest, and now I am so nice and slender, it would lap over all the way to the shoulder.

It has been so warm here this last week that it has seemed like summer. I suppose it must be Indian summer a week or so ahead of the usual time. October is usually cold, and we nearly always make a fire in the furnace about Oct. 15th. It was cold quite early, but we did not start the fire, and now it feels as if we would never need one.

I am glad you are having interesting work and are doing some riding around the country. All the towns you mention I look up on a fine map sent by the Girard Trust Co. You would be horrified to see said map hanging on the wall of your front office, but it is a handy place to see it, and it is war, so anything like that goes. When you come back, we will remove it.

It looks now as if the war would be over very soon, perhaps before you get this letter. Will the Kaiser abdicate, or will he be forced out? He has to get out some way. Will he possibly commit suicide? Well, soon we will know, for Germany cannot hold out much longer, and soon it will be all over but the shouting. That will keep up for years. Fancy the parades and things when you come home again.

Well, good-night, Major, my dear. You are only just my precious old Billy boy any way, and I wish you were home again.

Lots and lots of love to you from

Your own

Mabel

November 1918

Le Treport – November 11: Armistice

"I am detached and go to London [as] Consultant in Ophthalmology for England."

"On November 9, 1918, the Kaiser having fled, Germany declared herself a republic and two day later signed the Armistice in the Forest of Compiègne. The war cost the Central Powers three and a half million men. It had cost the Allies over five million."
–Paul Fussell, *The Great War and Modern Memory*, 20.

Nov. 2nd, 1918 [Saturday]*
[Le Treport] France[119]
My dear Mabel,

Three weeks ago, I told you the war was over. Well it is, you understand, but they do not know how to stop fighting. Germany evidently wants her Army destroyed, but if so, it is being rapidly attended to by Gen. Foch. This delay of a few weeks or months is really better, because the final exit of Germany will be the more complete.

You see what happened to Austria, had she stopped sooner, she could have saved her Army, but now it is routed & destroyed. The same thing will happen to Deutschland and how wonderful it will be. Just think of dethroning Wm. II. The erstwhile strongest & most securely entrenched monarch of the world. It is all very wonderful, and we who could not get to Europe and who would be very raw & poor troops – not to be taken seriously – made it possible. It's worth all the sacrifice so far and more indefinitely, if necessary.

The nervous tension caused by the rapidly happening events is considerable, and I heartily wish the end would come. The glooms are always with us, telling us with grave faces and signs of great wisdom that Germany is not broken yet – that she can carry on for many months yet – that she has lots of fight in her, etc., and that in 6 mos. or next year this time that the war might end. Charlie is back, but he has not much sense. Cloud is another. Harte of course has not enough of us in "unmarked graves" to satisfy him. He looks far into the future – sees nothing of course and thinks not of the end of the war. Taylor is apt to discourse on how, when the war is ended, it will be some great number of years – about 2 before anybody could go home. In other words, the great invading army is to remain on foreign soil with no enemies in sight – doing just what is not quite clear.

I suppose when Germany is beaten the 7 00 0000 men that beat her must remain in Germany as an "Army of Occupation" to make sure that she wasn't fooling us. By the way, Harte goes home Monday. He has been ordered to Washington. Mitchell will be C.O. At the last inspection this afternoon, the men gave him a very beautiful wrist watch. An awfully nice thing to do, and it pleased him very much naturally. He was rather liked by the boys. Everybody likes him in "spots." He is very much like the little girl with a little curl, that we used to read about. I am glad he is going home, because I feel that he has more than done his share, but I am sorry that he will not be able to take his Unit back as a "Crowning glory." And if I am right, he would only have to wait 3 or 4 mos. for that purpose, but if others are right, he would have to wait several years. It's very difficult to decide these things. However that may be, when this war is over, I look to your friend Lincoln Furbush and my friend Wallis Parker to get me home *"tout suite."* You will have to create a demand for such as I – Hard to do perhaps, but easier than licking the Hun, which will have been done by that time.

---

* WTS to MWS, ML, 12pp., on 6pp. 8x7in. note paper; opened by Censor 6848; passed by G.B. censor; stamped APO [3 NO 18]; received Nov. 27th.

Now some comments on your very nice letter of Oct. 6th. Have heard of the dreadful influenza epidemic in Phila. & elsewhere. Glad Ted escaped so easily, but have worried a good deal for fear you would get it, or fear of the children would get it in a severe form. It ploughed through this place, as I told you last week. Another of our boys [Pvt. John Wesley Thomas] died a few days ago. 5 or 6 nurses have it now, and of course a great many patients. I think though that the worst is over. One day this week we had over 20 funerals. They have averaged about 10-12 a day for some time (the top, I mean, not 16 alone). 16 has contributed 4-5 deaths a night, more than one lately.

Have not heard from Jesse yet. Have done everything I can think of. It is so likely now, that if he is not dead, he must be well, or invalided home. Received two nice letters from Warren this week. They came together. He is having a fine time. I suppose he told you how he went up to the Swiss border, to the entrance of the longest tunnel in the world – 7 miles. It was probably the Mount [Mont] Cenis tunnel. The two longest tunnels are the Mount Cenis & the St. Gotthard. One is 7 miles long and the other 9 miles long. I forget which is which but I have been though them both. The St. Gotthard is through the Alps. En route from Italy to Switzerland & the Mount Cenis you take in going from France into Italy. I hit it in going from Paris to Naples. That is the one that I presume Warren saw the entrance of. Just think, if I had staid [sic] there, I might have met him!! It would have been a long cold wait, and might have entangled his existence, because I was there in 1894.

I was rather surprised at the communications from the American College of Surgeons. But as I have intimated Francis [Franklin] H. Martin and his associates are a rather foxy bunch. For that reason only, I declined membership at the time the College was founded. I was asked three times to come in as an original member. My letter to him, a copy of which I have, was without equivocation. I told the Secretary (not Francis H.) that I would have to postpone my interest until after the war, as I was in no position at present to incur unnecessary expenses. His questions, I told him, I answered as a direct means of disposing of them. You were surely right in not paying him $100 – Nor will I pay $100 nor pay any attention to the thing until I am reestablished in my little home and office, and not then unless convenient. So now – [pause]

Sunday, Nov. 3. Was suddenly interrupted last night before I had time to add my good night kisses. Here they are

XXXXXXXXX

XXXXXXX

This like recent Sundays has been a long dull tiresome day. Unless something breaks soon, I will be a nervous wreck. Am becoming more and more nervous every day. Am thinking of running down to Paris this week just to get away from here. If I go, I will go back with Jed Fauce. I have no business in Paris, but my camouflage excuse will be to look up the whereabouts of my nephew. As Mitchell will be C.O., I should have no trouble in connecting. But I must get away from here.

Charlie is quite impossible and gets on my nerves very seriously. He is the most arrogant & ornery bully I have ever had the misfortune to live with. While he's sick, very lonely and childlike, he is presently less and less as time goes on. He is honestly fed up and as a consequence, his many rotten characteristics are rapidly gaining supremacy. Medea still devils him and insists on the great injustices and displeasures imposed upon her. I hope I will never have to see her when I return. The war has elevated and broadened most everybody. She has remained a fat peanut. However, for Charlie's sake, I suppose I will have to be nice. You know how nice I can be – well I just won't be that nice. Just a little nice for me – just enough – no more.

Will send you some money this week - $75.- if I don't go to Paris, maybe less if I do. I really guess I won't go, but you never can tell.

Good night my dear, please take especially good care of my Mabel. I will need her a lot after the war, to "stop doing what you're doing and pay attention to me." Does that sound familiar? Don't you wish you could hear it now, and put up your sewing after hustling to finish "just 3 more stitches" and pay attention to me. I will need lots & lots of attention after the war to make up for my 18 mos. absence.

By, By – Much love, your B.

Nov. 3rd. 1918 [Sunday]*
[Le Treport] France as usual[120]
My dear Barbara,

Thank you so much for your letter, but you should not frighten me by sending me such a picture of yourself. If I thought that "my Barbara" looked like that, I would have to go home and see her. Now the little pictures are very nice. You surprise me when you say that this is your last year of school. I thought you were just about starting. And your suggestion about the School of Industrial Art is a brand new one to me. I had no idea that you ever thought of taking up Art. Of course, it's very nice and you will do well at it, I am sure. Just what line are you going in for? Hope I can get home before you graduate.

You are quite right about your war work – You have plenty of time for the S.I.A. I hope though that you will associate yourself with a good organization. If there is an organization at home corresponding to one here (British) known as W.A.A.C. (Women's Army Auxiliary Corps) "Wacks" as they are called – keep away from it. "Doing anything at all" is not so good as doing some definite and particular thing for which you might be more or less suited. But you will of course look carefully into the thing before you sign up. Too bad about Bobby McAlbee or somebody like that, but have heard you speak of him. We also had influenza and a bad dose of it. It caused many deaths and has put thousands of troops out of action.

Warren told me about Aunt Florrie's "bit" in keeping our car – which "we had no right to have anyhow," and your displeasure. I was much amused, and have thought of it any times since. It must have sounded funny. But aside from being funny, it did not mean anything.

I guess you had a good time at Cape May this summer without the car.

This is the last letter you will ~~probably~~ get until after the war. (from me, I mean.) The war is ending very supidly [stupidly]. Unfortunately, all of the Germans are not dead, but enough are dead to discourage those left. If we had the time it would be nice to kill them all. As I can't think of anything of interest to tell you, I will ring off. Take good care of your mother.

Much love,
from
Dad

Nov. 7, 1918 [Thursday]†
Paris VIIIe / Hotel Alexandre III / Rond Point des Champs-Élysées[121]
My dear Mabel,

Am again in the great big city, where I investigated Jesse's whereabouts and condition and also to get away from Treport for a few days. Report has it, that the Armistice was signed at 2 o'clock today, but it has not been definitely confirmed yet. Paris is ready, and everybody is very cheerful despite the weather which is depressing. Have not seen the sun for 3 or 4 days; instead we have had nothing but rain. It is very interesting to be here just now, and I so planned it that I could be in Paris when the End came.

So here I am, and the End is also here. Along both sides of the Champs-Élysées are captured items of all kinds, and in the Place de la Concorde are thousands of things, trench mortars, machine guns, airplanes, balloons, rifles, Boche helmets, and every conceivable kind of captured material. It is a sight worth seeing. The French seem to be in a state of depression. Official confirmation of the end, and I think they will live. This beastly warm wet weather is also holding them back.

I came down with Ralph Straub. At the Red Cross H.Q. I learned that Jesse was evacuated Sept 22nd to a Hospital at Bordeaux. If he is still there, it is too far for me to go, so I can't see him. They will again today send particulars, but will not get an answer for 3 or 4 days. Whether or not I will stay over, I cannot say. If not, they will telegraph me at Treport. Just looked up an old school girl acquaintance and took her to dinner, so I dined all by my

* WTS to Barbara, ML, 4pp., on 2pp. 8x7in. note paper; opened by Censor 6848; passed by G.B. censor; stamped APO 4 NO 18.
† WTS to MWS, ML, 8pp., on 2pp. folded hotel letterhead paper; passed by U.S. Army censor; 2c U.S. stamp, cancelled U.S ARMY M.P.E.S. NOV 9 1918; received Nov. 26.

lonesome at Maxim's. It's just like the reproduction in the Merry Widow, but of course just now at the ordinary dining hour, not so gay & exciting. I had dinner there before and rather like it.

I had a check for $75. which I intended to send to you today, but when I learned that the war was over, I decided I had better wait a bit and see what is to happen. If there is any chance of getting out soon, I will need it to get home on. However, do not be discouraged. If possible, I will send it soon, or come myself with it. Which would you prefer? Have also put in for leave first week in December. But I won't have to take it when it comes along, and you can bet with the war over, and the prospect of getting home, I won't take that leave. Ed & I intended to go down to Pau – that is far in the Southwest of France – near Spain & the Pyrenees. It's a great place; we were going to take our golf sticks and have a bully time. But of course, the end of the war is better, so if we don't go, we will not complain too much. But I am afraid we left that a little too long. However, if the war were still in session, it would make a nice leave for us, wouldn't it? Ed & I could have a fine time playing golf – he plays so well, and I so poorly that it would just naturally even up, you know.

I wish I had you here to help select your Christmas present. It's awfully hard to do it without you. Another difficult thing to decide is whether to send it or bring it in person. Everything I look at that I think would be suitable for you costs about 1 000 000 Francs. The proximity of Christmas is another reason why you may not get your commutation of quarters this month. But what worries me is that you will only have your miserable little allotment of $180.- Now next month you will get $275.- Come to think of it, you will get that in Dec. Well, I feel better then.

Maybe you would be surprised to see the stupendous amount of war work carried on in Paris. The American Red Cross H.Q. is a bee hive – it is situated in the Regime Hotel – occupying all of it – and there are at work over 1100 offices, not officers. In a way which I had better not tell you. [pause]

Here it is:- It is now near 10:10 P.M. – the Armistice has not yet been signed because Marshal Foch wants it signed in Sedan. Sedan has just been taken, so there you are. You of course know the history of Sedan. It was there that the French Armies surrendered to the Boche in 1870. A very appropriate place for the present exchange of courtesies – is it not? Oh well, tomorrow will tell the story. I can hardly wait, but guess I'll have to.

This is rather a rambling letter my dear, but as there will be but few more, you will excuse it I know. I will probably return to Treport Sunday or Monday, depending upon the excitement here. If the sun does not come out some though, I will get very much depressed here, and may beat it back sooner. For the present, good night. You must love & miss me just the same, even if the war is over. XXXXXXXXXXXXXXXX

Much love from the end of the line.

from

B.

Nov. 10, 1918, Sunday*
[Le Treport] France
My dear Mabel,

Back from the Big City to the Little City of Treport. I wanted to be in Paris when the good news came through, but the 72 hour start made it a little more expensive than I cared for. So I left last night at 10:15, getting before leaving the joyous news of the Kaiser's acceptance. Arrived at Treport at 2:30 and climbed slowly up our 350 steps and got to the rim about 3 A.M. There were two nice letters for me; one from Owen, and the other from my very best girl. So I chatted a while, then went down the line and OKed a dead man, and saw several others who were getting ready to pass, thus saving Austin & Wray Stewart [?] the painful necessity of getting out of bed at that wee hour. Then I came back, got into my own little bed, and read my letters.

Had a good time in Paris, and feel much better for my 3 or 4 days' leave. Only met one man I know, and you do you think he was? Capt. Redrick – Warren's C.O. That was rather funny. I started out alone about 3 o'clock on a little tour of investigation. Poked around the way I want to, looked in shop windows and was having a fine time. I walked up to the Notre Dame, crossed the Seine, and started down the West Embankment, where for many years, street vendors have old books, prints, etc. The place was deserted so far as "pedestrians" were concerned. I was looking at the books & pictures when an officer passed & came back. I paid no attention to him until he called me by

* WTS to MWS, ML, 10pp., on 5pp. 8x7in. note paper; opened by Censor 3922; passed by G.B. censor; stamped APO 10 NO 18; received Nov. 27th.

name, and told me he was Capt. Redrick. Then of course I remembered him. He was on leave and had just 24 hrs. in Paris and was making a rapid tour on foot to at least see some of the places he had heard about.

As I knew them pretty well, I went with him and we walked & walked until I was foot sore as of old. I invited him to take dinner with me, reminding him that I took dinner with him in July. So we had dinner; I started him for his 8 o'clock train, went back to my hotel, gathered up Straub and we beat it. I was very glad to see him – he told me that Warren was fine and dandy. Won't Warren be surprised when his C.O. gets back next week and tells him he ran into me in Paris?

Well, we are all waiting anxiously for official information on the Armistice. Did you notice that the time set is the 11th day of the 11th month at 11 o'clock? Come 7 – Come 11!! I wish it would come. Of course, I feel that the Boche will sign up and surrender without condition. I am certain of it, but until it is done, there is a picking up of great suspense. If he does not, it will be a disappointment, but not for long. Marshal Foch is prepared to annihilate him, but it might take several months. It was very good of you to buy the Liberty Bond, but please don't do it again. Owen also bought one for me through the League. In fact, if my friends at home continue to buy me bonds, I will be a bloated or busted bond holder when I return. Did you ever get your bond? If not, [go] out and get it, or some one will have it. It has been paid for from morriches and should have been delivered to you at 109. When we get home, we can use my bonds to pay some debts. I don't like bonds much, never had any experience with such things. In large quantities, they are fine; in small quantities, such as we have, they are useless. Of course, I don't mean useless as subscriptions to Government funds. Owen tells me it was the biggest loan ever raised the <u>world</u>. It's perfectly wonderful $6 000 000 000, I think that has been subscribed.

I was sorry to read about a political mud slinging fight just at this time, but very glad that the Republicans were out [?]. In spite of the apparently able prosecution of the war, I am never the less more of a Republican than ever. But I am not disposed to know the administration until that unconditional surrender comes through, and I am out of the Army. I believe in winning the war, and it looks very much as if the war were won at this minute. Of course that fellow Roosevelt is at work for the Presidency next time, and I would be willing to bet right now that he will get it. Whether he gets it or not, we must get rid of the Democrats after the war, or life will not be worth living.

Now one more political view, and I will write about something else. I think the next worst thing to the war is national prohibition. I have no sympathy for it, and wish I had a million votes to defeat it. To my mind, it will be the worst possible thing for the country, and I am strongly against it.

Am very sorry to hear of the death of Harry Ingersoll. I think he called on you for a copy of his prescription about a year ago. He was a fine fellow. [WTS's writing becomes very difficult to read at this point, "sleepy"] There are 3 of them – Harry, Sturgis and Jarred – all 6 foolies + like Mr. Clement E. We also saw him or his father the night we saw the troops off for Mexico – from 20th & Chestnut. Do you remember? Nice picture of <u>you</u> and the rest. You look the best to me – great – natural – makes me home sick or Mabel sick. Have not seen pictures of Warren & children that you speak of – send them to me. Bought you a nice Christmas present in Paris, but don't know what to do with it. I don't know what is going to happen. The next couple of weeks will tell a lot. In case you do not receive the letter, be it known that I wrote you from the Big City.

Well my dear, I am getting sleepy and will have to stop. Will keep you posted on what is happening!! You will know when the war ends before I will. Good night. Be a good girl, and don't forget to love me along with all your other dates. XXXXXXXXXXXXXXXXXXXX

your <u>B</u>.

"The Hundred Days Offensive, fought from August 8 to November 11, 1918, was the final offensive of the First World War … Just after 5 a.m. on November 11, the armistice was signed between the Allies and Germany at Compiegne in France. On the eleventh hour of the eleventh day of the eleventh month of 1918, after more than four years of suffering, horror, and untold misery, bugles across Europe sounded the end of a devastating war that had left ten million soldiers dead, and the guns finally fell silent in Europe. At last, the 'war to end all wars' was over."[122]

Nov. 16, 1918, Saturday*
[Le Treport] France   Last letter from France
My dear Mabel,

At last this war has come to an end, and maybe we on this side of the Atlantic are now glad. And you on the Other side are probably more glad. I wish I could have seen you when the news burst upon you that the Armistice had been signed!! It reached us officially about 11 o'clock Monday [Nov. 11] morning, or a little before, and everyone with one accord stopped work and turned to jollyfurations. In 5 minutes, everything capable of producing noise was busily at it! A great gladark was spontaneously formed consisting of Officers – men – patients in blues, Ambulances, Chaufferines – lorries, etc., etc., and it marched headed by our band amid shouts – songs of men marching down to Treport & over to Mers. Then in the public square, the national airs were played, and the French people went wild. And so everyone carried on all day & all night for that matter. It was a wonderful piece of news and with it all it seemed like, and still seems like, a dream. It is hard to realize that the Great War is over and that the victory for the Allies has been so complete – hard to realize that the destruction of a great Empire has been so swift and thorough. But it is all true and is as it should be and as it was supposed to be. The delivering of the Hohenzollerns is the greatest achievement ever presented to the credit of civilization. And to think that you and I lived to see it! It has all been worthwhile, my dear.

I was very sorry to learn of Dr. Allen's death. The mail that brought your letter also brought newspaper clippings to our Mess. It's too bad and I feel so sorry for his family. In that family, the one lost his wife, and the one educated & trained for this purpose at the expense of the Government, probably got off with a Liberty Bond.

Our casualties have been unusually small. I am told that our total list is under 100 000. I mean battle casualties – Flu & pneumonia have raised havoc lately, but so did they with you. And I learned today an astounding fact. Our blind number only 100. Now, consider that we have over 2 000 000 men here and that 750 000 took part in the last show south of Verdun. If these figures hold as final, I think the record in that way is unprecedented.

Now I will tell you some local news that will interest you. I am detached and go to London in a few days – probably next Wednesday or Thursday. Allen Greenwood blew in today and offered me the position of Consultant in Ophthalmology for England, which I accepted. It's a good job, and carries with it Lt. Col. Which will probably not come through until after I am at home, but for which he will at once recommend me; I will travel to all the Hospitals (American) in England, and see their cases, etc., as Consultant, and that the younger men in local charge carry on right. As they must have a Consultant and have none, my orders will come by telegraph and I will move at once. So for me, good bye Treport. He also said, that if my Unit was ordered home, he would relieve me, so that I could go with it. Now "Ain't that grand?" In my opinion, our Unit will go home soon. Nothing to base it on, but opinion, unless you hear it from me to the contrary address future correspondence –

Major "Me." – American Officer's Club, 9 Chesterfield Gardens, London

When I get permanently located, I will notify you (naturally) after I strike a deal with Aunt Connie [Melhuish] – will see her anyway. Had a nice letter from Warren this morning. He also is glad that the war is over. Now I have told you all the important & interesting things that I can think of tonight. Will write now as usual tomorrow, but these XXXXXXXXXXXX and much love –

Sunday night. I have started packing – great experience after 17 months in my barracks. Of course, I have accumulated a lot of stuff, but not too much. It has gotten very cold here; we have ice and I do not like it. A telegram from the Red Cross today says, "Jesse Warren in Convalescent Camp near Bordeaux in good condition." Well, that is alright, but it did not come in time for me to be of any service to his mother. One might think that a fellow in Jesse's position would naturally communicate with a perfectly good family connection only a couple of hundred miles away. Had he done so, he could have saved his mother much unnecessary worry. But as I remember it, Jesse was never taught to think. Can't write any more now my dear, as I must do some packing. Will keep you posted on my movements. Good night – Cheery Ho – much love,

from B.

----

* WTS to MWS, ML, 8pp., on 4pp. 8x7in. note paper; opened by Censor 6435; passed by G.B. censor; stamped APO [date illeg.]; received Dec. 5.

Nov. 25, 1918, Monday*
London.W.14 / 17 St. Mary Abbots Terrace[123]
My dear Mabel,

No letter you Sunday did you get from Le Treport this week. [sic: this sentence is exactly as it was written] So much has happened that I don't know how to tell you about it. Well anyhow, I left Le Treport Thursday morning at 4 o'clock, dark, cold & lonely, bag & baggage. I left on the train which I thought was to take me to Boulogne. Had just curled up for a snooze before day light, when the guard opened the door and said *"finies"* –

I was in Abbeville, and the train for Boulogne was 12:30 P.M. So I had to spend the morning in Abbeville. At 12:30, I got the Boulogne train, but Alas! my trunk did not. So in Boulogne, I was seeing my precious trunk with everything of value, including all my instruments – <u>everything – lost</u>. Now that in France at this time is some mishap. The worst of it was that nobody was around the station when I left Treport, so I just put the trunk aboard, unregistered, intending to sit by it. The last I saw of it was on the station platform at Abbeville about 11 o'clock. Well, I was not going to leave France without my trunk if I could help it. So I made repeated personal searches of every place & platform around the Boulogne station, hoping to spot it. Nothing doing [on] Friday or Saturday morning, so I hopped a train & returned to Abbeville. When I likewise made my personal investigation, it was not there. As I was overstopping my orders, I decided that I would have to leave Sunday afternoon, trunk or no trunk.

But early Sunday morning, I returned to the baggage rooms, platforms, galore, sidings, etc., etc., of the Boulogne station. There were <u>millions</u> of baggage crates, everything. The goods transportation [sic] is enormous. I had reached my last hunting section, when something just led me on in despair as it were, a little farther, <u>where</u> there stood my little trunk. I could have kissed it & hugged it. I would not go ten feet from it, but called a porter, put it on his "wheel barrow," took it right down to the boat and checked it for London. It was personally conducted. Then promptly at 2:30 Sunday afternoon I left France, for good and all, after a stay of just 18 mos. less a few days.

Arrived in London, in the dark & rain. – Impossible to get a taxi – had no place to go, and just $7- to carry on with. Could not see anything. 6 or 7 million people here – no accommodations for anybody and nobody cared a damn. I started for the American Officers' Club – and by means of the Underground & my two legs – arrived safely. Could I have a room? Not one in the place. Well then, I can have something to eat – so I had a *bon* eats, and then called up "Connie" on the phone. She was lovely, said come right out, your room is ready. Well, as I had not enough money to travel "my style" *a'la* Ritz, or the like, and waited until the following day to gently explain my problem to her. I took bus No. 9 on [sic] Piccadilly and found my way out here. She's fine and her family is fine and I am for the immediate present living in luxury. She insists that I shall remain here while in London, but I can't do that you know. I am trying to get located, but it is very difficult. Every place is full to the roof. My trunk is still at Victoria Station, and I must get some place to live in a day or two. Reported this morning – was sent to visit a hospital in the north of London. Tomorrow or next day will go to Poynton [Paignton] down in the Southwest. Then I will have to go to Liverpool and every place where we have American hospitals. I do not think I will have time to make more than one or two rounds, as I believe our Unit will be sent home. But I am here for the finish, and it can't be long.

Unfortunately, Allen Greenwood was too slow in reaching me. He said he would put me in for a Lt. Col. at once, but the Gov[ernment] has decided that there shall be no more promotions after Nov. 11th Armistice day. My recommendation will therefore be about 10 days too late. That's hard luck, but it's <u>my</u> luck so what's the odds. But it's kind of annoying because Geo. Norris, and John Gibbon, who had similar jobs (Consultants) – but got them sooner, are both <u>full Colonels</u>. Had Greenwood sent me over her a month or 6 weeks ago, I could have been one too, or at all events a Lt. Col. However, if I can get home, I won't bother about eagles and silver oak leaves –

Called on [Maj. Robert] Stuart Smith, and was darn glad to see him. Will dine with him tomorrow. He is doing good work, and is very happy and looks fine. Unfortunately, I must cable home tomorrow for money. I am sorry that I can't live in London with $4 in my pocket, and that's about what I have. I will draw on J.L.L. [Lippincott] & Co., and you can give them two of my "Liberty bonds." Will ask for $100.-

I think I did pretty well to get here so unexpectedly. You know, I had just been to Paris. I paid my Mess bill at Treport and cleaned everything up- and started out with $35.- to be a Consultant in London, and so far, I am as good as the next fellow <u>so far as he knows</u>. Stuart Smith is going to arrange with Morgan Grenfell so that I can draw at once. Now my dear, I must stop. Everybody is in bed and asleep. I am in a beautiful library with everything so

---

* WTS to MWS, ML, 12pp., on 3pp. 9x7in. folded embossed stationery; "passed as censored" signed by WTS; received Dec. 13.

different from my shack at Treport. Beautiful. Couches, wall paper, pictures, pianos, chushy chairs – the whole show, just about like Chestnut Hill or Greenbush. Your people are very good to me, and appreciate it. Mr. M. [Melhuish] is fine and Betty [Marjorie] is lovely.

Good night, my dear.

Much love,

B.

Written but not read

Nov. 27, 1918, Wednesday*
Torquay, Grand Hotel[124]
My dear Mabel,

Devonshire is Lorna Doone country – it is very picturesque. This is a beautiful place down in Devonshire, far down that long tongue of England, ending in Land's End. It is 201 miles from London and at Poynton [correction: it was Paiginton], 3miles away – connected by tram, is an American Red Cross Hospital which I visited today in my official capacity as Consultant. Return to London tomorrow. But it is real civilization, and the hotel is a swanky affair such as we have at home along Long Island towns, or on the New England coast. The place is much like New England. It's not cold, and palms are palming in the beautiful gardens. And the cushy trains running fast are perfectly smooth trains, and on time are such a contrast to the trains in France. It would take a French train several days to run 200 miles, and I came down in 5 hrs. Possibly the war did not get down here; if it did, you would scarcely notice it.

I think I told you that I landed in London with very little money. Well, as it was impossible for me to carry on, I drew on J.L.L.&Co. for $100 – which I got at once from Morgan Grenfell. It was a war necessity – Sorry. Reimburse them at once by selling my Liberty Bonds. – Not the one that Owen put up, because he expects me to pay for that one. He told me about it, and said he took the liberty of buying it for me, and I could etc. – when I returned. That's all right, and I can handle it when I return.

Return! It won't be long now; my work is done, I am trying to make at least one round of the American Hospitals, just to see them. As I told you, Greenwood discovered me too late and thereby deprives me of promotion. It is impossible to get suitable quarters in London now, and I am still with Connie, and my trunk is still at Victoria Station. She says she wants me to remain with her, but you know how I feel about it. Of course, I don't bother her during the day, but I occupy a large spare room, eat a wonderful breakfast, and sometimes a wonderful dinner. And she may have, and in fact has, other guests. But I can't find any other place, so there you are. Am a member of the American Officers Club, 9 Chesterfield Gardens, which is my official and permanent address for mail, or messages, etc. Remember, I have no "Unit" – just "Me" – Am. Officers' Club, London. Of course, I would get anything addressed to Connie's house.

Yesterday, I went to the Washington Inn to try to get a room. The lady in charge, when she heard my name, said, that sounds like a good Germantown name. I replied that Pastorius and myself founded Germantown. She knew all about the Germantown Hospital, and why not, - she was Mrs. Vann, our old Superintendent – now Mrs. Eager. However, she had no recollection of me. She would have known Theo. Le Boutillier better, I fancy. She has run Washington Inn since 1914. Maybe not so long, but since it started. But you can only stay there two weeks, so that won't do. Then I tried Cavendish Square, another Officers' Club run by the Y.M.C.A. Nothing doing there. Every place is full. You see, the Officers' Clubs are very inexpensive; no tips – the best of things and the lowest possible prices. Hotels, regular apartments, are way up in G.- and you get stuck. For example, at the Officers' Club in Boulogne, my room – first night with another fellow in it was 2 francs – (38 cts.) – next night, a single room, 3 francs. Meals, and good ones, 3 or 4 francs – At the Folkstone Hotel where I have eaten, they would be as many dollars.

Now in London of course it is a little more expensive – but the same proposition holds. – and The American Officers' Club is almost as finely furnished and as handsome as the [Union] League or Art Club. It's the private

---

* WTS to MWS, ML, 8pp., on 2pp. 8x5in. folded hotel letterhead stationery; "passed as censored" signed by WTS; rec'd Dec. 13.

mansion of some wealthy Lord – turned over as is. Holesbury has nothing on the hotel. I can do everything there but sleep – so I guess I will have to sleep at Connie's. The travelling stunt is a little expensive on account of hotels, meals, etc., but I get 4 cts. a mile which helps some. Of course, I pay no R.R. fare. Government transportation and 4 cts. a mile. Many hotels like this, and I would have to apply for a higher mileage rate. They charge me for everything. I had to pay one cent for this sheet of paper, and one cent for the envelope. Our hotels supply stationery ad lib. Good night my dear. Keep a loving, a missing, and expecting me. I will drop in on you some day; don't be surprised – too much.

Edwin was in Paris when I was there. Sorry I did not write her [Martha].

Good night.

Much love, B.

Chesterfield House – London
9 Chesterfield Gardens in 2019

American Officers' Club in World War I

Cavendish Square – London – 2019

Y.M.C.A. American Officers' Club
Was located here in World War I

Belgrave Hotel – London – 2019

Belgrave Mansion Hotel was
U.S. Military Headquarters
in World War I

December 1918

From London to Philadelphia

"Will you be glad to see me?"

Some Never Returned

"The cross on the grave would be inscribed 'Kenneth MacLeish, Born the 9th September 1894, Glencoe, Ill. Found the 26th of December 1918'." MacLeish wrote, "The Gates of Honor are opened to us, the lucky ones of us who are over here." He took off in his Sopwith Camel biplane fighter on the morning of October 14. "In a wild melee over Belgium he scored his first aeriel victory of the war. MacLeish joined a second patrol later that afternoon, a flight from which he never returned."[125]

Dec. 7th, 1918 [Saturday]*
London.W.14 / 17 St. Mary Abbots Terrace[126]
My dear Mabel,

I have not forgotten you but you will doubtless think that I have neglected you. But I have been so much on the go since coming here that I have had little opportunity, and I am pretty tired. We have in England 10 Base & Camp Hospitals and in less than two weeks I have been to all once and some twice, travelling in all about 1500 miles.

Wrote you last as I remember from Torquay (Paignton) [Dorsetshire]. Then I took in Portsmouth [Dorset], Sarisbury Court [Hampshire], Hurlsey Park [Hampshire], Romsey [Hampshire] and Winchester [Hampshire] in one trip of two days. These five hospitals are all more or less close together – or at least in the same section. I then went to Liverpool and visited Knotty Ash and Mossley Hill [both suburbs of Liverpool]. My next trip was to Dartford [Kent], which is not far from London. Then I thought I was through, but the Col. thought I was a little "hazy" on the conditions at Sarisbury Court, so he said you had better run down there again. Alright Sir! said I, and started off determined to get back to London the same day. Took the 1 o'clock train for Portsmouth. Got there in 3 hrs – had telegraphed for an auto, or H.Q. did – took a wild ride of some 20 miles in the dark to Sarisbury Court – found out in a few minutes what I wanted and beat it back to Portsmouth in time for the 7 o'clock train for London. Arrived just after 10 – tired but all here. Next morning, I reported to Col. Washburne. He said, "Have you been down there already." "Certainly," said I. "You are some fast traveler." I allowed I was.

Well my dear, here is a bit of news for you. This letter will be in the nature of an anticlimax, because I think it will not beat me home by much. I hope to sail for the U.S. on the *Saxonia* about the 15th of this month. Now, I am not sure and while I rather expect to, I may not connect. The *Saxonia* will go as a hospital ship and will probably carry a thousand or more of our sick & wounded. So I will travel on duty and work my way as it were. Norris & Gibbons are going on her [for] sure, and I am living in hopes. But even if I miss the *Saxonia,* I will be coming along before many weeks. Will you be glad to see me?

Am living a high life here, and I have always been so fond of London. It's a wonderful city. It suits me all the way down to the ground to finish here. And now that I have finished my work, I can visit my old friends. Yesterday, I went far out to the north of London – Golders Green and called on Mrs. Morgan. She was so glad to see me and so nice. Had tea with Mr. & Mrs. Melhuish. They are quite rich and live in a lovely place. She invited me for dinner tomorrow, but I turned it down as I am so tired of chasing around that I decided I would take a Sunday and do nothing.

---

* WTS to MWS, ML, 12pp., on 3pp. 9x7in. folded embossed stationery; "passed as censored" signed by WTS; no postage stamp.

A V.A.D. [Voluntary Aid Detachment worker] that used to be at our hospital in Treport and whom I liked, has been here at H.Q. for some time. She heard that I had come over and wanted to see me. Well she saw me alright and was so pleased that she threw her arms around me right before everybody and did all but kiss me. She is a nice little girl, so I took her to dinner last night and we talked old times on the top. Night before last I took Connie & Betsey [Melhuish] to the theatre – the first chance I had to do anything of that kind. We had a fine time. This afternoon Connie & Betsey took me to a play – "The Chinese Puzzle" – very good. I blew us all off to lunch at the Piccadilly Restaurant first. So you see my dear, I am ending the war just like a swell tourist. But there is never a dull moment in London, unless you want one.

But I am so tired of it all. If it does not stop soon, my clothes will possibly fall off. I have lost 15-18 lbs. in the last year, and I naturally do not care about buying new uniforms. Am alright for looks so far, but I would not give this uniform more than two months. It has been mighty serviceable. Do you remember that when I was abroad studying, I was away just 19 months, and ended in London! Same this time – Funny, isn't it? Here's hoping for the *Saxonia* my dear. I am dead tired, and must stop. Maybe, more tomorrow. Good night. Much love

XXXXXXXXXX

<u>Monday</u> – Loafed all day yesterday. Did not leave the house until 6:30 when I took Mr. M. down to the Club to dinner. Nice quiet evening. No real news yet. Goodbye – Much love

from

<u>B</u>.

Dec. 14th, 1918, Saturday*
London.W.1 / American Officers' Club / 9, Chesterfield Gardens
My dear Mabel,

This is another dull day. As you will know when you receive this, I did not get on the *Saxonia,* which is a disappointment, for I rather thought I would and that I would be home by Christmas day. Norris & Gibbon are on her of course – they get everything. Now I have no idea when I will get home. Maybe, not until January, or February or March or April. I am out of luck and have found it necessary to revise my opinion of the Colonel, who I am sure could have sent me home had he wished to. My now revised opinion is that he is a stuffed shirt. You understand, I have nothing to do in London. I drop in at H.Q. about 10 o'clock every morning to see if there are any orders for me. Of course, there are none, so I spend the ensuing day & night as best I can. I have shopped, walked around, ridden around & loafed until I am sick of it. And yet the Government pays me $4500 – per year for my services. But yet, it is better than being at Treport & I am fond of London. But I thought I would be with you Christmas, and am disappointed. Instead, I accepted an invitation from Stuart Smith to dine with him and Col. Washburne (<u>the</u> Colonel) and others at their house Christmas night. About half a dozen of them have an apartment in Half Moon St. Connie will also have a Christmas dinner for everyone in sight, but I don't know her time – if at night, I won't be there, and she will be sore. She has been awfully nice to me, but I try to be out as much as possible as every good guest should. Mr. Melhuish met Gibbon here last week and took a great shine to him. So he wanted me to take him out home to dinner. This I did Wednesday night, and maybe Connie did not give us a great dinner. She spread herself and we had a delightful time. They can entertain alright.

One thing today helped a little. I got some belated mail. A letter from you dated No. 19th and 3 or 4 other letters. Have already gotten letters of a much later date from you, but that didn't make any difference, you see, all of my mail is still coming through Treport, so it is quite late. Herbert seems to be out of luck with his girls. I am afraid he does not select the right kind. There is a whole lot in getting the right kind as I did. It was very nice and thoughtful. Bill Drayton should call you up. I would if he called up "Medea." Gibbon is also going to call you up. He tried to get me on the *Saxonia,* but could not work it. Base 10 is sitting at Treport, awaiting orders, and dead tired of the job. I guess "Chawoles" is boiling with rage and indignation. Am glad to be out of that anyway.

---

* WTS to MWS, ML, 5pp., on 3pp. 9x7in. letterhead stationery; "A.E.F. passed as censored" signed by WTS; EXAMINED BY L 11; no postage stamp.

You speak of Christmas presents. I purposely did not send you any because I expected to be home or nearly so. You need not worry about it, because I do not need Christmas presents this year. As for your presents from me, I did not send them, because I expected to bring them. But I have them, and will deliver them just as soon as possible. So don't worry about that. Sorry I miscalculated, but it is difficult to do anything else in the Army. Just remember that you will all get them, but just a little late. So now – Geo. Ross, Jim Tally & Bob Le Conte have gone home, and will be there by Christmas. Others who know how to land right XXXXXXXXXX

Monday morning. Dec. 16th. Well my dear, things are moving a little. The Colonel is better this morning and is trying to get me home. In the meantime, I am off for another junket – Cambridge, Oxford, north of England, etc. Must gather up American patients in British hospitals, so that they can be sent home. Well, if we must stay here, it's better to travel and see something. Will write you soon again. Goodbye. Much love.

yours as ever

B.

Dec. 17, 1918 [Tuesday]*
London, W.1 / Regent Palace Hotel / Piccadilly Circus[127]
Dear major Shoemaker,

I want to congratulate you on returning Saturday on the *Mauretania* and to beg you to bear me in mind and if there are any more doctors needed on board or if there is room for one more to remember that I would love to be that one. I am enclosing a small photo of my wife and baby which will give you a good idea of why I want to return at once if at all possible. Wishing you all the best

Very sincerely yours
Robert Lockhart
Capt. M.C.
U.S.A Base Hospital No. 40
Sarisbury Court,
Hants, Eng.

"Mrs. Lockhart and Susanna" enclosed in letter from Captain Robert Lockhart

----

29th December 1918 [Sunday]†
London.W.14 / 17 St. Mary Abbots Terrace[128]
My dear Major,

Santa Claus presented to me a very fine cigarette case, handsomely initialed, and told me it came from you.

Very many thanks for this gift, which I have mobilised at once and I think of many pleasant smokes I have had in your genial company as the aroma ascents into the blue.

I hope you have had a pleasant voyage and found your family all well. It would give us great pleasure if the opportunity comes for you to bring Mrs. Shoemaker over here to visit us.

I am one of the few Londoners who has not seen President Wilson, at least not yet! I intended going to my Club to see him drive up St. James' Street with the King, but Jack had a seat for Chu Chin Chow "thrown on his hands" at the last moment so I went with him. I hope the President will overlook this omission on his programme!

As he has had the biggest & most enthusiastic crowd London has ever turned out, to greet him, I guess my absence was not noticed! I am afraid you gave me the *Mauretania* in mistake, to cable. I see she was, & perhaps still is at Southampton.

---

* WTS from Robert Lockhart, ML, 2pp, on 8x5in. hotel letterhead, enclosing two 1in. photos; addressed to "Major Shoemaker, M.C. / The Ophthalmologist / Office of the Chief Surgeon / Belgrave Mansions Hotel / London S.W.1 / postmarked 18 Dec 18.

† Two letters from 17, St. Mary Abbot's Terrace in one envelope, postmarked London 29 Dec 1918, with 1½ d. British stamp. The first letter is signed G. Melhuish, with salutation to WTS; the other is signed Constance Melhuish. Letter #1 is on 1p. of folded 9x7in. embossed stationery, showing Kensington as the address. Letter #2 is on 2pp. of 10x8 embossed thin weight stationery, showing London.W.14 as the address.

I discovered this too late to correct in a later cable. Anyway you doubtless arrived! Constance is writing her memo. With best wishes for the New Year to you + your family,

Sincerely yours,

G. Melhuish

My dear Major – & Mabel –

He has been a dear good boy! – & we just loved having him here – and when he went off – he left things to fill all our stockings – & we are all most grateful & love our gifts – knowing they are "Ultimates" from affection! That bringing forth into something tangible – of our love for each other! –

Well – George had his gift forcibly removed from his desk, before he had a chance to see it – & and he found it in his stocking on Christmas morning – as I found my adorable little "Georgian" picture – & Betty her beautiful silk stockings – She is so pleased – & so am I – & we all send our best thanks to the dear "Major", & to Mabel, for having brought such a very dear person into the family!

Oh, isn't it good to get him back – Mabel Dear – & I hope soon, you will have your big son – too – but I'll not give up until he is in America again.

Thank you, "Will" – for the good letter from the *Mauretania* – It was good of you to write – We just loved having you here – & wish we might have seen more of you! –

Jack has gone to Brighton to visit Irene & her parents – She & he return here tomorrow & we are all going to the Drury Lane Pantomime in the evening – Betty has gone to Essex – to visit friends who have a most beautiful place – with a deer park – & everything lovely - She is to be gone about six days. –

This reminds me of Kipling's description of a Lie: –

"– Not a private hansom Lie, but a pair & brougham Lie; –

"Not as little place at Tooting, but a country-house with shooting

"And a ring-fence deer-park Lie"

The house she is visiting is the "Ring-fence deer-park" kind of a house! – I hope she will have a happy time –

George is writing a line or two – & I will close – as I have been writing a lot of letters – & must write to Mother now – My dearest love to you & yours – And all good wishes for a Happy New Year – and years to come!

Always your devoted old sister-in-law – or some-kind-of relation –

Constance Melhuish –

"The effect of the war in Britain was catastrophic: a whole generation was destroyed that might have furnished the country's jurists, scholars, administrations, and political leaders. … America's only comparable catastrophe would seem to be the collapse of the stock market in 1929 and the ensuing depression. … Americans, while they fought enthusiastically (some said well), were on the front lines for only about six months. … British dead finally amounted to almost a million, American to a 'mere' 48,000."[129]

"Over the course of the war, nations developed effective military medical systems that instituted preventive changes like vaccination, established efficient evacuation chains, and pioneered novel therapeutic interventions to treat the sick and injured. … The military experience in World War I profoundly shaped the medicine practiced on the battlefield … By 1917, surgery became increasing aggressive across multiple specialties as laparotomies and craniotomies emerged as standard of care."[130]

History of the Pennsylvania Hospital Unit

Part Three

After the War

Epilogue

After World War I, William Shoemaker became Professor and Chairman of Ophthalmology at the Graduate Hospital of the University of Pennsylvania. He eschewed garlic and onions, saying that an ophthalmologist must not have bad breath, yet his portrait shows him with a cigar in hand. The portrait, which is now in the possession of his great-granddaughter Barbara (Johnson) Riley, shows him seated, wearing his characteristic brown suit. He has a scar on his left cheek, the result of an abscessed tooth in World War I. The portrait is rather dark, as is his visage. His daughter Barbara referred to his picture in jest as "The Big Bad Wolf," for she used to creep past it on the hall stairway when she came in at night. His children teased him about his gruff appearance, and he took this teasing with good humor. His portrait also appears in a mural in the Lankenau Hospital of Philadelphia.

Dr. and Mrs. Shoemaker's other five children grew to adulthood, were married and produced a total of 16 grandchildren. Dr. Shoemaker was the "eye doctor" for all of his children and grandchildren, including Barbara's daughter, Helene, whose strabismus ("squint") he treated. It is ironic that this prominent ophthalmologist carried the genes for green-brown color blindness, which passed to at least two of his great-grandchildren, Helene's sons James and David.

Mabel (Warren) Shoemaker died of recurrent carcinoma of the ovary in 1932 at the age of 62, ten years before her husband's death. She was cremated. Mabel died before all of her grandchildren were born, and she was not well known to many of the others, for they were young at the time. From photographs and the recollections of her grandchildren, she appears to have been a quiet, calm person. She dressed conservatively, and when she was young her hair was dark, with a small bun at the back. Mabel was in her 20's during the last decade of the nineteenth century – the decade known as "The Gay Nineties," when "Casey would waltz with the Strawberry Blonde, and the band played on." This was also the height of the Victorian age, and the conservative manners and dress of Queen Victoria were taken as a model by many women on both sides of the Atlantic Ocean. Mabel Warren may have been intrigued by the "Gay Nineties" excitement that was reflected in song and dance – even if it was only an ankle showing below a long, tightly girdled dress. However, growing up as an orphan in the home of a very wealthy and strongly religious family, her personal life must have been greatly constrained, and thoroughly proper. The swift and glamorous victory of the United States in the Spanish-American War occurred when she was a young mother with two children. When she saw her husband off in uniform twenty years later, it was again to be followed by a triumphant conclusion: The Great War was the "war to end all wars." Mabel died in 1932 during the Great Depression, and before the world was again at war. Her husband lived to see it coming, and to see some of its impact on their children and grandchildren, although when he died, in 1942, the Second World War was far from over, and the Atomic Age had not yet begun. Winston Churchill wrote in 1948, that from the start of the Great War in August 1914 until the end of World War II in August 1945, it was "another Thirty Years' War."

Maps of France with Locations Mentioned in Letters

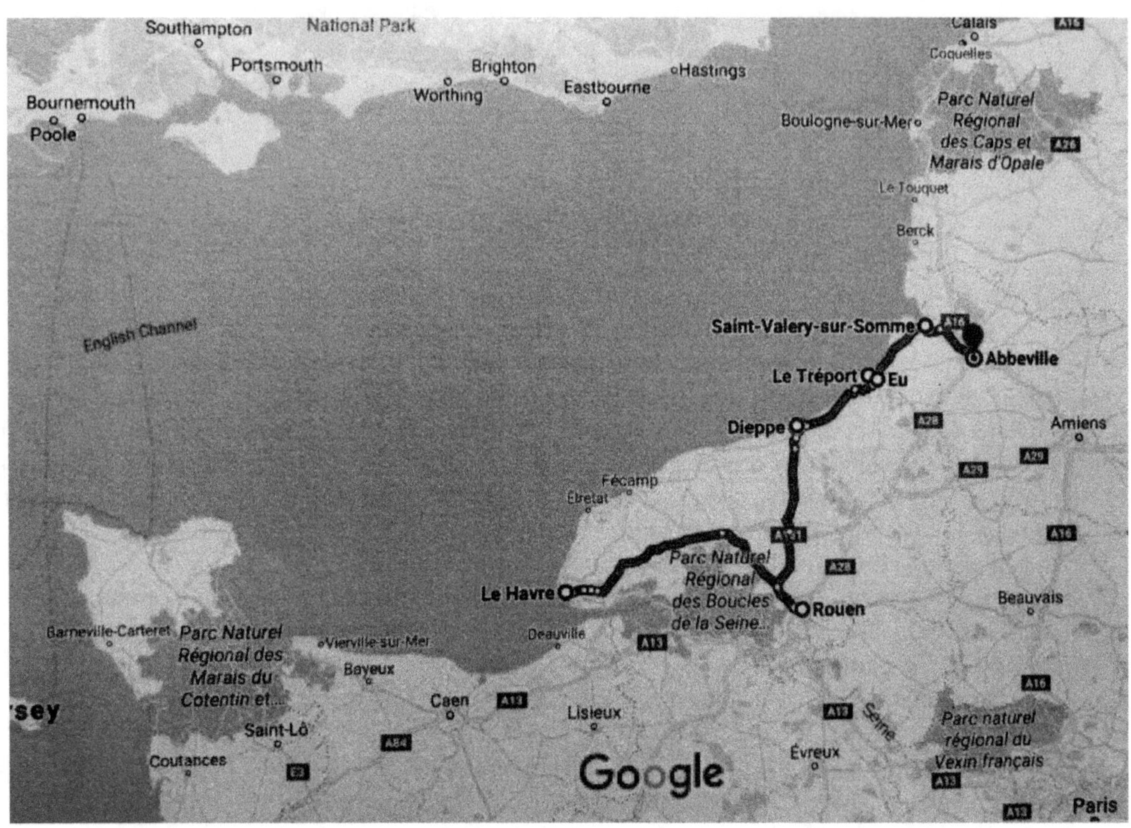

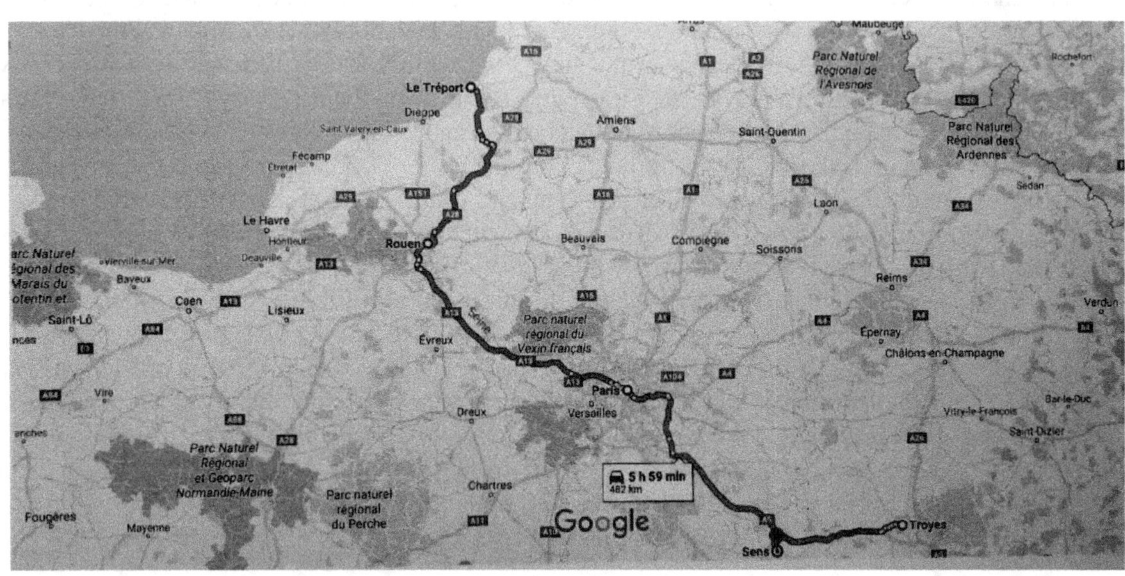

Notes

[1] Winston Churchill, *The World Crisis*, vol. 1, *1911-1914*, 2-4.
[2] George J. Hill, *Quakers and Puritans: The Shoemaker, Warren and Allied Families. Ancestors and Descendants of William Toy Shoemaker and Mabel Warren, Who Were Married in Philadelphia in 1895* (Berwyn Heights, Md.; Heritage Books, 2014), with additional notes about research on the Warren and Shoemaker families; Betsey Warren Davis, *The Warren, Jackson, and Allied Families* (Philadelphia, Pa.: J. B. Lippincott, 1903); and Benjamin H Shoemaker, *Genealogy of the Shoemaker Family of Cheltenham, Pennsylvania* (Philadelphia, Pa.: J. B. Lippincott, 1903).
[3] *Science News* 143 (19 June 1993):390: "It starts as an inability to distinguish features at night. Before long, RP leaves its victims with tunnel vision or no sight at all. Though a cure for this inherited disease remains elusive, Boston-area researchers reported this week what appears to be the first successful treatment. Large daily supplements of a particularly stable form of vitamin A slowed the inexorable degeneration of retinal function that characterizes RP." The research was led by Dr. Eliot L. Berson of the Harvard Medical School.
[4] Warren Brothers became the largest company of its type in the United States, and it established subsidiaries in many other countries. One subsidiary, Kaiser-Bechtel-Warren, was one of the six companies that built Hoover Dam on the Colorado River in the 1930s. Warren Brothers merged with Ashland Oil in 1966.
[5] *World War I: A Captivating Guide* (2019), 6-7.
[6] Coningbsy Dawson, *Carry On: Letters in War-Time* (1917; New York: Wallachia, 2020).
[7] S.S. *Westernland*, Capt. Mills commanding, on which WTS had sailed to England many years before, was built for the Red Star line in 1883. She was Red Star's first steel-hulled ship.

Emanuel Swedenborg, *The Path of Life* (Philadelphia: J. B. Lippincott, 1913) is the most important writing in the Church of the New Jerusalem, which was founded by Swedenborg, and to which MWS's family belonged.

WTS's end note, "Written but not read," and his very poor penmanship, suggests that this letter was written in the dark and could not be read by him.
[8] WTS indicates in this letter that his son Warren would have his permission to go into military service, and "the sooner the better," if his mother approved. WTS believed that he might be drafted into service before the end of the war. Warren did enlist, however, and so did his brother, Ted. Warren served in Europe, but Ted had surgery and was discharged from service.
[9] The Mersey River is the gateway to the port of Liverpool. Maj. Richard H. Harte, MRC, USA, was Director of the Administrative Staff of B.H. No. 10. See Roster, *History of the Pennsylvania Hospital Unit (Base Hospital No. 10, U.S.A.) in the Great War* (New York: Paul B. Hoeber, 1921), 204. The S.S. *Mongolia* sailed from New York with the Chicago Unit at the same time that B.H. No. 10 left New York. A practice firing of its guns accidentally killed two nurses. For "vigilant watch," the periscope, and the destroyer, see *History*, pp. 35-6.
[10] The *Saint Paul* had been launched to serve as an auxiliary cruiser in the Spanish-American war, and was in heavy action. After that war, it was a trans-Atlantic cruise ship. The ship was re-commissioned as a troop carrier in April 1918 as the *St. Paul* (ID-1643); it was scrapped in 1923. The St. Louis Unit was B.H. No. 21 under Dr. Frederick Murphy; the orthopedic surgeons were led by Dr. Joel Goldthwaite; and the civilian passengers included Dr. Thomas W. Salmon, a neurologist (*History*, p.35).
[11] *History*, 43: The men who were ambulatory from hospitals wore distinctive blue uniforms; and the service at St. Paul's Cathedral was followed by "an open air investure by the King [George V] at Hyde Park." The U.S. Ambassador was Walter Hines Page. The program for the event was saved and is in WTS's wartime letters. It is 8 double-sided pages, 8x6in.
[12] MWS, perhaps somewhat reluctantly, approved of her boys' applications to go into military service, but WTS chided her for believing all that the recruiter told her about their prospects.
[13] WTS writes to his son Ted, saying that Ted may have to go into service, if the war doesn't end. And if so, Ted must be prepared to do his duty. A "Great Game" is mentioned by WTS. At that time, the Great Game meant the conflict between two Empires, Russia and Britain, in central Asia, but we see here that it had a wider use. In this instance, the Great Game is a conflict between a German submarine and a slow-moving passenger liner, with American destroyers spoiling the Game.
[14] The final letter written on the night of May 31 to his daughter Barbara is nearly illegible, with nonsensical language and many spelling errors. He was probably very fatigued by the time he wrote it.
[15] Mabel and the children wrote letters to WTS in May, not knowing of the events that took place on the trip across the Atlantic. However, passenger ships were at risk, and of course they worried about him. In contrast to the letters of WTS, the excellent penmanship of Mabel and the children make their letters easy to read. Some of the names mentioned cannot be identified, but I will transcribe them as written.
[16] The Americans soon began to wear "Sam Browne" belts, and WTS would wear one on his formal picture in uniform. The belt was designed to allow a heavy pistol to be carried at the waist, with a strap that went over the shoulder to support it.
[17] St. Dunstan's Hostel for Blind Soldiers and Sailors was founded by Sir Arthur Pearson in 1914; now is "Blind Veterans UK."

[18] Mr. Andrew Stanford Morton, F.R.C.S.Eng., was "in his day, unsurpassed as an operator on the eye." He died in 1927. Hospital mergers in London have eliminated the Royal Westminster Ophthalmic and Moorfield's as independent institutions.
[19] "Mr. Lot" is another one of WTS's nicknames for German submarines; it refers to Lot, who lived in the evil city of Sodom (Genesis 19). The story of the Unit's crossing the Channel, the submarine, and the destroyer are also in *History*, 45.
[20] The numbers, 10,000,000 and 2.8 million, look like wild estimates. "Crazy Bill" was his derogatory nickname for Kaiser Wilhelm II, the Emperor of Germany. A popular song in 1914 began with "Kaiser Bill went up the hill to take a shot at France." "Boche/s" = Derogatory slang for Germans, from French *Alboche*, contracted from *al(lemand)* German + *ca(boche)* blockhead.
[21] The operation in which the gall bladder is completely removed (cholecystectomy) was rarely performed in those days; Burnham, who was Mabel's cousin, probably had a safer procedure (cholecystostomy), in which the gall stones were removed but the gall bladder was left in place. The physicians mentioned were notable: Dr. John B. Deaver (1855-1931) was the John Rhea Barton Professor of Surgery at the University of Pennsylvania; he invented a surgical retractor which is still in use. Dr. John Hinchman Stokes (1885-1961) was the revered Head of the Department of Dermatology at the University of Pennsylvania. Dr. J. Lincoln Furbush, noted for his work on yellow fever, was originally from Philadelphia; in 1917 he was a colonel with the Medical Department of the Army in Washington.
[22] The Aldine was a Philadelphia hotel located at 1910-1922 Chestnut Street; its principal client was J. B. Lippincott Co., the publishers, who were related to the Shoemakers.
[23] Their friend, Walter Clarke Rodman, was a well-known poet at that time; his *Twilight and Other Verse* (1915) is still available as a reprint in hardcover and paperback. It was dedicated to L.N.R.: "To one whose gentle spirit is to mine / companion still, though sight and touch no more / May recognize her present, I devote / and dedicate this record of my thoughts."
[24] Dr. John H. Gibbon, Sr., was B.H. No. 10's Chief of Surgery. Dr. John H. Gibbon, Jr., designed the first heart-lung machine.
[25] This letter shows that Will's son, J. Warren Shoemaker, known as Warren, had the mechanical aptitude that had been in the Warren family for several generations. "Uncle Ralph" was his mother's youngest brother; she was her parents' 8th child; he was the 10th, a bachelor. "Uncle Josh" was Joshua Lippincott Shoemaker, one of his father's older siblings. Charles (mentioned previously) was the first of the nine children of Julien and Hannah Shoemaker; Josh was the second; and William was the eighth.
[26] Camp Becket, a YMCA summer camp for boys in the Berkshire Mountains of western Massachusetts, was founded in 1903. The second director, Henry Gibson, is mentioned by Ted in this letter. He espoused values of hard work that are still taught there.
Harriet Beecher Stowe's pre-Civil War novel, *Uncle Tom's Cabin*, was later parodied in crude minstrel shows.
[27] Mentioned in this letter: The dentists, Charlie and Ed, are his relatives: 1st Lieutenants Charles S. Jack (a cousin) and Edwin Shoemaker (a nephew). The other "dandy fellows" are Captains Arthur Newlin, Edward B. Hodge, and J. E. Sweet; and 1st Lieutenants William B. Cadwallader and Norris W. Vaux. Mrs. Packard was probably the wife of Capt. Francis R. Packard.
[28] The first paragraph is scrawled, with incomplete, illogical sentences; he is obviously upset and worried. The rest of the letter continues with much of the same nonsense. Nothing more is known about those named as patients and others who owe him, or who he believes are dodging the "need" to serve on duty in the war. Near the end of his letter, he wonders if he will ever come home, and writes about the "gloomies," but he then says that "I feel better, after writing all of this."
[29] "Fridays" refers to Friday Evenings, an invitation-only social event in Philadelphia.
[30] Ed and Charlie = Edwin Shoemaker was his nephew and Charles Jack was his cousin. Henry and Anne cannot be identified.
[31] https://www.poconorecord.com/article/20080704/LOCALENT/807040321 (posted 4 July 2008): "Camp Owaissa was on the northern shore of Lake Naomi and was started in 1916 and continued as a premier camp for girls until 1966. In addition to water sports, crafts and general camp activities, it had its own horse stables and featured equestrian activities. Today, the former camp is part of the Lake Naomi/Timber Trails community." Billy was killed in a tragic accident in 1916 while crossing train tracks with his brother Bob. Cousin Ethyl was the daughter of Barbara's uncle Charles Shoemaker.
[32] Names in this letter: *History*, 205, shows the names of Amina Furhmann (Nurse) and 1st Lt. Arthur H. Gerhard, Medical Reserve Corps, U.S. Army. WTS's brother Owen Shoemake, his cousin Burnham, and Dr. Deaver are mentioned previously. Nothing more is known of Mr. Rutan, Miss Hale, or Mr. Levin.
[33] Lightly edited with a few spelling and punctuation errors corrected. Bob became an inventor, with a hobby as a printer.
[34] 48 years = WTS is making a pun; he was born in January 1869, and was 48 years old at the time that England's King, Queen, and Prince of Wales were scheduled to visit his hospital in France. King George V and Queen Mary were the parents of the debonair Prince of Wales, who succeeded his father in 1936 and became King Edward VIII. Determined to marry Wallis Simpson, he abdicated and became the Duke of Windsor. He was succeeded by his brother as King George VI (of "The King's Speech"). The latter was the father of Queen Elizabeth II. The sequence will be familiar to those who view recent English films, such as "The Crown."
[35] Already depressed by what he has seen, WTS refers to Civil War General William T. Sherman's comment, "War is Hell."
[36] Mentioned previously: Goldensky's photo studio; Walter Burgess Warren, Mabel's brother; Florence Shoemaker (b. 1895) is the daughter of WTS's brother Joshua; Ethyl Shoemaker (b. 1892) is the youngest daughter of WTS's brother Charles; her cousins Ed and Charlie are dentists with B.H. No. 10. *History* shows that Dorothy's friend Lawrence Jones is one of the young men in B.H. No. 10. He was named Pvt. Herbert L. Jones on the Unit's Roster, and as Pvt. L. H. Jones, when he played a mandolin and was "Marjorie Goodly" in the cast of "What Happened to Jones" – a humorous comedy, which was performed on 13 March 1918 in Dieppe. The Lankenau Hospital is still in existence, and WTS is shown there in a painting of a group of the hospital's physicians. Dr. Le B. = Dr. Theo. Le Boutillier. Nothing more is known about Dodner's photography studio.

[37] *History* shows that Major Matthew A. DeLaney, Medical Corps, U.S. Army, was Commanding Officer of Base Hospital No. 10. Will Downs is Pvt. Stephen W. Downs; Tom Barratt is Pvt. Thomas L. Barratt. The Wharton School is the business school of the University of Pennsylvania. The "Main Line" is the Pennsylvania Railroad's commuter line that runs from Philadelphia to west to Paoli; the term also refers to the wealthy suburbs and its resident who live in towns along that branch of the railroad. It is not clear what WTS mean by the "pot of Orangia." It is certainly not Orangina, a bottled beverage which was first made in 1935. "Charlie" is his cousin and roommate, Dr. Charles Jack, a dentist in Base Hospital No. 10. "When It's Apple Blossom Time in Normandie" (1912) was a popular song in World War I; it was re-recorded and was popular in World War II. WTS used shorthand for Love You: ly, ly.

[38] The Stetson Hospital is still in existence in Philadelphia. Creighton = William J. Creighton. Nothing more is known about Dr. McGinn. The clipping that Miles mentioned was not preserved. She was a woman, although she used Ray as a given name.

[39] The crowd of Dorothy's aunts and cousins at Cape May includes: "Aunt Florie" is Florence (Hey) Shoemaker, the wife of WTS's brother Joshua; "Florence" Shoemaker (b. 1895) is her daughter and "Cousin Low" is probably her older brother, Louis Jack Shoemaker. "Lucretia" is her cousin, the daughter of WTS's brother Charles and his wife Lucretia (Hay) Shoemaker. *History*, p. 208, shows Pvt. Fenno Hoffman, and also (probably related) Pvt. William A. Hoffman. By the end of the war (pp. 232-3), Fenno had been promoted to Private, 1st class, and William to Sergeant.

[40] F.S.S = Friends Select School. Its website says it is "a college-preparatory, Quaker school for pre-kindergarten through 12th grade located at 1651 Benjamin Franklin Parkway ... in Center City, Philadelphia."

[41] Harvey, Owen and Josh are WTS's brothers. Caroline is Harvey's daughter, named for their sister Caroline (1867-1906).

[42] *History* shows Privates Samuel H. Chapman, Jr., and Frank T. Chalk were promoted to Private First Class by the end of the war. Bob's "Seen and Heard" was not preserved. The Girls Scouts formerly had a summer camp on Lake Yokum. No trace of it remains on the internet. Two postal cards were also mentioned by Barbara in her letter of 15 August.

[43] Mentioned in this paragraph: WTS's son Bob; Catherine Warren, who is the wife of Mabel's brother, Herbert; WTS's daughter Dorothy; Mabel's brothers Ralph and Walter; and WTS's children Barbara, Ted, and Warren.

[44] Major Richard Harte, Director of Base Hosp. 10; Capt. William J. Taylor, MC; Chaplain Edward M. Jeffreys, American Red Cross. Mary is the wife of Charles Jack. None of the other names in this letter can be identified, except for WTS's family.

[45] What WTS called the "Great Drive" was the Third Battle of Ypres. "The British and French attack on a 15-mile front in Flanders; take 12 villages and claim 5,000 prisoners." See: https://www.firstworldwar.com/onthisday/1917_07_31.htm. However, the next day, a German counter-attack took place. See: https://www.firstworldwar.com/onthisday/1917_08_01.htm, "Germans counter-attack, retake St. Julien and regain some positions in Ypres-Roulers railway district."

The phrase "inhumanity to man" appeared in Robert Burns' poem in 1794. It appears again in *Grapes of Wrath* (1939).

[46] Hunter Watt Scarlett, M.D. (1885-1954) was an All-American football player at Penn, and he graduated from the University of Pennsylvania School of Medicine in 1911. He became an ophthalmologist, and volunteered for service in 1914 with American Hospitals in France. He served in heavy combat, on the front lines. He was repatriated for illness in 1916, thought at first to be meningitis, but it was probably sleeping sickness. He later served with U.S. Hospital No. 11 in New Jersey. In 1920, he was awarded the *Médaille de la Reconnaissance française* from France, and he was an ophthalmologist at the Pennsylvania and Bryn Mawr Hospitals. In this letter, Scarlett tells of his discovery of the organism that was named for him: "Diplobacilli scarletti."

The subsequent history of Scarlett's discovery is told in Sverre Dick Henriksen, "*Moxarella, Acinetobacte*r, and the *Mimeeae*," *Bacteriological Reviews* 37 (no. 4, Dec. 1973), 522-561; 522-3: "The original culture has been lost, but "Bovre studied a collection of strains believed to represent Scarlett's species, now known as *Moraxella nonliquefaciens*."

[47] Morgan Harjes was in Paris. It was the French banking firm associated with the New York bank, J. P. Morgan. The letter that WTS refers to, written on 8 August, was not preserved. Many letters from WWI were lost – some because they were mis-addressed, and others because they were lost at sea. WTS refers to his colleague, Captain Arthur Newlin. His deposit of 342 francs is equivalent to $68, if the exchange rate is calculated at 1 franc = .20 US cents. In July, WTS had requested that his wife make a transfer of $75 in Philadelphia to his nurse's account, and she would lend him the same amount in France.

[48] MWS writes of many visitors to her summer home. Those who are members of her extended family are WTS's brother Owen (and his wife Margareta "Greta") and his brother Harvey (and his children Caroline, Katherine, Alice, Margery, and Arthur). Ann and Marion are also implied to be his children, although I cannot confirm this. Nothing is known about the other guests and business associates. J.L.&Co. is the Lippincott publishing company. *Jaeger* = hunter (German). Formerly stout trousers, used for hunting; now a brand name.

[49] *History*, 69-70: "In reply to Colonel DeLaney's urgent request, a reinforcement comprising 8 officers, 47 enlisted men, and 30 nurses, was sent from Philadelphia. The men under command of Lieutenant H. B. Wilmer, sailed on August 18, 191 7, on the S. S. *Aurania*, but the nurses were delayed and did not sail until August, 1917, when they embarked under command of Captain J. Paul Austin on the S. S. *Baltic*. Lieutenant Wilmer and his detachment had an uneventful voyage, landing at Liverpool, thence proceeding by train to Southampton, from which port they crossed the Channel to Havre, and from there were sent to Dieppe, where they were met by ambulances which took them to Le Treport, arriving on September 7. Captain Austin and the nurses arrived on September 21." Ralph Straub went to France as a Private; he returned as a Sergeant at the end of the war.

[50] Sackaley Hall has disappeared from Ocean City, N.J., without a trace. The postcards sent by WTS were not preserved.

[51] Dr. Milton Howard Fussell (1855-1921) was an acknowledged expert on tuberculosis at that time; his wife was Isabell.

[52] Born 4 October 1899, Ted was 17 years old at this time. Charlie and Ed were his cousins with B.H. No. 10. Hudson Chapman was Pvt. Samuel H. Chapman, Jr.; he was promoted to Pvt. 1st Class at the end of the war.
[53] Ralph Warren was Mabel's youngest brother; he had no children. Catherine (née Winter) was his sister-in-law, married to Herbert Marshall Warren, Jr. Jesse Warren was a brother; his wife was Margaret (née Schlegel); their son Jesse was drafted.
[54] Names in this letter = George Warren is Mabel's brother. Lt. Edwin Shoemaker is the son of WTS's brother, Owen. He was very popular with the younger officers and nurses, but WTS thought he was not a serious professional. WTS, however, admired Lt. Charles Jack, a nephew of Owen's wife, Margareta "Greta" (nee) Jack. Lt. Harry Bond Wilmer was in charge of the support unit that arrived; he was promoted to captain by the end of the war. Henry W. LeBoutillier came as a private and was promoted to sergeant at war's end. Majors Matthew A. DeLaney, commanding officer, and Richard H. Harte, director, of B.H. No. 10.
[55] WTS is sad and sarcastic in this letter, returning many times to the disaster of war, especially this one; and of the mis-information, the slackers, and those who profit from the service of others. Many of those named in this letter cannot be identified, but George Warren (Mabel's brother) is the "publicity bureau"; Charlie Jack's wife is the anxious "Mother"; and WTS's brother Harvey is "Sir Harvey."

Sir William Osler's son Revere, a 2nd Lt. in the Royal Field Artillery, was wounded in France and died the next day, 28 August 1917, in spite of surgery performed by the famous Boston surgeon, Dr. Harvey Cushing, and a blood transfusion given by Dr. George Crile, co-founder of the Cleveland Clinic. The "Boche" (singular) referred to the German Army (plural). WTS paraphrases his own previous comment, when he wonders, "who said 'This is an awful war'."
[56] Hun = Disparaging term for Germans, derived from the Turkish people who settled east of the Rhine. The most famous was Attila the Hun in the 5th Century.
[57] Mentioned in this letter: Eu, France, is about 7 km inland from Le Treport.
[58] Some corrections were made in punctuation and spelling. Ted mentions Privates Samuel H. Hudson, Jr., and Frank T. Chalk, who were in B.H. No. 10.
[59] WTS is sarcastic about Mary Jack. She is the wife of Dr. Charles Jack, half-brother of his brother Owen's wife, Margareta "Greta (née Jack). WTS refers to her as Lamenting Mary (from the Frankfurt folk tale about the Mother of Christ) and as Medea (a half-goddess Sorcerer and wife of Jason, who killed his children). He mis-spelled it as "Media," a town near Philadelphia; I have corrected the spelling. WTS was fond of Charles Jack at this time, who e was his roommate at Le Treporte, but he was skeptical about Owen's son, Dr. Ed Shoemaker, who he thought was a rich, light-headed playboy. As the war continued, Charles Jack also got on his nerves, too. *History,* 249-50: excerpts, humorously mentioning WTS ("Bill") and the dentists Charlie and Ed. Told in verse, with each letter of the alphabet about a member of the Unit. At the end of the war, WTS and Ed were promoted to Major, and Charlie to Lt. Colonel.

AN INDEX OF THE UNIT
OR WHO'S WHO IN THE OFFICERS' MESS
A Poetical Effusion read at the Nurses' Thanksgiving Day Party, November 1918

B stands for Bill, "Uncle Bill" being meant,
Who holds down Ward 12 with terrific intent.
Tea is served at all hours; drop in, you'll be glad
You'll find in addition good grub there, by Gad . . .

E stands for Eddie, the slippery, the beau,
The pride of our Unit, whom all of us know.
He's a sight for the Gods when he's dressed up to "kill,"
When he goes out to dance, or to dine "Uncle Bill" [Shoemaker] ...

Also **J** stands for Jack, of Chivalrous renown,
Who takes up the gauntlet ere it is thrown down.
Who don't like remarks made by bachelors bold,
—But the rest of this story had best not be told.

"Miss Hill" = Barbara graduated from high school at Miss Hill's School.
[60] The censor redacted the name of "our good ship" (probably the *St. Paul,* on which the Unit arrived). If so, the rumor was wrong; the *St. Paul* survived the war. "God's will be done" is a paraphrase of Matthew 6:10 (KJV)
[61] The "Mutt and Jeff "cartoon (1907-1983) is said to have been the first daily comic strip; the piece that WTS mentioned was not preserved. The Penn Charter School was founded in 1689. It is still in existence at 3000 School House Lane in Philadelphia.
[62] In this letter, WTS mentions their maid, Emeline, a fair skinned African American woman, who was a part of their family. He wants MWS to explain her parentage to Barbara, who has apparently wondered about it.
[63] Mentioned in this letter: 1st Lieut. William B. Cadwallader; 1st Lieut. William Drayton, Jr.; Consulting Ophthalmic Surgeon Colonel William Lister (1868-1944), Lord Lister's nephew.
"Fritz" = Derogatory slang for Germans, derived from the nickname for Friedrich. Boche and Fritz are nicknames for German Army soldiers.

[64] Aunt Lulu is probably Laura Shoemaker, wife of his uncle Harvey Shoemaker.
[65] *History*, 204: 1st Lieutenant Edward Bell Krumbhaar, Medical Reserve Corps, U.S. Army; 231: he returned with the Unit as a Major; 34: "We were also fortunate in having the services of Mrs. Edward B. Krumbhaar as a laboratory technician, who was granted permission, by the Secretary of State, to accompany her husband Capt. Krumbhaar, the pathologist"; 206: Secretaries ... Krumbhaar, Helen D., American Red Cross; 240-6: "Laboratory of Base Hospital No. 10" shows the work that they did, including the diphtheria epidemic, preparation for evacuation during the German offensive in the Spring of 1918, and the six publications in scientific journals that resulted from this work; 251:

K stands for Krumbhaar; we've two in our store
And I'm sure you'll agree, that we wish we had more.
For they've done their duty, done more than they should,
Or to put it quite tersely, they've more than "made good."

Also see "Pathologist Edward Krumbhaar Dies at 83," *JAMA* 196 (no. 2, 1966): 39. He was a founder of the American Association for the History of Medicine and was president in 1934; a pathologist, he held the presidencies of the American Society for Cancer Research and the Philadelphia College of Physicians, and several other professional groups.

WTS comments on MWS's reference to "Warriors of Hell," who were perhaps Sikhs. WTS suggests that his family read Arthur Guy Empey, *"Over the Top" by an American Soldier Who Went* (New York: G. P. Putnam's Sons, 1917). He refers to the Red Cross armband on his formal uniform (cover photo), which must have been taken before he went to Europe. However, the Sam Browne belt was not worn until the Unit arrived in Europe, so this is puzzling.
[66] The Allies were finally victorious at Passchendaele on November 10, 1917, one day before this letter was written.

WTS mentions several people in this letter who are apparently from Philadelphia, but who are not in B.H. No. 10: Jones, Langdon, "R. & C." and Isaac. "Medea" is his derogatory nickname for Charles Jack's wife, Mary. The British, French, and Americans had serious problems with their Russian and Italian Allies in World War I, as alluded to in this letter.

On WTS's comments on Russia and Italy: "In late October 1917, Germany intervened to help Austro-Hungary, by moving seven divisions from the Eastern Front when Russia withdrew from the war. This resulted in a victory over the Italians in the Battle of Caporetto (otherwise known as the Twelfth Battle of the Isonzo). When the battle had run its course, 11,000 Italians were dead, more than a quarter-million had been taken prisoner and Italy had retreated well behind their original lines. Caporetto was an unmitigated disaster and the whole Italian front along the Isonzo disintegrated, sparking a crisis in Italy. With the Central Powers now threatening Italy's territory, the Government changed tactics and implemented more defensive military strategies, replaced the Chief-of-Staff, improved soldier morale and Allied troops arrived (mainly British and French) to reinforce the front." (https://www.history.co.uk/italy-in-wwi)
[67] Ruth Chatterton (1892-1961) was a well-known actress, and one of the first women to become a pilot.
[68] Charlie = Dr. Charles Jack, the step-brother of Greta (Jack) Shoemaker, wife of WTS's brother Owen. 5 fr = ca. $1 (at what was then the current exchange rate, shown in the previous letter to be 5.4 fr. = $1.00). The British government paid "2/6" = a "half-crown," 2 shillings + 6 pence (2s6d). There were 20 shillings per pound and 12 pence per shilling. The value of a pound was set at £1 = $5.00. "Josh" = WTS's older brother Joshua Lippincott Shoemaker. His "mother" = Hannah (Hester) Shoemaker. 109 = his home, 109 So. 20th St., Philadelphia. Something was redacted at the end of his letter on Sunday, with only parentheses remaining ( ). WTS's word *ratiner* is probably French, but cannot be translated.
[69] The Hindenburg Line was a 140-mile segment of the line of battle that extended from the North Sea to Switzerland. The battle at Cambrai in November 1917 produced a temporary break in the Line, but the Line was re-established by the Germans and remained intact until October 1918. WTS refers sarcastically to several people whose names are unknown, including the American "Prince." Martha is the wife of his nephew Ed Shoemaker. Jane, who wrote the enclosure, appears to have been the governess for WTS and his siblings. Her handwriting is awkward, and I have corrected the capitalization, but I left her sentences intact. WTS recommends the book *German Spy in America* by John Price Jones.
[70] *History*, 88. Photo on unnumbered page, following p.86.
[71] Padre Jeffreys = Chaplain Edward M. Jeffreys, American Red Cross. Les Norris = Maj. George W. Norris. Aunt Caroline = WTS's sister. Ralph = Pvt. Ralph Straub, who is probably related by marriage in some way to the Shoemaker family. Kidwell = Capt. H. L. Kidwell, Quartermaster Corps.
[72] WTS continues to be angry about the stay-at-homes and their spurious reasons to avoid the war, mentioning "wobbly spine" and Quaker mentality. He is not optimistic, saying that the war may go on for four or five years. The First Casualty Clearing Company completed its work in December, having been deployed to the front from B.H. No. 10. Nurse Helen Fairchild became ill near the end of December at the Clearing Company and died in January after being evacuated back for surgery at Le Treport. The enormous explosion on December 6 in Halifax, Nova Scotia, was the result of a collision of the *Mont-Blanc*, a French cargo ship carrying munitions, with a Norwegian ship, *Imo*, not German sabotage. Blame was placed on the *Mont-Blanc*. The success that was hoped for at Cambrai was dashed when the battle was called off on December 7.
[73] "Charlie Jack ... is not my kind" shows that WTS has now soured on both Charlie Jack and Ed Shoemaker.
[74] Their son WTS, Jr. (Billie) was hit by a train and killed while walking with his brother Bob in 1916.
[75] Mentioned in this letter: The Westcotts are distant cousins of WTS. Edwin B. = probably Edmund Boeriche, who later married Dorothy. The first name is difficult to read, but first letter of the surname is clearly "B." Edmund Boeriche and his sister Margaret were orphaned before he was six. They grew up in their grandparents' home, with their five uncles and three aunts. The uncles married, so Ed Boeriche had three Aunts and five Aunts-in-law in Philadelphia.

[76] Miss Dunlop had been Directress of Nursing at the Pennsylvania Hospital; she was Matron in charge of Nurses at B.H. No.10. Sad to say, the letters from MWS do not show the numbers that are referred to, and many were not preserved. There is now a long gap in the letters that WTS saved and/or brought back to Pennsylvania from MWS and the children.

MWS's 48th birthday was on January 5, 1918.

[77] WTS's 49th birthday was on January 22, 1918. The actor John Martin Harvey was knighted in 1921; he was then known as Sir John Martin-Harvey. Dr. Burton Chance was another ophthalmologist in Philadelphia, and a competitor of WTS. The *Pennsylvania Medical Journal* 22 (Nov. 1918), 103, shows that he discussed a paper on war injuries to the face and orbit in September 1918.

[78] President Wilson delivered a message to Congress on January 8, laying down his "Fourteen Points"; four more would be added on February 11. William Taft and General Leonard Wood were Republicans; Wilson was a Democrat.

The comments about Ed and Martha and "the Pope" are facetious. Martha is the wife of WTS's nephew Ed Jack; 2011 is the address of WTS's older brother, Joshua, so he is "the Pope."

[79] "Greta" is the wife of WTS's brother Owen; she was the mother of the dentist, Ed Shoemaker, in B.H. No. 10. Josh is sarcastically referred to as WTS's "esteemed" brother, Joshua Lippincott Shoemaker. "2011" is the address of Josh's house in Philadelphia. WTS was especially scornful of Josh's wife, Florence, known as "Aunt Florrie."

Harry Augustus Garfield, son of President James A. Garfield, was Woodrow Wilson's Food Administrator and Fuel Administrator. He faced difficult decisions. In January 1918, he instituted what were known as "idle" Mondays.

WTS's comments about "our convoy" being "very lucky" undoubtedly refers to the sinking on February 5, off of the coast of Ireland, of the British ship, S.S. *Tuscania*, carrying U.S. troops.

The term "Veil of Life" is associated with the writings of Swedenborg. I cannot locate the precise book that WTS refers to.

[80] By "working women" WTS probably means those who work on the streets at night.

[81] The statues of Adam and Eve, naked, are shown in a postcard that WTS obtained on a visit on December 6 to the Rouen cathedral (see above, for that date). *The Music Master* opened in New York in 1904, and ran for two seasons, and was in revivals and on tour for two decades. David Warfield (1866-1951) originated the title role of "Anton von Barwig," whose character was modeled on Beethoven. He later starred in *The Grand Army Man* and *The Return of Peter Grimm*, and in his most famous role, in *The Auctioneer*. Maude Adams (1872-1953) was arguably the most famous American actress of her time. She was best known for playing *Peter Pan* when it opened in New York. John Willard (1885-1942), a playwriter and actor, wrote many successful melodramas for the theatre, some of which became movies. He was a Captain in the U.S Army Air Corps in World War I and flew bombing raids over Germany. John Mason (1858-1919) was an American actor whose most successful role is said on IMDB to have been as "Jack Brookfield" in *The Witching Hour.*

Dan Dietz, *Off Broadway Musicals, 1910-2007: Casts, Credits, Songs, Critical Reception and Performance Data of More than 1,800 Shows* (McFarland, Inc., Publishers, 2010), 411 (#1467), notes that the 1912 play *Peg O' My Heart* was recreated in 1967 as *Sing Me Sunshine* by Johnny Brandon. Could he have been a descendant of the Brandon who was mentioned by WTS?

"Captain Vernon Castle, famous dancer and member of the Royal Flying Corps, fell to his death at Benbrook [Texas] flying field February 15 [1918], sacrificing his own life to save that of a fellow aviator." *Sausalito News* 34 (no. 8, 23 Feb 1918).

[82] *History*, 250-2:

H stands for Harte, whom we know and revere,
To whose Labor we owe it, that we're over here.
Not back in "the States," tuning up for the dance
Instead of at Treport in warm, sunny France.

N stands for Norris, of whom it is said
The "Division Chief" business has quite turned his head.
He's so feeble of mind that he cannot decide
To wear his hair parted, at middle, or side.

S stands for Sweet, whom we all like to see,
Making rainbows and halos with Chloramine-T.
He's been made a Major by those who know best
As a fitting reward for or was it a jest?

WTS comments sarcastically as he says, "go home to read a paper before a medical society," knowing that MWS will recognize that he is referring to his competitor, Dr. Burton Chance (see January 26, 1918).

The phrase, "tears cannot restore thee" may be a paraphrase from an earlier poem by Tennyson, but the first time it appears in a Google search for precisely those words was in a magazine published in Philadelphia some twenty years before WTS mentioned the quote. In *Success with Flowers* v.3 (no.1, 1892), 146: "This inscription appears in a Wisconsin cemetery, over the grave of a woman, written by her husband: 'Tears cannot restore thee, therefore I weep'."

[83] 21March-05 April 1918 First Battles of the Somme. The German Army launched a major offensive against the Allied Front Line on 21 March 1918. After a successful advance of several miles into Allied-held territory the German attackers became exhausted after several days without relief, they were experiencing difficulties of supply to the forward lines and German Supreme Command terminated the offensive on 5 April.

History, 91: "SECOND CASUALTY CLEARING TEAM. On March 21st another Casualty Clearing team was called for and Major Hodge, Captain Dillard and Miss Stambaugh with two orderlies were sent up the line. Their experience was crushing. During the heavy retreat they were forced from one station to another rapidly until they reached Amiens. A few hours after reaching Amiens a bombing raid occurred and all members of the unit were injured—Miss Stambaugh receiving a very bad injury in the calf of the leg. She was carried immediately back to Abbeville, Amiens, being in a distressing condition. At Abbeville she was cared for, and as soon as able when traffic was allowed, she was transferred to the Queen Alexandra Hospital in London where every attention was given to her by the British authorities and she was well looked after by the American colony."

History, 249:

D stands for Dillard, who's known far and near
For his fine manners, tennis and his ways with a "dear;"
But please to remember, tho it may seem hard,
That at good Lady Murray's, he known as D*illard.*

[84] On April 7, as WTS was writing this letter, the German Spring Offensive began with the Battle of the Lys, also known as the Fourth Battle of Ypres. The German offensive was unsuccessful and they withdrew on April 29.

The phrase, "back bone of winter is broken" is in the first line of the poem "Savage Thoughts of a Winter Fiend on the Passing of the Last Winter Month" by Henry Harrison, *Infunitive* (Melomime Publications, 1923), 152. It must have appeared earlier and was reprinted in this collection.

[85] WTS enclosed a clipping: E.H. Alden (letter to the Editor), "Worshipping with Unitarians" *Messenger* (27 February 1918). "I think there may have been many of your readers who felt, as I did, a distinct shock at the announcement … that on account of the coal shortage the Brookline Society would join with the Unitarian Society in Sunday morning worship. It is inconceivable to me that there can be any sphere of true worship in such a service."

Herbert Warren Arnold is the son of MWS's sister Ella, who m. Dr. William Franklin Arnold. Mary Kille Warren is the daughter of MWS's brother Henry Jackson Warren.

WTS asks, "Was it the Pharisees who thanked the Lord that he was not like other men?" This is a paraphrase of the parable of the Pharisee and the Publican (Luke 18:9-14).

The phrase "inhumanity to man" first appeared in WTS letter of August 2, 1917. See above for comment on that note.

[86] Dr. Ralph Butler was a distinguished surgeon in Philadelphia at that time. He succeeded Dr. Burton Potts, a distant cousin of WTS, whose great-grandfather Isaac Shoemaker married Elizabeth Potts.

Edward Treacher Collins (1862-1932) was a well-known British ophthalmologist, remembered now for the rare eponymic Treacher Collins syndrome, with cranio-facial abnormalities.

Mount Kemmel, near Ypres, was first attacked unsuccessfully by the Germans on April 17-19, 1918. The Germans attacked again and captured the peak on April 24-25, but with no apparent change in the war's outcome. The mountain was recaptured by Americans in the summer of 1918, and it is now the site of a U.S. cemetery.

"Bunny" appears to have been WTS's nickname for his brother Owen.

[87] WTS is making what he intends to be a humorous comment about dates; he married MWS on June 6, 1895, not May 6. In this war, everything seems to be on the wrong date to him. WTS refers to the poem by Richard Harris Barham, "Jackdaw of Rhiems" (originally published in 1870); see Edmund Clarence Stedman, ed, *A Victorian Anthology*, 1837-1895: "The Jackdaw sat on the Cardinal's chair! / Bishop and abbot and prior were there. …"

[88] The CC of TN dated May 13 is transcribed in this text. The other CCs of TNs show transactions as follows:
Feb. 15: "We … hand you herewith Frs. 424.05 … with Frs. 1.20 insurance and postage" / Feb. 21: "received for your account from: Messrs. Drexel & Co. as from Mabel W. Shoemaker ... Fr. 568." / April 23: "We beg to hand you … Fcs. 300., … with Fcs 1.10 insurance and postage." / May 6: "We beg to inform you … from Messers. Drexel … by order of Mrs. Mabel Shoemaker … Frs. 140 – duly credited."

[89] The menu was not preserved, nor were any of MWS's numbered letters mentioned in this letter from WTS. The only scrapbook that was preserved was of Shoemaker family photos with dates from 1898-1906. The wartime scrapbook mentioned in this letter has not been located. It may have passed to one of William and Mabel's other children and was lost. "Uncle Jake" was his mother's oldest brother. Born in 1818, he would have been about 100 years old at this time.

[90] The "Chinks" were members of the Chinese Labour Corps. Chinese laborers were also recruited by the French. Some 120,000 served in Europe in World War I, with at least 2,000 dying – mostly of influenza – though Chinese sources contend that many more died, perhaps 20,000. None are recorded to have died at Le Treport, but there are Chinese graves in 40 other cemeteries in France. See also WTS's comment on June 15, 1918.

V.A.D. = Voluntary Aid Department. Organization of British nurses who provided aid in the combat zone in France. WAAC = Women's Army Auxiliary Corps. Organization of British women volunteers who served in Britain and in the combat zone, without military ranks but with distinctive insignia for leaders (Controllers and Administrators) and followers (Workers).

[91] http://www.historyofwar.org/articles/battles_aisne3.html:

"The Third Battle of the Aisne was the third major offensive launched by the Germans of the Western Front in the summer of 1918. … The battle began with one of the most intense artillery bombardments of the war. The Germans fired some two million shells in four hours on the morning of 27 May and then launched their attack with seventeen divisions. The Allied lines on the Chemin des Dames were shattered.

The Germans were able to advance thirteen miles on the first day of the battle, the single biggest advance since the beginning of trench warfare in 1914. The bridges across the Aisne were captured intact and the Germans began an advance towards the Marne.

"The Allied response to this crisis was rather better organized than during the attack on the Somme. Twenty seven divisions were fed into the line between 28 May and 3 June. The German advance continued throughout May. Soissons was captured on 28 May. German troops crossed the Marne around Jaulgonne and on 30 May reached Château Thierry. German troops were now only thirty-seven miles from Paris.

"This was as close as they would get. The advancing German troops had outrun their supplies. They had lost over 100,000 men in the battle (as did the Allies), and by the start of June had lost much of their numerical advantage. The battle also saw the appearance of large numbers of American troops in the front line for the first time. On 1 June the American 3nd Division had taken up the defense of Château Thierry, before launched a counterattack that forced the Germans back across the Marne. They were followed into action by the 2nd Division, who on 6 June attacked the German positions at Belleau Wood, to the north west of Château Thierry."

[92] http://www.historyofwar.org/articles/battles_cantigny.html:

"The battle of Cantigny, 28 May 1918, was the first American offensive of the First World War. Cantigny had been captured during the Second Battle of the Somme (21 March-5 April 1918), the first of Ludendorff's series of major offensives during the spring and summer of 1918. The village had then been fortified and turned into a German observation point. It was defended by veteran troops of General Oskar von Hutier's Eighteenth Army.

"The American attack was made by the American First Division under Major General Robert Lee Bullard. The village was captured, and then held against repeated German counterattacks on 28 and 29 May. American losses were 100 dead and 1,500 wounded, out of an initial force 4,000 strong (one infantry regiment), later increased to 8,000. German casualties are unknown, but around 200 men were captured during the battle.

"In the context of the Western Front, the battle of Cantigny was little more than a skirmish. However, it gained great significance part because it was the first combat success of the American army, after nearly a year of preparation in France, and partly because it took place on the second day of the Third Battle of the Aisne (27 May-3 June 1918). The first day of that battle had seen the Germans advance thirteen miles, the greatest distance achieved in a single day since the start of trench warfare. The American victory at Cantigny was therefore a valuable boost to Allied morale."

[93] WTS is paraphrasing a sentence from Washington Irving satire, *A History of New York* (1809), "a dark and stormy night."

[94] Excerpted from https://www.history.com/this-day-in-history/battle-of-belleau-wood-begins

[95] WTS paraphrases Matthew 25:35-45: "I was hungry and you fed me …" when he wrote, "Can't say I am hungry, nor thirsty, nor without clothes." His word "baksheed" is a memory of his time spent in Egypt in the 1890s; in Egypt and to the east in Southern Asia, *baksheeh* refers to a tip, or something extorted, such as extra rations.

[96] Gilbert, *First World War*, 434.

[97] Herbert = probably MWS's nephew, Herbert Marshall Warren (b. 1891), son of her brother George Copp Warren. He would have been 25 years old at that time. For Chinks, see previous end-notes. I presume "Hutchinseme" was a nickname given by WTS to Edward Hodge, for he is the only other Lt. Col. in the Unit. Brother Alfred Shoemaker was seven years older than WTS, born in 1862. It is not clear what he could do for the war effort, but WTS is obviously scornful of him. Syphons = Sparking or carbonated water; and the equipment for making it. Shumway and Zentmeyer can be identified as Philadelphia ophthalmologists.

[98] Harry = perhaps Howard Hey Shoemaker (b. 1885), the son of WTS's brother Charles.

[99] WTS predicted that "Barbara will get married to some very rich man." She did, when she married Albert Zimmermann.

[100] Excerpted from https://www.thoughtco.com/second-battle-of-the-marne-2361412:

"In the opening days of the fighting, German forces made only minor gains before being halted by a constellation of Allied troops. Due to intelligence gathering, the Allies were largely aware of German intentions and had prepared a sizable counter-offensive. This moved forward on July 18 and quickly shattered German resistance. After two days of fighting, the Germans commenced a retreat back to trenches between the Aisne and Vesle Rivers. The Allied attack was the first in a series of sustained offensives that would bring the war to an end that November.

"In early 1918, Generalquartiermeister Erich Ludendorff commenced a series of attacks known as the Spring Offensives with the goal of defeating the Allies before American troops arrived on the Western Front in large numbers. Though the Germans scored some early successes, these offensives were contained and halted. Seeking to continue pushing, Ludendorff planned for additional operations that summer.

"Believing that the decisive blow should come in Flanders, Ludendorff planned a diversionary offensive at the Marne. With this attack, the hoped to pull Allied troops south from his intended target. This plan called for an offensive south through the salient caused by the Aisne Offensive of late May and early June as well as a second assault to the east of Reims.

"In the west, Ludendorff assembled seventeen divisions of General Max von Boehm's Seventh Army and additional troops from Ninth Army to strike at the French Sixth Army led by General Jean Degoutte. While Boehm's troops drove south to the Marne River to capture Epernay, twenty-three divisions from Generals Bruno von Mudra and Karl von Einem's First and Third Armies were poised to attack General Henri Gouraud's French Fourth Army in Champagne. In advancing on both sides of Reims, Ludendorff hoped to split the French forces in the area. Supporting the troops in the lines, French forces in the area were buttressed by approximately 85,000 Americans as well as the British XXII Corps. As July passed, intelligence gleaned from prisoners, deserters, and aerial reconnaissance provided the Allied leadership with a solid understanding of German intentions. This included learning the date and hour that Ludendorff's offensive was set to commence. To counter the enemy, Marshal Ferdinand Foch, Supreme Commander of the Allied forces, had French artillery strike the opposing lines as German forces were forming for the assault. He also made plans for a large-scale counter-offensive which was set to launch on July 18."

[101] WTS and his companions planned to visit "Marché Eglies" and "Chateau d' Orge," ("English Market" and "House of Barley"). Touche d'Argues ("a touch of seaweed") was probably in a beautiful forest near the still-extant Chateau d'Argues. The scrapbook which WTS mentions has disappeared. WTS refers to the genealogy of MWS, which suggests (without proof) that her emigrant ancestor, Arthur Warren, was a descendant of the Duke of Warenne, and also of William the Conqueror. William's uncle, mentioned by WTS, is Richard III, Duke of Normandy, who reigned over Dieppe.

The words, "are numbered" is from the Bible, I Chronicles 23:1; "God gavest" is from Nehemiah 9:35 (KJV). WTS coined an expression "*pas du trip*" to mean "no trip." "Franco-American Soup" is a brand name of Campbell's Soup Co., which was located in Camden, N.J.

[102] Troyes is 80 miles SE of Paris; Subigny 40 miles W of Troyes; Villeroy is a street in Subigny; Sens is 5 miles W of Subigny.

[103] Excerpted from: https://www.britannica.com/event/Battle-of-Amiens:

"By late July 1918 Allied forces held a superior position on the Western Front; troops from the United States were pouring in to reinforce the war effort, and German soldiers were exhausted in the wake of a stalled offensive on the Marne. Having gained the initiative, Allied commanders had hoped to launch a limited offensive to secure a series of strategic transit hubs. As part of this, French General Ferdinand Foch planned an attack in the Amiens region of northern France that would protect the vital Paris-Amiens railway.

"The attacking force comprised the Canadian Corps, the British 4th Army, the French 1st Army, the Australian Corps, and others. In early August, the Allies made a show of weakening their front line so that German officers expected no assault. In reality, troops were being moved to the front at night, while bogus radio communication reinforced the deception. The Allied offensive would be supported by thousands of heavy and super-heavy field guns, more than 600 tanks, and 2,000 aircraft. The Germans were greatly outnumbered and, in the words of German military chief Erich Ludendorff, "depressed down to Hell." The Germans were protected by three lines of trenches, which were poorly wired for communications and without good dugout shelters. Unlike earlier offensives, the Amiens assault would not be preceded by bombardment so as to preserve the element of surprise. …

"The battle ended on August 11 as German resistance stiffened and Canadian commander Sir Arthur Currie urged the Allied leadership to consolidate the gains they had made thus far. In three days, the Allies had advanced some 8 miles (13 km), a huge achievement in a war characterized by minute gains at enormous cost. More than 19,000 Allied soldiers were killed or injured, while the Germans lost more than 26,000, including some 12,000 prisoners. Also captured by the Allies was the "Amiens gun," a 280-millimetre (11-inch) Krupp naval gun that had been mounted on a railway carriage.

"The 'Amiens gun' had been shelling the city of Amiens throughout the summer, and previous attempts to disable it had been unsuccessful, but an enterprising Australian sapper commandeered the train's engine and drove it back to Allied lines.

"Ludendorff described the opening day of the battle as 'the black day of the German Army in the history of this war…Everything I had feared, and of which I had so often given warning, had here, in one place, become a reality.' When Ludendorff informed German emperor William II of the disaster at Amiens, William replied, 'We have reached the limits of our capacity. The war must be terminated.' Indeed, Amiens sparked the 'hundred days' campaign, the successful Allied push that would drive the Germans backwards until their ultimate defeat and the signing of the armistice on November 11, 1918."

[104] John B. Murphy, M.D., F.A.C.S. (1857-1916) was the revered founder of the American College of Surgeons. In this letter, WTS continues his rant against his contemporary Philadelphia ophthalmologists, Drs. Burton Chance and H. Maxwell Langdon.

[105] HMAT *Warilda* (Australia), was torpedoed by German U-boat *UC-49* on August 3, 1918, in the English Channel, and sunk with the loss of 123 lives.

Mrs. Constance (née Gibbens) Melhuish, wife of George J. Melhuish, was the sister of Frances Vaughn Gibbens, who married George Copp Warren, brother of MWS. See letters from WTS (below) in England for his visits to the Melhuish family.

[106] The operation performed on Ted was not defined, but I think it was probably a herniorraphy. The operation that WTS helped with may be described in part in the enclosure, transcribed at the end of this letter. Montmorillon is about 300 miles SW of Paris, and La Courtine is about 150 miles SE of Montmorillon. Neither town is near Switzerland or the foothills of the Alps.

[107] Martin Gilbert, *The First World War: A Complete History* (New York: Henry Holt, 1994), 457.

[108] WTS is referring sarcastically to senior officers in the Army Medical Department. Col. Furbush was from Philadelphia, now serving as the principal assignment officer in Washington. U. S Army Surgeon General Gorgas was famous as the officer who gave permission for the experiments carried out by Walter Reed in Cuba that proved the mosquito was the route of transmission of yellow fever. Gorgas used this information to suppress both yellow fever and malaria in Panama. He thus enabled the Canal to be constructed, after the French had failed.

[109] Maj. General Merritte W. Ireland was Surgeon General of the Army from 1918-1931, succeeding William C. Gorgas. Ireland was one of the most effective officers ever to hold that position.
See: https://www.worldwar1centennial.org/index.php/practice-of-medicine-in-ww1.html

[110] WTS opines, "the East was pretty good," which was but a limited view of the situation. It was more complicated. The Russian Empire had collapsed, and the Allies sent troops into Russia from the east. American troops landed at Vladivostok on August 16. French troops had arrived eight days earlier. And then, on August 25, Germany and Bolshevik Russia signed an agreement to give Germany control of Red Navy vessels and facilities on the Black Sea. "Lenin and the Kaiser were making common cause, as Stalin and Hitler were to do twenty-one years later, to the day" (Gilbert, *First World War*, 452-3).

WTS says "the reconstruction was as bad as the war," based on the belief expressed by most historians at that time that Reconstruction following the Civil War was a disaster. They opined and taught that the South was flooded with opportunistic Northern "carpetbaggers," and that uneducated former slaves were elected to high positions in government. Reconstruction is now considered to have been a valiant attempt to treat African Americans fairly. That is to say, the disaster in America was not due to Reconstruction; but instead, what "was as bad as the war" was caused by the termination of Reconstruction in 1877.

[111] Gilbert, *First World War*, 464-5.

[112] In commenting on Foch's age, WTS may be referring to the opinion of Sir William Osler, expressed in *Aequanimitas* (1914):400, that, "The teacher's life should have three periods, study until twenty-five, investigation until forty, profession until sixty, at which time I would have him retired on a double allowance."

[113] Gilbert, *First World War*, 472.

[114] Re "Jesse" in this letter: WTS's letter on October 25 says his name is Jesse Warren. Jesse Warren (b. 1894) was the son of MWS's brother Jesse Warren, and his wife Margaret. "R.a L.M. and M.M." = "Remember a' Lovin' Me and a' Missin' Me."

[115] MWS must have started over in her numbering of letters, for WTS mentioned much higher numbers in his letters to her.
F.S.S. = Friends Select School. MWS mentions "Academy of Surgery" = American College of Surgeons.

MWS says the influenza epidemic of 1918 was the "scourge, worst the country has ever known." She was writing on October 25, 1918, a little less than a month after the epidemic swept across Philadelphia. The city was beginning to recover at that time. 12,000 died of the flu in Philadelphia, with a population of 1.7 million, plus other transients. Every hospital bed was filled soon after the Liberty Loan Parade on September 28, 1918, which was enthusiastically attended by 200,000 people. The health director, Dr. William Krusen, warned that the parade could place the city at risk, because influenza had already appeared in sailors at the Navy Yard. But the *Philadelphia Inquirer* and *Evening Bulletin* put the story on back pages, and city officials proceeded with the parade. The event rapidly spread the disease. In St. Louis, in contrast, the parade was cancelled and only 600 deaths from influenza were recorded. Mabel didn't mention the yellow fever epidemic of 1793, which was the subject of Longfellow's poem, "Evangeline." 2,000 died of a population of 20,000 in 1793, a mortality rate of 10.00 percent. It was more than ten times higher than the mortality rate from influenza in Philadelphia in 1918, which was 0.07 percent.

[116] Words for casualties in World War I: "Blighty" = wounded and sent back to England; St. Dunstan's is the hospital for blinded soldiers. "go [or Gone] West" = died.

[117] *History*, 198: Pvt. Frank X. Dochney d. 10/25/18. He was known affectionately as "Doc" as his name was pronounced, but probably also because of his skill as an orderly and caregiver.

[118] Aunt Linie = Caroline Shoemaker (b. 1838), sister of Julien Shoemaker (father of WTS). She m. Louis Jack (1831-1914) and was mother of Dr. Charles "Charlie" Shoemaker Jack (b. 1874) of B.H. #10. "Charlie" Jack m. Mary Miller Lewis, who WTS called by various derogatory nicknames such as "Bleeding Mary" and "Media" (the spelling of which I corrected to the ancient temptress, Medea).

[119] *History*, 198: John Wesley Thomas, died, October 30, 1918.

WTS quotes the beginning of a nursery poem about the girl with a curl. It continues, "When she was good, she was very, very good. / And when she was bad, she was horrid." WTS misspelled the founder of the American College of Surgeons as Francis H. Martin; his name was Franklin.

I cannot determine if he had a son, Franklin H. Martin, Jr., who may have been Ted's surgeon in Cape May, N.J., when he was in Army basic training. WTS's handwriting in this letter was unusually difficult to read. There are also many grammatical errors and words left out, some of which I have corrected, but in other cases, I left the text as he wrote it.

[120] Aunt Florrie = Florence (Hey) Shoemaker, wife of WTS's brother Joshua Shoemaker.

[121] I suppose WTS imagined that he took Mabel to dinner at Maxim's when he wrote, "Just [imagined that I] looked up an old school girl acquaintance and took her to dinner, so I dined all by my lonesome at Maxim's."

[122] *World War I: A Captivating Guide to the First World War, Including Battle Stories from the Eastern and Western front and How the Treaty of Versailles in 1919 Impacted the Rise of Nazi Germany* (Middletown, Del.: Captivating History, 2019, 82-3. I have added quotation marks around "war to end all wars."

[123] Lorna Doone country = from the novel, *Lorna Doone: A Romance of Exmoor*, by R. D. Blackmore (Sampson Low, Son, & Marston, 1869).

Tracking down clues about other names and places mentioned in this letter:

Allen Greenwood, "The Organization and Activities of the Ophthalmic Service in the American Expeditionary Forces," *Transactions of the Section of Ophthalmology of the American Medical Association at the Seventieth Annual Session, Held at Atlantic City, N.J., June 9 to 13, 1919* (Chicago: American Medical Association, 1919), 87-158. Lt.-Col. Greenwood was Chief Consultant in Ophthalmology to the American Expeditionary Force. He appears first as a Major, and later as a Lt.-Col., and he appears never to have risen above that rank. Approximately 150 men are listed by Greenwood as having been ophthalmologists with the American Army in Europe. None ranked above Lt. Col. WTS appears only once, as last on the list, on page 138: "Chief Surgeon's Office, Base Section 3 / Major William T. Shoemaker." Dr. Burton Chance, who is mentioned with scorn by WTS because of his short term of service overseas, is not mentioned.

The Red Cross Bulletin v.3 (No. 1, December 30, 1918), 80: "R. Stuart Smith, American Red Cross Commissioner for Great Britain, recently was paid the very high honor of being called to the Bar of the Middle Bench. He was specially admitted to the Honorable Society of the Middle Temple as a student on December 5, 1918. ... On the same day as the call, Major Smith also signed the Roll of Barristers of the High Court of Justice." He was Robert Stuart Smith, L.L.B., of Philadelphia.

Calendar: Register of Alumnae and Former Students: 1922. v.40, Part I (Bryn Mawr, Pa.: Bryn Mawr College, 1922), 157: Peckham, Emilie Comstock, . . . 132 East Market Street, Bethlehem, Pa.

Graduate Student in French, 1901-03. Married, 1906, Mr. Frank Stuart Smith. One daughter (one child [d?] 1908).

It is apparent from this entry that some members of the Stuart Smith (two words) family had come to Pennsylvania. This provides a possible connection with Maj. R. Stuart Smith, who certainly had prominent British connections, yet was an American.

----

https://www.victorianresearch.org/atcl/index.php: Constance Isabella Stuart Smith was born in 1859 in London, the daughter of the Rev. Hinton C. Smith, the curate of Chipperfield. She was educated in Belgium and Germany and attended King's College, London. In the 1890s, she began writing novels beginning with *The Repentance of Paul Wentworth* (1889). In the twentieth century, her interests turned to labour law and she wrote books and articles on the subject, including T*he Case for Wages Boards* (1905).

Smith served in a number of official capacities: senior woman inspector of factories (1913–1921), joint-secretary of the Women's Employment Committee (1917–1919), and deputy chief inspector of factories (1921–1925). For her social work, she was awarded an O.B.E. She died in 1930, unmarried. Perhaps there was a family connection to Major R. Stuart Smith.

----

124 *The Bulletin of the Geographical Society of Philadelphia*, Volume 7., p.66: Dr. Theo. Le Boutillier, member; 216 So. 20th St., Philadelphia. He lived near the Shoemakers, who lived at 109. He was well-known in his time as a physician, with publications related to childhood diseases and unpasteurized milk.

Franz Daniel Pastorius was the principal founder of Germantown, Pa. He was not related to WTS, although WTS was related to several other original settlers of Germantown.

125 Geoffrey L. Rossano (ed.), *The Price of Honor*, 1, 230, dust jacket. The letters of Kenneth MacLeish to his fiancé are cited here as another example of the pathos of the Great War.

126 "Mr. and Mrs. M.", and "Connie and Betsey" = Mr. and Mrs. George Melhuish and their daughter Marjorie. They lived at 17, St. Mary Abbott's Terrace, London. Constance "Connie" (née Gibbens) Melhuish was the sister-in-law of Mabel's brother, George Copp Warren.

Melhuish is said to be one of the oldest family names in English history. Its origin traces back to 501 A.D., from Maegla. All Melhuishes are said to be descended from Melewis, who saved the life of Edward I in the 1272, by pulling a poisoned dagger from his arm. They are thus entitled to use the crest and coat of arms awarded to him ("Argent on band engrave Sable with three Fleur de Lis' in middle chief point is a dagger hilted or blad azur"). This is unusual for England, for the armorial bearings are ordinarily awarded only to an individual, not to the entire family and descendants.

George John Melhuish was born in Croydon, Surrey, 19 November 1857, youngest of the four children of John Melhuish and his wife Ann; he died in Switzerland, 3 July 1924. When he was 23, he was a "clerk tea broker" in Japan, and it was there that he married, eleven years later, in 1891, Constance "Connie" Gibbens, younger daughter of Edwin Augustus Gibbens and his wife Mary Elizabeth (née) Chandler; born 3 September 1864 in Waltham, Mass.; died in the U.S. in 1928. How Constance Gibbens happened to be in Japan in 1891 is unknown; at the time of their marriage, he was 34 and she was 27.

All their three of children were born in Japan; Marjorie "Betsy" or "Betsey" (b. 1894), Charles John "Jack" (b. 1896), and Mary (b. 1900). They returned to England and sometime after 1911 they moved to the posh neighborhood of St. Mary Abbott's Terrace in Kensington. The six residents of that house in the census of 1911 were apparently unrelated (one was a medical student), but it gives an indication of the size of the house.

It was typical: a ground floor with a kitchen, living room, and dining room; a second floor with bedrooms; and a third floor with closets and servants' quarters. As the wife of an Englishman, Constance Melhuish was registered to vote in England, and she appears on the Electoral Register in 1918 and 1919 as a resident of 17, St. Mary Abbot's Terrace. They travelled widely before and after the war; one of their cruises was on the S.S. *Governor Cobb*, which returned to Key West from Cuba in 1920; at that time, they were living at Notting Hill in Kensington. After George died, Connie lived at 29 Bedford Gardens in Kensington. On 27 June 1927, she arrived back in the U.S. on the *Lancastria*. Giving her occupation as "housekeeper," she applied successfully to become a Naturalized U.S. Citizen. Her obituary, which was published in Boston on 20 December 1928, shows that she resided in New York. Her parents' names were said to be Edward A. and Mary Chandler "Gillem" but the given names of her parents allowed me to find the correct spelling of her maiden surname; it was Gibbens. Constance Gibbens was the younger sister of Frances Vaughn Gibbens, who was the wife of Mabel Warren's brother George Copp Warren. Mabel and George Warren's parents lived in West Newton, Mass., not far from Waltham, the birth place of Constance and Frances Gibbens.

"The Chinese Puzzle" played at the New Theatre in London, starting on July 11, 1918. It opened in Liverpool on July 1.

127 S.S. *Mauretania* was a fabulously luxurious Cunard ship, launched in 1906. It was the largest ship in the world when it was built, and it held the "Blue Riband" record for the fastest transit across the Atlantic for 22 years. It usually took about 5 days. It was a troop carrier and hospital ship in World War I. Captain Lockhart though the ship was due to sail on it on December 21, but in a photo, it is seen in New York harbor on December 23. WTS was still in England on Christmas day. I cannot find a record of the passage of *Mauretania* shortly after Christmas 1918, but the letters at the end of this chapter imply that WTS returned on it. He was in London for Christmas, and left soon after that. It is likely that the ship returned to England and WTS was aboard for its next trip back. I have silently corrected the mis-spelling of the ship's name in all three letters.

128 "Chu Chin Chow" was a musical based on the legend of Ali Baba and the 40 thieves. It premiered at His Majesty's Theatre in London in 1916, and it ran for five years – a record that stood for forty years. It opened in New York in 1917, starring Tyrone Power. A silent movie version was produced in 1925, and it has often been revived; the last one was 2008 in London.

There is actually no "St. James Street" in London, though St. James Square is surrounded by a street to which the writer surely refers. Many of the private clubs in London are located along St. James Square and the adjacent street known as Pall Mall. Also, within two or three blocks are His Majesty's Theatre (now known as Her Majesty's Theatre), Piccadilly Circus, and Trafalgar Square. The Travellers Club, which was featured in "Around the World in 80 Days" (movie, 1956) is at 106 Pall Mall.

129 Paul Fussell, *The Great War and Modern Memory* (New York: Sterling, 1974), 422.

130 Justin Barr, Leopoldo C. Cancio, David J. Smith, Matthew J. Bradley, and Eric A. Elster, "From Trench to Bedside: Military Surgery During World War I Upon Its Centennial" *Military Medicine* 184 (11/12:214, 2019).

THE PHOTO ALBUM

Part One

Before the War

Photo courtesy Mrs. Mary K. W. Rouillard

THE WARREN FAMILY. Seven founding brothers together with brother Jesse (deceased before the Company was organized) and two sisters. Photograph was taken about 1886. Seated, from left, Walter B., Ella, Albert C., Mabel and Ralph L. Warren. Standing, from left, Herbert M., Frederick J., Jesse, Henry J. and George C. Warren.

The Warren Family in 1886
Young Mabel is seated, 2d from Right

William T. Shoemaker's parents
Hannah Hester (1826-1914) and Julien Shoemaker (1825-1909)

1888 ***Mabel Warren*** ***1884***

RETINITIS PIGMENTOSA

WITH AN

ANALYSIS OF SEVENTEEN CASES OCCURRING IN DEAF-MUTES

BEING AN ESSAY FOR WHICH WAS AWARDED
THE ALVARENGA PRIZE OF THE COLLEGE OF
PHYSICIANS OF PHILADELPHIA, JULY, 1908

BY
WILLIAM T. SHOEMAKER, M.D.
PHILADELPHIA

LABORATORY EXAMINATIONS OF THE BLOOD AND URINE
BY
JOHN M. SWAN, M.D.
PHILADELPHIA

WITH ILLUSTRATIONS AND THREE COLORED PLATES

"Valeat Quantum Valere Potest"

PHILADELPHIA
J. B. LIPPINCOTT COMPANY

PLATE I.

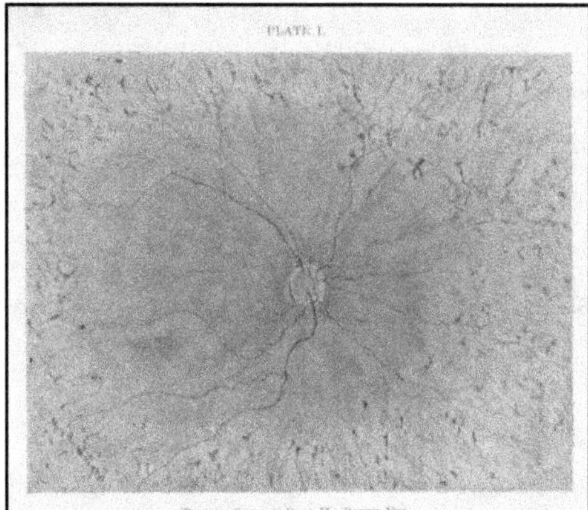

FUNDUS SKETCH, CASE II—RIGHT EYE.

The disk is well defined and atrophied. The vessels are very small but well distributed. The veins are dark and without light streaks. The pigmentation is of the classical variety, i. e., feathery and branching, and completely encircles the equator. Although very evenly deposited, the pigment is rather densely packed around the vessels in a number of places.

Both retina and choroid are markedly [illegible]. The case is thirteen years of age, has good central vision and a fair field.

Will and Mabel's first four children (ca. 1902): Barbara, Ted, Warren, Dot

STUDEBAKER SIX *Touring* CAR—ED

STUDEBAKER CORPORATION OF AMERICA, Detroit, Michigan

COLOR, Body, blue; chassis, black
SEATING CAPACITY, Seven persons
POSITION OF DRIVER, Left side
WHEELBASE, 122 inches
GAUGE, 56 inches
WHEELS, Wood
FRONT TIRES, 34 x 4 inches
REAR TIRES, 34 x 4 inches, anti-skid
SERVICE BRAKE, Contracting on rear wheels
EMERGENCY BRAKE, Expanding on rear wheels
CYLINDERS, Six
ARRANGED, Vertically
CAST, En bloc
HORSEPOWER, 36.04 (N. A. C. C. Rating)
BORE AND STROKE, 3⅞ x 5 inches

LUBRICATION, Splash with circulating pump
RADIATOR, Cellular
COOLING, Water pump
IGNITION, Storage battery
STARTING SYSTEM, Two unit
STARTER OPERATED, Chain to crank shaft
LIGHTING SYSTEM, Electric
VOLTAGES, Six
WIRING SYSTEM, Single
GASOLINE SYSTEM, Vacuum
CLUTCH, Cone
TRANSMISSION, Selective sliding
GEAR CHANGES, Three forward, one reverse
DRIVE, Plain bevel
REAR AXLE, Full floating
STEERING GEAR, Worm and gear

PRICE INCLUDES top, top hood, windshield, speedometer, battery indicator and demountable rims

Price $1085
3-Passenger Roadster. 1060
3-Passenger Landau Roadster $1350
4-Passenger Coupe... 1600
7-Passenger Sedan...$1675
7-Passenger Limousine 2500

1916 "Studebaker Six" advertisement

The Shoemakers had a family of 7 and a Studebaker. It was probably a Limousine

Shoemaker family photograph, undated

Dr. William T. Shoemaker's Home and Office
109 So. 20th Street, Philadelphia, Pennsylvania

Part Two

In Time of War

HISTORY OF THE
PENNSYLVANIA HOSPITAL UNIT
(BASE HOSPITAL No. 10, U. S. A.)

IN THE GREAT WAR

I am a soldier and now bound to France.
KING JOHN: I. I.

Then did Sir Knight abandon dwelling
And out he rode a coloneling.
BUTLER'S "HUDIBRAS."

NEW YORK
PAUL B. HOEBER
1921

The Pennsylvania Hospital Unit, Base Hospital No. 10, U. S. A., leaving Philadelphia, May 18, 1917.

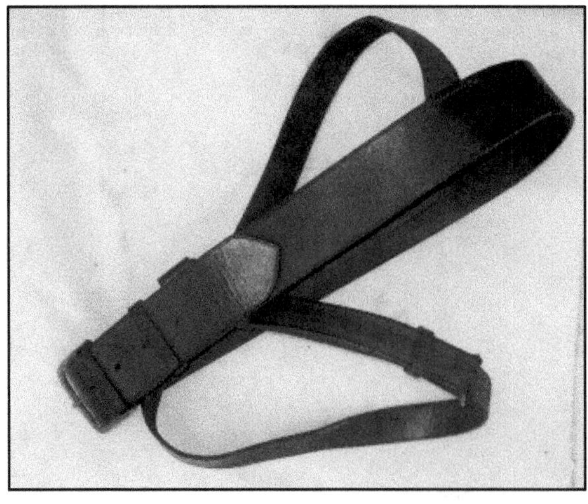

Captain Shoemaker's Sam Browne Belt

USS St Paul (above and R)

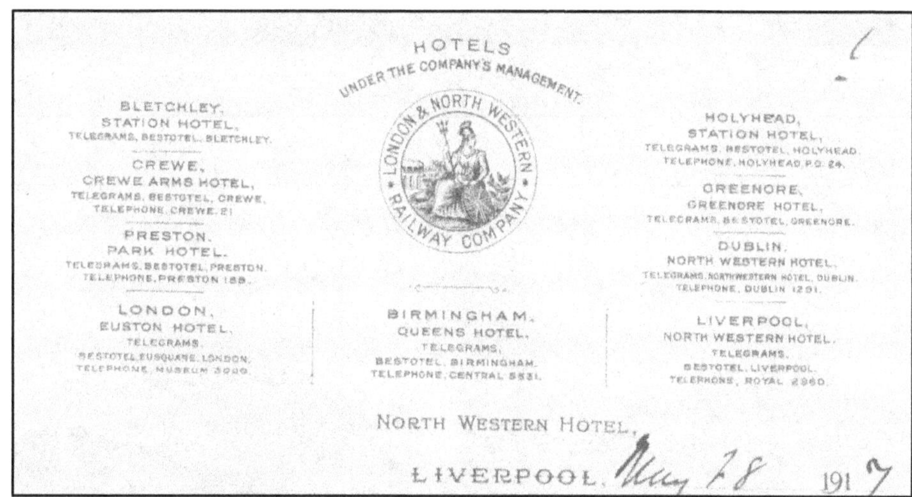

HOTELS
UNDER THE COMPANY'S MANAGEMENT.

LONDON & NORTH WESTERN RAILWAY COMPANY

BLETCHLEY,
STATION HOTEL,
TELEGRAMS, BESTOTEL, BLETCHLEY.

CREWE,
CREWE ARMS HOTEL,
TELEGRAMS, BESTOTEL, CREWE.
TELEPHONE, CREWE 21

PRESTON.
PARK HOTEL.
TELEGRAMS, BESTOTEL, PRESTON.
TELEPHONE, PRESTON 188.

LONDON,
EUSTON HOTEL,
TELEGRAMS,
BESTOTEL EUSQUARE, LONDON.
TELEPHONE, MUSEUM 3000.

BIRMINGHAM,
QUEENS HOTEL,
TELEGRAMS,
BESTOTEL, BIRMINGHAM.
TELEPHONE, CENTRAL 5531.

HOLYHEAD,
STATION HOTEL,
TELEGRAMS, BESTOTEL, HOLYHEAD.
TELEPHONE, HOLYHEAD P.O. 24.

GREENORE,
GREENORE HOTEL,
TELEGRAMS, BESTOTEL, GREENORE.

DUBLIN,
NORTH WESTERN HOTEL,
TELEGRAMS, NORTHWESTERN HOTEL, DUBLIN.
TELEPHONE, DUBLIN 1291.

LIVERPOOL,
NORTH WESTERN HOTEL,
TELEGRAMS,
BESTOTEL, LIVERPOOL.
TELEPHONE, ROYAL 2960.

NORTH WESTERN HOTEL,
LIVERPOOL, May 28 1917

Examples of WTS's stationery in London and envelope from France

Panoramic view of the hospital area at Le Treport.

Officers of Pennsylvania Hospital Unit, Base Hospital No. 10, U. S. A., in charge of British General Hospital No. 16, B. E. F., Treport, France.

Le Treport

Cliff - about 350 feet above the beach

Funicular from town to the top

Le Treport - France
6/18/1917

My Dear Barbara,

Your nice letter came through a friend of others, and tells me things that I did not know. That for instance not heard of Dorothy's scholarship. You seem to hurry to grow up and I trust it will continue. I am not so much in love with army life yet that I care to continue beyond the necessity. Today I am Officer of the Day, and have many duties aside from the treating of my patients. At 12 o'clock I inspected the food in the kitchen, that was to be served to the patients; then passed through the several rooms of patients (walking patients) and also orderlies etc. and gave the one over asking in each place if there were any complaints. At 4 o'clock the [illegible] is inspected - and passed on

Example of WTS's handwriting - Letter to Barbara, 6/18/17

Le Treport, France.
June 18, 1917.

My dear Barbara;

Your nice letter came through ahead of the others and told me things I did not know. You seem to be having a good time and I trust it will continue. I am not so much in love with camp life yet that I care to continue beyond the necessity. Today I am officer of the day and have many duties aside from the treatment of my patients. At 12 o'clock I inspect the food in the kitchen that is to be served to the patients. Then pass through the mess room of the patients (walking patients) and also orderlies, etc. and give the once over, asking at each place if there are any complaints. At 4 o'clock the Tea is inspected and after supper I must again go to the kitchen, see that everything is clean and in order. At 10 or 11 to night I must visit the night superintendent, medical and surgical and get complaints if any. If anything happens during the night I will be routed out. At 6 o'clock tomorrow must go to the parade grounds and see that the men going back to the front at that time have their equipment and rations. About 8 o'clock inspect the meat going by transfer. Visit the dead and sign their certificates, mkaing sure that they are dead. In bewtween times I take care of my patients just as usual. We take turns at this. The Majors are exempt. It will come around about once in 2 weeks. The day preceding the one referred to I am supernumerary in waiting. I was that yesterday. My only duty yesterday was to attend the funerals at 2.30 P. M. , that was very sad. The Padre, (minister) walked ahead of the Lowry, the ambulance or truck containing the coffin wrapped in the British flag. There is a double file of about 15 men and at the rear 2 officers, an English officer and myself. We walked over to the mortuary and there joined the procession. Then we marched at a measured tread to the cemetery about 1 1/2 miles. There the English burial service was read, bugle sounded, final taps, and we saluted, and then came home in the Lowry. I nearly collapsed on account of the heat, boiling sun about 100°, and of course I could not open my coat. When on duty, every button must be buttoned. Two heart-broken women were at the grave, probably a mother and wife or daughter. Another duty that I have for a week starting last Saturday is censoring letters. This I do immediately after breakfast, glance over them, stamp them and sign my name. It is more or less perfunctory. This morning I read one from a chap in one of our wards to his mother. He asked for money to buy cigarettes and I jotted down his name and location and when I found time took him some cigarettes. He was surprised, but I did not tell him how I knew he wanted them. This is the end of the paper. Tell your Mother not to be jealous, I will write to her next. With love to all.

June 18, 1917

My dear Dorothy;

Your letter arrived today after being opened, censored and marked insufficient address and numerous things. A proper and safe address would seem to be Base 10, American Expeditionary Force, France might be added, but is not necessary, but LeTreport must not be in evidence. We are kept very busy and have not enough doctors, nurses or enlisted men for 2000 beds, which we have when full. We are far short of a sufficient staff. At the present time we have about 1000, but a

Transcription of the letter to Barbara 6/18/17

Entrance to British General Hospital No. 16, Le Treport.

Convalescent patients leaving the admission and discharge hut at British General Hospital No. 16, Le Treport.

This picture represents some of
the horrors of war.
By it the High Commissioner of
the American Expeditionary Force
visualizes one of his Ophthalmologists drawing government money somewhere
in France.
It behooves the original
to look just like this
for duration, otherwise he
may have difficulty in
proving that he is himself
and therefore may never
get home.
Time:- About January 1918-
Place:- Operating Room, 16 Gen.
Operator:- Frank Knowles.
Victim:- Me.

Photograph of WTS, January 1918, sent to MWS with a humorous note on the back:

"This picture represents some of the horrors of war. By it the High Commissioner of the American Expeditionary Force visualizes one of his Opthalmologists drawing government money somewhere in France. It behooves the original to look just like this for the duration, otherwise he may have difficulty in proving that he is himself and therefore may never get home. Time: About January 1918. Place: Operating Room, 16 Gen. Operator: Frank Knowles. Victim: Me."

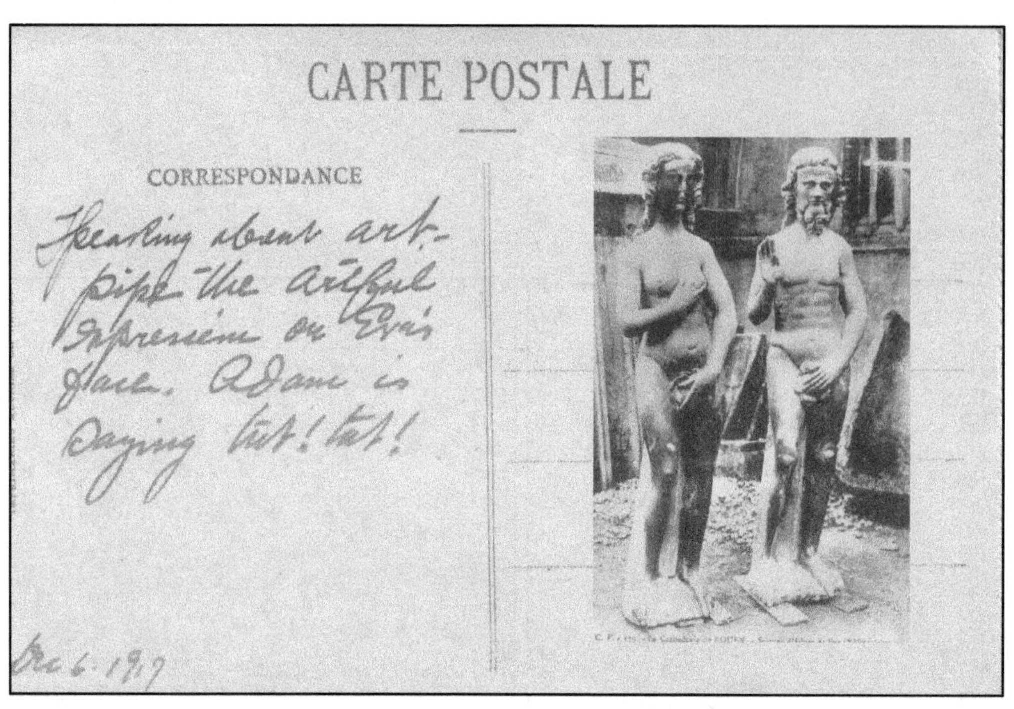

WTS wrote to BWS from Rouen, France, on Dec. 6, 1917, "Thinking about art - pipe the artful expression on Eve's face. Adam is saying tut! tut!

Ambulance, Base Hospital No. 10

Helen Fairchild, Nurse at Base Hospital No. 10, died January 18, 1918, following emergency surgery for a gastric ulcer complicated by liver failure, probably from exposure to poison gas and/or chloroform

Originally buried at the Base Hospital, her grave is now in Somme American Cemetery and Memorial, Bony, France. She is buried at Plot A, Row 15, Grave 13.

Funeral of Miss Helen Fairchild.

The grave of Miss Helen E. Fairchild in Mont Huon Cemetery, B. E. F., Le Treport.

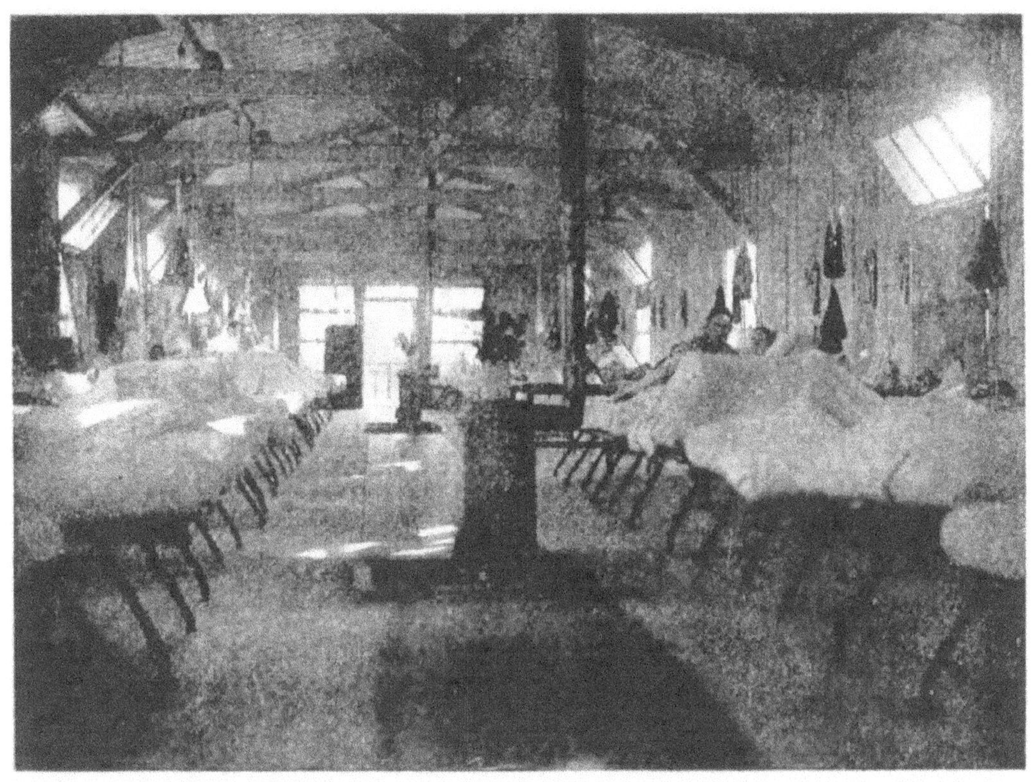

A surgical ward in No. 16, General Hospital, Le Treport.

Military Cemetery in Le Treport

St. Mary Abbot's Terrace, Kensington
A very fashionable district in London

Major William T. Shoemaker stayed with Mabel's in-laws, George and Constance Melhuish, in their house at #17 when he was detached from Le Treport to work as a Consultant after Armistice Day in 1918

From #17, it is 3 miles to George's Club at St. James Square, going east through Hyde Park and in front of Buckingham Palace. The Club is near Pall Mall, Piccadilly Circus, and Trafalgar Square

American Troops marching in London, as they returned home after World War I

SS* Mauretania *in "dazzle" camouflage, photographed in December 1918
Maj. William T. Shoemaker probably returned to America on this ship sometime after Christmas 1918 – as explained in the letters of that month

William T. Shoemaker's Insignia – World War I

Chevrons for overseas service

Worn on lower left sleeve of coat (blouse)
One chevron for each 6 months of overseas duty

World War I Victory Medal

William T. Shoemaker's advancement from 1st Lieutenant to Captain to Major is shown in the badges from one silver bar, to two silver bars, and the gold maple leaf

APPROVED

APPLICATION No. 298907

OFFICER—ORC

Shoemaker, William Toy (Surname) (Christian name) White PHILADELPHIA 1278

Residence 109 So 20 St (Street and house number) Philadelphia (Town or city) (County) PENNSYLVANIA (State)

Born in Philadelphia Pa Jan 22/. 69

Called into active service as 1 Lt MC May 15/17 fr ORC XXXXXXXX Training Camp.

Promotions: Capt Sept 28/17; Maj Oct 1/18

Organizations and staff assignments: MC Cas to disch

Principal stations: Philadelphia Pa; AEF

Engagements: none

Wounds received in action: None.

Served overseas May 9/17 to Dec 30/18.

Hon. disch. Dec 30/18 for convenience of the Government, services no longer required.

Was reported 0 per cent disabled on date of discharge in view of occupation.

Remarks: XX

MAY 29 1934 AUD. GEN.

Form No. 84c-1 A. G. O. Oct. 18, 1922

* Give place and date. † Insert (a) grade; (b) arm or staff corps or department; (c) date; (d) source, civil life (CL), RA, NG, ORC, NA; and (e) designation of training camp attended, if any. ‡ Strike out if he did not attend a training camp. § Give dates of departure from and return to the United States. ‖ Give date.

Dates as per Card Record From 5/15/17 to 12/30/18

Commonwealth of Pennsylvania

1278

VETERAN'S COMPENSATION APPLICATION

NUMBER 298907(1)

Read instructions and then type or print answers to the following questions striking out words that do not apply. Numerals in parenthesis refer to corresponding numerals in the instructions.

COMMONWEALTH OF PENNSYLVANIA
COUNTY OF Philadelphia } William Toy Shoemaker

being duly sworn, deposes and says: I am the veteran named in the following descriptive information; I served in the military or naval forces of the United States from March 17, 1917 to December 30, 1918; I claim veteran's compensation as authorized by Act No. 53, approved January 5, 1934; I was a resident of Pennsylvania at time of entry into service; I make this claim with full knowledge of the penalty for making a false statement relative to a material fact concerning this claim and the answers to the following questions and descriptive information are true:

(2) Shoemaker, (Last name) William Toy (First and middle name) (Serial number) White (White or colored)

(3) Legal residence at entry into service 109 S. 20th St., Philadelphia, Pa. (Street and number, City, County and State)

(4) Present residence: 2017 Locust Street, Philadelphia, Pa. (Street and number, City, County and State)

(5) Enlisted, commissioned or inducted in RA; ~~ANC~~; NA; NG; ERC; Navy; NRF; MC; ORC; Pa. Vols; US Vols; at Philadelphia, Penna on March 17, 1917

(6) On active duty in Navy or in Reserve from to

(7) Place and date of birth Philadelphia, Pa., January 22, 1869

(8) Names and addresses of dependents: Wife (full name) Deceased, Mabel Warren Shoemaker
Minor children grand children- two
Mother (full name) None
Father None

(9) Service in organizations, at stations or on vessels in the order named as follows:
Penna.Hospital-Base 10 from Mar.17,1917 to Dec.30,1918
from to
from to

(10) Grades or ratings with dates of appointments or promotions: 1st Lt Mar. 17, 1917, Captain Sept 4 1917, Major

(11) Engagements: None

(12) Wounds or other injuries received with dates: None

(13) Served overseas from May 17, 1917 to December 30, 1918

(14) Honorably discharged on December 30, 1918 at Hoboken N. J.

(15) Did you ever apply for or receive a bonus or veteran's compensation from any other State? No

(16) Did you ever refuse on conscientious, political or other grounds to perform full military duty or to render unqualified service? No

Subscribed and sworn to before me this 19th day of March 1934.

(17) Elizabeth K. Shearer (Signature and seal of officer administering oath) NOTARY PUBLIC My Commission expires Mar. 9, 1935

(18) William Toy Shoemaker (Signature of Applicant)

CERTIFICATE OF IDENTIFICATION

March 19 1934 (Date)

(19) I, Elizabeth K. Shearer (Name of person certifying to), do hereby certify that I am Notary Public (Title of officer and position); that William Toy Shoemaker (Name of applicant) applying for Veteran's Compensation is known to me to be the applicant named in the application; that he is mentally competent; that the signature of the affidavit is the signature of the applicant.

Signature and impression of seal or stamp indicating office or position held.

Elizabeth K. Shearer (Signature of person identifying applicant)

Read instructions (19) before signing.

(1) 20 Months at $10.00 per month; Total $ 200

(1) Computed by S. Miller Reviewed by Bo Stine

(1) Checked and approved by F. Slonaker For The Adjutant General.

Dorothy Shoemaker, Petty Officer 3rd Class, USNR

Part Three

After the War

William Toy Shoemaker, 1924, by Elias Goldensky of Philadelphia

Mabel Warren Shoemaker by Elias Goldensky of Philadelphia

Dorothy Shoemaker

Warren Shoemaker

Warren and Ted Shoemaker, ca. 1923

Barbara Shoemaker (R) with her future sister-in-law, Kate Wheelock, as bridesmaids, 1930

Ted Shoemaker

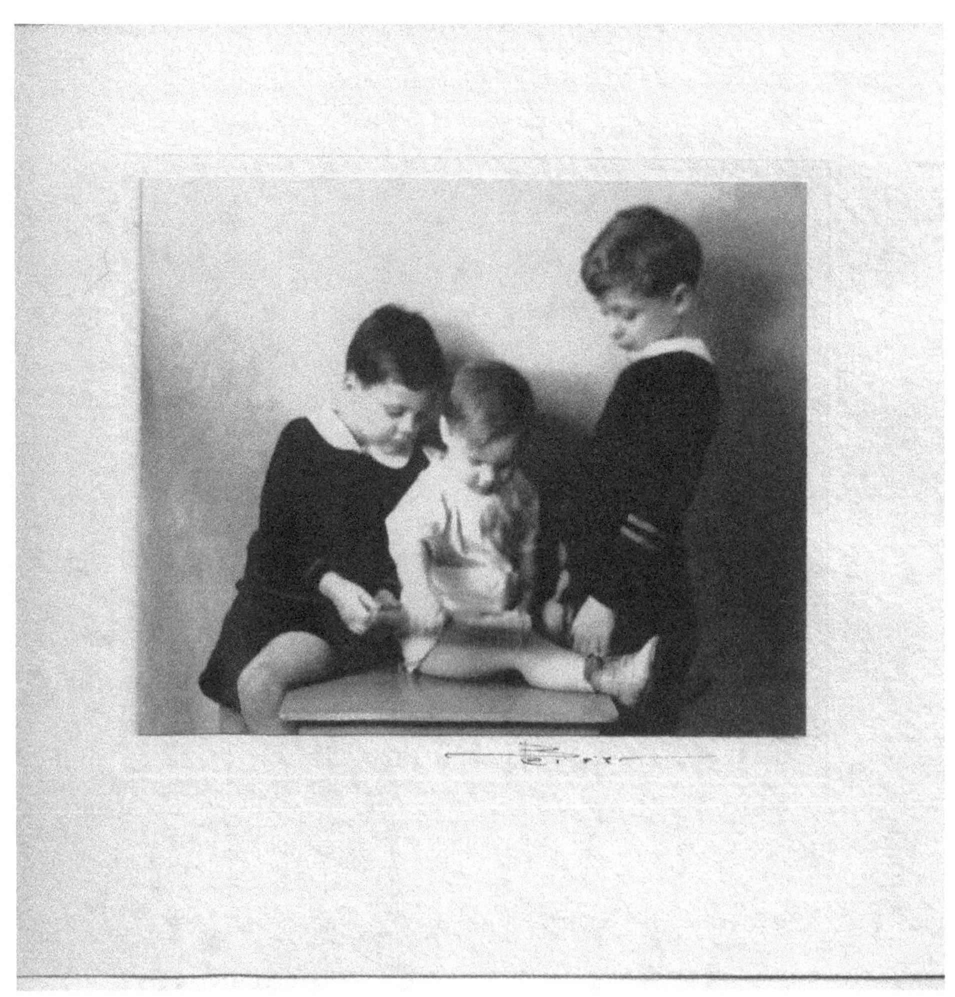

Ted Shoemaker's children in 1931 – Bix, Julien "Shoe," and Ted, Jr.

Barbara Shoemaker in 1924

Barbara (Shoemaker) Zimmermann in 1931, with her daughters Helene and Barbara

The First Seven Grandchildren of William and Mabel Shoemaker – 1930

Top: Bill Boericke, Lanie Zimmermann, Bill Shoemaker, Dorothy Boericke
Floor: Babs Zimmermann, Bix Shoemaker, Ted Shoemaker

Mabel Warren Shoemaker by Elias Goldensky

William Toy Shoemaker by Elias Goldensky
American Legion badge on lapel

Mabel Warren Shoemaker by Elias Goldensky
Died July 8, 1932

William Toy Shoemaker
Wearing American Legion lapel button
Died November 26, 1942

J. Warren Shoemaker wearing American Legion lapel button

Barbara's daughter Helene and her family
The family gathering in Northampton, Mass., after the ceremony when Helene "Lanie" Zimmermann Hill was presented with the Smith College Medal, 19 February 1997. Between Lanie and George is their granddaughter, Heather. In back row, left to right, are David, Sarah, Jim, Helena

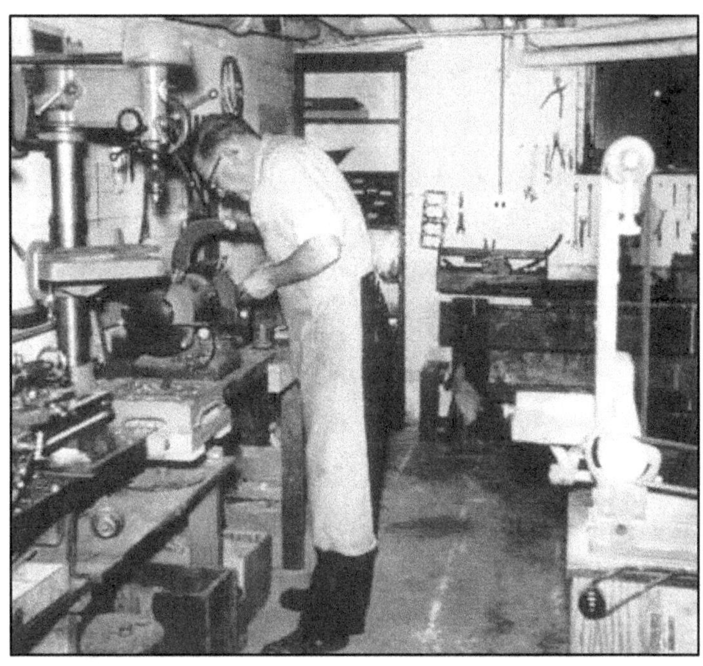

Bob Shoemaker in his shop in Portland, Oregon

Portrait by Boychuk, Philadelphia

Children of Bob and Nancy Shoemaker
Dick, Ann, and Bob, Jr. – December 1938

Bibliography

Books

- Aminoff, Michael J., and Robert J. Daroff, eds. *Encyclopedia of the Neurological Sciences*, 2nd ed. Oxford: Elsevier, 2014. Col. William Lister, p.606
- Bacon, John U. *The Great Halifax Explosion: A World War I Story of Treachery, Tragedy, and Extraordinary Heroism*. New York: HarperCollins, 2017.
- Bible. King James Version
- Blackmore, R. D. *Lorna Doone: A Romance of Exmoor*. 2 vols. Philadelphia: Porter & Coates, 1882.
- Churchill, Winston. *The World Crisis*. Vol. 1, *1911-1914*. 1938. London: Bloomsbury, 2015. In this edition, Churchill wrote on page xv (unnumbered), "Foreword to New Edition," dated November 22, 1938, "This book was originally published in four large volumes … I write this new Preface in a day of extraordinary difficulty and danger."
- _____. *The World Crisis, 1911-1914*. 5 vols. in one. 1931. New introduction by Martin Gilbert, abridged and revised. New York: Simon and Schuster, 2005.
- _____. *The Gathering Storm*. Vol. 1 of the series, *The Second World War*. Cambridge, Mass.: Riverside Press, 1948. Churchill began the Preface, page iii: "I must regard these volumes of *The Second World War* as a continuation of the story of the First World War … Together, if the present work is completed, they will cover an account of another Thirty Years' War."
- Curtin, D. Thomas. *The Land of Deepening Shadow: Germany at War*. New York: George H. Doran Co., 1917.
- Cushing, Harvey. *From a Surgeon's Journal: 1915-1918*. Boston: Little, Brown and Co., 1936.
- _____. *The Life of Sir William Osler*. London: Oxford University Press, 1925.
- Davis, Betsy Warren. *The Warren, Jackson, and Allied Families*. Philadelphia, Penna.: J. B. Lippincott, 1903.
- Dawson, Coningsby. *Carry On: Letters in War-Time*. New York: John Lane Company, 1917.
- Dietz, Dan. *Off Broadway Musicals, 1910-2007: Casts, Credits, Songs, Critical Reception and Performance Data of More than 1,800 Shows*. McFarland, Inc., Publishers, 2010.
- Empey, Arthur Guy. *"Over the Top" by an American Soldier Who Went*. New York: G. P. Putnam's Sons, 1917.
- Enders, Gordon. *Foreign Devil: An American Kim in Modern Asia*. New York: Simon and Schuster, 1942.
- Enders, Gordon B[andy], and Edward Anthony. *No Where Else in the World*. New York: Farrar & Rinehart, 1935.
- Fay, Sidney Bradshaw. *The Origins of the World War*. 2d ed. 1928, 1930. New York: Macmillan Co., 1964.
- Fife, George Buchanan. *The Passing Legion: How the American Red Cross Met the American Army in Great Britain, the Gateway to France*. (London: McMillan, 1920). Names many hospitals in England with American patients, including 10 on p.70; and Red Cross Commissioner in England, Maj. R. Stuart Smith.
- Fulton, John Farquhar. *Harvey Cushing*.
- Fussell, Paul. *The Great War and Modern Memory*. 1974. New York: Sterling, 2009.
- *Gibson, Henry William.* "Twenty-five Years of Organized Boys Work in Massachusetts and Rhode Island." In Cheley & Baker, *Camp and Outing Activities*. Introduction by H. W. Gibson. Assoc. Press, 1915.
- Gilbert, Martin. *The First World War: A Complete History*. New York: Henry Holt, 1994.
- Harrison, Henry. *Infunitive*. Melomime Publications, 1923.
- Harte, Richard H., and others. *History of the Pennsylvania Hospital Unit (Base Hospital No. 10, U.S.A.) in the Great War*. New York: Paul B. Hoeber, 1921.
- Hemingway, Ernest. *A Farewell to Arms*. New York: Scribner's, 1929.
- Hill, George J. *Quakers and Puritans: The Shoemaker, Warren and Allied Families. Ancestors and Descendants of William Toy Shoemaker and Mabel Warren, Who Were Married in Philadelphia in 1895*. Westminster, Md.: Heritage Books, 2015.
- _____. *Liberia and the U.S., 1917-1947. Race, Religion, Rubber, and Politics in the U.S.A. and Liberia*, 2d ed. Saarbrucken: Scholars Press, 2014. For biography of Henry Serrano Villard, who is portrayed in the movie, *In Love and War*.

- _____. *"Dearest Barb" from Karachi, 1943-1945: Letters and Photographs in the World War II Papers of a Naval Intelligence Officer, Lieutenant Albert Zimmermann.* Berwyn Heights, Md., 2018. For biography of Gordon Bandy Enders.
- _____. *Proceed to Peshawar: The Story of a U.S. Navy Intelligence Mission on the Afghan Border, 1943.* Annapolis, Md.: Naval Institute Press, 2013. Biographies of Gordon Enders and Lt.-Col. Rex Benson.
- Hillcourt, William. *Baden-Powell: The Two Lives of a Hero.* Boy Scouts of America, 1937. Reprinted, 1981.
- Homer. *The Odyssey.* Translated by Robert Fitzgerald. New York: Doubleday Anchor Book, 1931.
- Hopkirk, Peter. *The Great Game: The Struggle for Empire in Central Asia.* 1990. New York: Kodansha International, 1994.
- Kipling, Rudyard. *Kim.* London: Macmillan, 1901. Republished by Dover Publications, Mineola, N.Y., 2005.
- _____. *Something of Myself and Other Autobiographical Writings.* Edited by Thomas Pinney. Cambridge: Cambridge University Press, 1990.
- Jones, John Price. *The German Spy in America: The Secret Plotting of German Spies in the United States and the Inside Story of the Sinking of the Lusitania.* London: Hutchinson & Co., 1917.
- Litoff, Judy Barrett, and David C. Smith. *Since You Went Away: World War II Letters from American Women on the Home Front.* Lawrence, Ks.: University Press of Kansas, 1991.
- Rossano, Geoffrey, ed. *The Price of Honor: The World War One Letters of Naval Aviator Kenneth MacLeish.* Annapolis: Naval Academy Press, 1991.
- *Obituary Record of Graduates and Non-graduates of Amherst College for the Academic Year Ending June 18, 1919.* Amherst, Mass.: Published by the College, 1919). Col. Robert Stuart Smith, 258-9.
- Osler, William. *Aequanimitas.* Philadelphia: P. Blakiston's Sons & Co., 1914.
- Remarque, Eric Maria. *All's Quiet on the Western Front.* 1929. Translated by A. W. Wheen. Boston: Little, Brown, and Co., 1930.
- Rodman, Walter Clarke. *Twilight and Other Verse.* Philadelphia: J. B. Lippincott Co., 1915.
- Service, Robert W. *Rhymes of a Red Cross Man.* Toronto: Barse & Co., 1916.
- Shoemaker, Benjamin H. *Genealogy of the Shoemaker Family of Cheltenham, Pennsylvania.* Philadelphia, Penna.: J. B. Lippincott, 1903.
- Stedman, Edmund Clarence, ed. *A Victorian Anthology*, 1837-1895.
- Swedenborg, Emanuel. *The Path of Life.* Philadelphia: J. B. Lippincott, 1913.
- Tuchman, Barbara W. *The Guns of August.* New York: Macmillan, 1962
- Villard, Henry Serrano, and James Nagel. *Hemingway, In Love and War: The Lost Diary of Agnes von Kurowsky.* 1989. London: Sceptre, 1997.
- *World War 1: A Captivating Guide to the First World War, Including Battle Stories from the Eastern and Western Front and How the Treaty of Versailles in 1919 Impacted the Rise of Nazi Germany*. Middletown, Del.: Captivating History, 2019.

Periodicals

- Alden, E. H. "Worshipping with Unitarians" *Messenger* (27 February 1918).
- Anon. "Pathologist Edward Krumbhaar Dies at 83," *JAMA* 196 (no. 2, 1966): 39.
- Anon. *Success with Flowers* v.3 (no.1, 1892), 146: "*Tears cannot restore thee*, therefore I weep."
- Barr, Justin, Leopoldo C. Cancio, David J. Smith, and others. "From Trench to Bedside: Military Surgery During World War I Upon Its Centennial. *Military Medicine* 184 (Nov.-Dec 2019), 214.
- Butler, Ralph, "Infective Sinus Thrombosis," *Pennsylvania Medical Journal* 20 (Nov. 1916), 277.
- Chance, Burton, ophthalmologist in Philadelphia. The *Pennsylvania Medical Journal* 22 (Nov. 1918), 103.
- Henriksen, Sverre Dick. "*Moxarella, Acinetobacte*r, and the *Mimeeae*," *Bacteriological Reviews* 37 (no. 4, Dec. 1973), 522-561. [About Diplobacillus scarletti, discovered by Dr. Hunter Scarlett (letter, 7 Aug 1917)]
- Jopson, John H., in *Transactions of the Philadelphia Academy of Surgery* 22 (1922), 166: "Dr. John H. Jopson, … spoke of the use of the Carrell-Dakin method in the pericardium … Presbyterian Hospital."
- Shoemaker, William T., in "Medical Reserves," *Pennsylvania Medical Journal* 20 (June-July 1917), 675, 736.
- *Transactions of the College of Physicians of Philadelphia*, vol. 27 (1905), edited by William Zentmeyer, 238-9.

William T. Shoemaker, "Binasal Hemianopsia" (December 1906), 238; Edward A. Shumway, 239-40.
Ibid, v. 15 (No. 11, 1932), p. 1060. H. Maxwell Langdon, chairman, Section on Ophthalmology, presided when a paper was presented on "Removal of oxidized tissue remaining after foreign body in cornea."
Ibid. v. 45 (1923), xlv: Henry F. Page, M.D. was elected Fellow of the College in 1898.

Ibid. v. 41 (1919), xxxiii: Page was Assistant Physician to the Lankenau Hospital, and Clinical Professor of Medicine at Women's Medical College.

Ibid. v. 44 (1922), xxi: William J. Creighton, 1905 Chestnut St., was elected Fellow in 1920. He gave a paper on Herpes Zoster Ophthalmica (pp. 307-8).

Movies and Plays

1917
A Farewell to Arms
All's Quiet on the Western Front
Chu Chin Chow
In Love and War
It's a Long Way to Tipperary
Journey's End
Last Post
Lawrence of Arabia
Paths of Glory
Reds
Roses of Picardy
Sergeant York
The Chinese Puzzle
The Fighting 69th
The Hunt for Red October
There's a Long, Long Trail A-Winding
War Horse

Poetry

Brooke, Rupert. "The Soldier"
Graves, Robert. "Two Fusilliers"
Liep, Hans. "Lilli Marleen" (1915)
It was set to music in 1938
Kilmer, [Alfred] Joyce. "Trees"
McRae, John. "In Flanders Fields"
Owen, Wilfred. "Anthem for Doomed Youth"
Sassoon, Siegfried. "The Kiss"
Seeger, Alan. "I have a Rendezvous with Death"
Yeats, W. B. "An Irish Airman Foresees His Death

Songs and Hymns

America
God Save the King
It's a Long Way to Tipperary
Keep the Home Fires Burning
K-K-K-Katy
Last Post
O Holy Night
Old Virginny
Onward Christian Soldiers
Over There
Pack Up Your Troubles
Battle Hymn of the Republic
Star Spangled Banner
Taps
When It's Apple Blossom Time in Normandy

Internet

Army Medical Department, U.S.

- The Base Hospitals of the AEF - Worldwar1.com www.worldwar1.com›dbc›basehosp
- Ford, Joseph H. *The Medical Department of the United States Army in the World War*. United States Government Printing Office, 1927. Chapter 24, "Base Hospitals." p. 638: Base Hospital #10

https://history.amedd.army.mil/booksdocs/wwi/adminamerexp/chapter24.html

BASE HOSPITAL NO. 10

Base Hospital No. 10 was organized at the Pennsylvania Hospital, Philadelphia, Pa., during February, 1917. It was mobilized at Philadelphia early in May, 1917, and on May 19 sailed from the United States on the *St. Paul*, arriving in England on May 28, 1917. After a few days' delay in England the unit was assigned to station at Le Treport (Seine Inferieure), France, arriving at that station on June 12, 1917. It was one of the original six hospitals assigned to duty with the British and operated No. 16 General Hospital, British Expeditionary Force. It remained at Le Treport, attached to the British during its entire overseas existence. It ceased to function about February 27, 1919; sailed from Brest, France, on the *Kaiserine Augusta Victoria* April 8, arrived in the United States April 17, 1919, and was demobilized shortly thereafter.

PERSONNEL

COMMANDING OFFICER

Col. M. A. Delaney, M. C., May, 1917, to March 11, 1918.
Lieut. Col. Richard A. Harte, M. C., March 12, 1918, to November 3, 1918.
Lieut. Col. William J. Taylor, M. C., November 4, 1918, to December 24, 1918.
Lieut. Col. Charles F. Mitchell, M. C., December 25, 1918, to demobilization.

British Hospitals

• https://www.moorfields.nhs.uk/content/our-history
In 1837 Queen Victoria gave the hospital in Lower Moorfields its Royal Charter. Although it was renamed the Royal London Ophthalmic Hospital, everyone continued to call it Moorfields.

• https://www.londonremembers.com/subjects/royal-westminster-ophthalmic-hospital
Royal Westminster Ophthalmic Hospital and Lost Hospitals of London

• https://bjo.bmj.com/content/11/6/319 (accessed 1/15/20)
Morton, Andrew Stanford, F.R.C.S.Eng.: Obituary of Mr. Andrew Stanford Morton, F.R.C.S.Eng., who died in his seventy-ninth year. "There can be no question that Morton was, in his day, unsurpassed as an operator on the eye." He was on the staff at Moorfield's. ("ANDREW STANFORD MORTON" *British Journal of Ophthalmology* 1927;11:319-320.)

• http://www.vlib.us/medical/hspindex.htm -> A comprehensive list of British Military Hospitals in the United Kingdom during WW1. The list cannot be searched but it is indexed by county; e.g., Dorsetshire: Paignton.

• https://vad.redcross.org.uk/en/ List of auxiliary hospitals in the UK during the First World War.pdf. This list is also shown by county names; e.g., Hampshire County: Portsmouth and Romsey both have more than one hospital.

Camps

• Camp Becket: See Gibson, in Books.
• Camp Owaissa: https://www.poconorecord.com/article/20080704/LOCALENT/807040321 (posted 4 July 2008).

History of World War I

• Battles of the Somme: http://www.greatwar.co.uk/battles/somme/somme- battles.htm#firstsomme1918
• Italy and Russia in World War I (October 1917): https://www.history.co.uk/italy-in-wwi

People

• Beckman, Capt. Elizabeth. http://www.usmilitariaforum.com/forums/index.php?/topic/265351-captain-elizabeth-b-eckman-base-hospital-no-10-bef-ww1/
• Cavell, Edith. https://www.telegraph.co.uk/news/bbc/11861398/Revealed-New-evidence-that-executed-wartime-nurse-Edith-Cavells-network-was-spying.html. She was shot at dawn by a German firing squad on October 12, 1915.
• Cornell, Sgt. Horace H. https://www.classicbooksandephemera.com/pages/books/004340/horace-h-cornell/letter-from-sergeant-horace-h-cornell-to-stacy-b-brown-may-31-1917-from-base-hospital-no-10-on-its-way
• Fairchild, Helen. http://www.vlib.us/medical/MaMh/MyAunt.htm
• Garfield, Harry Augustus. https://encyclopedia.1914-1918-online.net/article/garfield_harry_augustus
• Lister, Col. William. See Michael J. Aminoff and Robert J. Daroff, eds., *Encyclopedia of the Neurological Sciences*.
• Melhuish. https://www.genealogy.com/forum/surnames/topics/melhuish/65/
• Osler, Revere. https://www.westernfrontassociation.com/on-this-day/28-august-1917-the-life-and-death-of-edward-revere-osler/
• Stuart Smith, Constance. https://www.victorianresearch.org/atcl/index.php: Perhaps there was a family connection to Major Robert Stuart Smith.

Schools

• Friends Select School: https://www.friends-select.org/ (accessed 30 January 2020).

Ships

• R.M.S. *Mauretania* (1907) See: HMS Tuberose aka RMS Mauretania enters New York harbor in full dazzle Dec. 23, 1918, carrying returning troops after the Armistice.
• S.S. *Saint Paul* https://www.navsource.org/archives/04/stpaul2/stpaul2.htm

Appendix – Names
Families – Shoemaker, Warren, Jacks, Melhuish
Personnel of Base Hospital #10

Families

Shoemaker Family

Charles SHOEMAKER (1786-1847) m. Rachel COMLY (1793-1867); 10 children: Jonathan, Alfred, Eliza Lukens, Martha Comly, Hannah m. Owen [1] SHOEMAKER (her cousin), Julien m. Hannah HESTER (see next), George, Henry Charles, Harriet Mary, Caroline (1838-1870) m., as 2d wife, Dr. Louis JACK (1831-1914) (Jack Family, see below)

Julien SHOEMAKER (1825-1909) m. Hannah HESTER (1826-1914); 9 children:

1 Charles John (1856-1923) m. Lucretia HEY (1856-1927); 5 children: Howard, Lucretia, Anna Caroline, Orlando (b. 1891), Ethyl

2 Joshua "Josh" Lippincott (1858-1926), m. Florence "Florrie" HEY (1861-1922); 5 children: Louis Jack, Dudley, Emmanuel, Herbert, Florence

3 Henry (1859-1865).

4 Owen [2] (1860-1944) m. Mary Margaret ("Greta") JACK (b.1864); 5 children: Walter, Edwin "Ed" (b. 1888) m. Martha Clauson REED, Margaret, Rachel, Elizabeth Jack

5 Alfred (1862-1935)

6 Harvey (1864-1931) m. Laura Montgomery LANE (b. 1863); 6 children: Caroline, Katharine, Alice, Margery, Laura, Arthur Meigs

7 Caroline "Linie" (b. 1867)

8 William Toy "Bill" Shoemaker (1869-1942) m. Mabel WARREN (1870-1932); 6 children: Dorothy (1896-1992), Jesse Warren "Warren" (1898-1964), Theodore "Ted" (1899-1962), Barbara (1902-1985), Robert Comly "Bob" (1903-1976), William Toy, Jr. "Willie" (1904-1916)

9 Mary Westcott (b. & d. 1870)

Warren Family

Jesse WARREN m. Betsey JACKSON; 11 children: Joseph, Mary Ann, Joseph, Betsey, Elvira, John, Samuel, Cyrus, Herbert Marshall (1827-1880), Harriet, Ebenezer Burgess.

Herbert Marshall WARREN (1827-1880) m. Eliza Caroline COPP (1828-1879); 10 children: Albert Cyrus, Ella, Jesse m. Margaret SCHLEGEL, Herbert Marshall m. Catherine WINTER, Henry Jackson, George Copp (b. 1863) m. Frances GIBBENS, Frederick John, Mabel, m. William T. SHOEMAKER (see above for their 6 children), Walter Burgess, Ralph Lambert

Jack Family

Louis JACK (1831-1914), son of Josiah JACK, m. (1st) Thankful CORBUS; child: Mary Margaret "Greta" (b. 1864) m. Owen SHOEMAKER (1860-1944); 5 children (see above)

Louis JACK m. (2d) 1870, Caroline SHOEMAKER (1838-1870) (see above); 4 children: Caroline, William, Charles Shoemaker JACK (dentist with B.H. No. 10) m. Mary Miller LEWIS (2 children: Sarah m. John LOBER, Mary), Howard Foster

Melhuish Family

John MELHUISH and his wife Ann had four children. Their youngest child, George John MELHUISH (1857-1924) m. Constance "Connie" GIBBENS (1864-1928), daughter of Edwin Augustus and Mary Elizabeth (née CHANDLER) GIBBENS. Edwin Augustus and Mary Elizabeth GIBBENS had two children: Frances "Fanny" Vaughn GIBBENS, m. George Copp WARREN; and Constance, m. George John MELHUISH, known as George J. Melhuish; children: Marjorie "Betty" (1894), Charles John "Jack" (1896), Mary (1900).

Base Hospital #10

Medical Staff (as they first appear in *History of the Pennsylvania Hospital Unit*)

Administrative Staff

Matthew A. DeLaney, Major, Commanding; Medical Corps, U.S. Army
Richard H. Harte, Major, Director; Medical Reserve Corps, U.S. Army
N. L. McDiarmid, Captain, Adjutant; Medical Corps, U.S. Army
H. L. Kidwell, Captain, Quartermaster; Quartermaster's Reserve Corps, U.S. Army
John H. Gibbon, Major, Chief of Surgical Service; Medical Reserve Corps, U.S. Army
George W. Norris, Major, Chief of Medical Service; Medical Reserve Corps, U.S. Army

Captains – Medical Reserve Corps

Wm. J. Taylor
Francis R. Packard
J. E. "Ted" Sweet
Wm. T. Shoemaker
Arthur Newlin
Charles F. Mitchell
Edward B. Hodge
Henry C. Earnshaw
John M. Cruise

1st Lieutenants – Medical Reserve Corps

Edward Bell Krumbhaar
Norris W. Vaux
William Drayton, Jr.
J. Howard Cloud
Arthur H. Gerhard
Frank C. Knowles
Henry K. Dillard, Jr.
William B. Cadwallader
John B. Flick

Charles S. "Charlie" Jack – Dental Surgeon Reserve Corps, U.S. Army

Charles Shoemaker Jack, D.D.S., was a 1st cousin of WTS. His mother, Caroline (née Shoemaker) Jack was the youngest of the ten children of WTS's parents, Julien and Hannah Shoemaker. Charles was also the half-brother of Margareta "Greta" Jack, who was married to WTS's brother Owen. Charles' wife was Mary "Medea" Lewis. He called Ed "brother-in-law"

Edwin "Ed" or "Eddie" Shoemaker – Dental Surgeon Reserve Corps, U.S. Army

Edwin Shoemaker, D.D.S., was a nephew of WTS. He was the 2d child of Owen and Margareta "Greta" (née Jack) Shoemaker. Edwin's wife was Martha Clawson Reed; they had 2 children.

Edward M. Jefferys – Chaplain, American Red Cross. "Padre Jefferys"
Helen D. Krumbhaar, Secretary, American Red Cross

Medical Staff (arrived later)

Capt. J. Paul Austin. 1st Lieutenants: Richard C. Beebe, Wm. L. Cunningham, Michael M. Nolan, George W. Outerbridge, Isaac B. Roberts, William Whitaker, H. B. Wilmer, Hershey E. Orndoff

Nurses who were mentioned in WTS's letters

Margaret A. Dunlop – Chief Nurse
Eva Gerhard – Assist. Chief Nurse
Helen Fairchild
Bertha Elliott
Jessie Gault
Sara S. Hopkins
Elizabeth Volz
Amanda D. Faunce
Sara A. Fidler
Amina Fuhrmann
Bessie "The Little One" Metz
Isabella Stambaugh

Privates who were mentioned in WTS's letters

Ralph Straub
Stephen W. Downs
Thomas L. Barratt
Herbert L. Jones
Samuel H. Chapman, Jr.
Henry W. LeBoutillier
William A. Hoffman
Frank T. Chalk
Gerald J. Sullivan
Irwin Hamilton
Frank X. Dochney
John Wesley Thomas

Acknowledgements

I begin this page on March 25, 2020, as the coronavirus pandemic is rising, and I am unsure of the future of *War Letters* and of my participation in it. I therefore am writing this page, in case someone else may have to pick up the work and complete it. The first draft was completed on April 7. With the assistance of Debbie Riley of Heritage Books, the covers have been drafted. On January 17, 2019, I signed a contract to publish *War Letters*, and also my two previous books, *Rolling with Patton* and *The Home Front*, with Leslie Wolfinger, the publisher of Heritage Books. The latter two books have already been delivered by me, and the publisher has an ISBN for *Patton*. I therefore give thanks to all who complete this work and bring it to the public, and to my family for its support. Especially, thank you Lanie.

I have long been fascinated with World War I, partly because of my interest in history, which I learned to appreciate as a school boy. And also because of the visage of my uncle Capt. Manly Thompson, in uniform, when he visited us while on leave in World War II. I was told by Mother that he had also served in the American Army in World War I. I learned that my uncle Howard Woodin was a veteran of that war, and that my mother's cousin, Private Joe Thompson, had died of influenza in 1919.

In high school, I was introduced vicariously to the horrors of that war, when I played the role of a British officer who left his trench, presumably to die, in *Journey's End.* I have recalled that haunting introduction to the grim realities of the Great War many times since then, as I read about it in poetry and books, and saw plays and movies about the war. I mention those in the Bibliography.

In recent years, I was reintroduced to events of the Great War as I worked on the papers that my wife's parents preserved from World War II, and which I began to examine in 2007. These papers led me to the involvement of her family in World War I. I discovered that one of the characters that my father-in-law met in India was a spy, Major Gordon Enders – secretive, adventuresome, and literate – whose life might be re-imagined for a movie. Like *Lawrence of Arabia*, but in the Far East. Enders' career began as pilot in France in World War I. He fell from an airplane and was nursed back to health. Enders' story was true, yet it was strikingly similar to the *Roman à clef* that Ernest Hemingway fictionalized in *Farewell to Arms*. The idea of a movie was suggested to me in a cold call by Harvey Rochman, and by chance at about the same time, I was contacted by Maynard Creel, who as a boy had known Enders; and by Patty Hoenigman, a great-niece of Enders' wife. I learned more about Enders from his niece, Dr. Trudy (Enders) Huntington. Rochman suggested that the movie be called *Khyber Pass*, and with that title, and with his persuasiveness, others in Hollywood have taken interest in the movie. I thank them all.

I regret that I never talked about World War I with any of the others that I met who had been in that war, though thinking about their experiences has helped me to write about that war. Most notable was Vice Admiral Dwight Dickinson, MC, USN (ret). He was one of my sponsors for a commission in the Navy. Admiral Dickinson was said to have been the second-most highly decorated Navy doctor in World War I. On his tuxedo's lapel, he wore a miniature of the Navy Cross, the Distinguished Service Cross, and two French decorations. His rank was the result of what was called a "tombstone promotion."

At Yale, some of the older faculty members had been in the Great War, and I learned later that several of the younger ones had been in World War II. Fred Kilgour was my most history teacher, and I later learned that he had been in the O.S.S. The Yale Aviation Unit was active in World War I, and many of its leaders were members of Skull and Bones. Robert Dudley French, who was Master of my College, was said to be the senior advisor for S&B. We never talked about that, of course. It was in Political Science 101 at Yale that I read Paul Fay, *The Origins of the World War.* At the Harvard Medical School, several of my professors told stories about Dr. Harvey Cushing, who operated on Sir William Osler's son, Revere Osler, in France. We walked every day across "Oscar C. Tugo Circle," which was named in memory of the private in the Harvard unit who was killed in France, September 5, 1917.

Heritage Books by George J. Hill:

American Dreams: Ancestors and Descendants of John Zimmermann and Eva Katherine Kellenbenz, Who Were Married in Philadelphia in 1885

"Dearest Barb": From Karachi, 1943–1945, Letters and Photographs in the World War II Papers of a Naval Intelligence Officer, Lieutenant Albert Zimmermann, USNR

Edison's Environment: The Great Inventor Was Also a Great Polluter

Four Families: A Tetralogy Reader's Guide to Western Pilgrims, Quakers and Puritans, Fundy to Chesapeake, *and* American Dreams*; Synopsis of 481 Immigrants and First Known Ancestors in America from Northern Europe in the Families of George J. Hill and Jessie F. Stockwell, William T. Shoemaker and Mabel Warren, William H. Thompson and Sarah D. Rundall, John Zimmermann and Eva K. Kellenbenz, with Outlines of Their Descent from the Immigrants*

Fundy to Chesapeake: The Thompson, Rundall and Allied Families; Ancestors and Descendants of William Henry Thompson and Sarah D. Rundall, Who Were Married in Linn County, Iowa, in 1889

Hill: The Ferry Keeper's Family, Luke Hill and Mary Hout, Who Were Married in Windsor, Connecticut, in 1651 and Fourteen Generations of Their Known and Possible Descendants

John Saxe, Loyalist (1732–1808) and His Descendants for Five Generations

Prairie Daughter: Stories and Poems from Iowa by Essie Mae Thompson Hill

Quakers and Puritans: The Shoemaker, Warren and Allied Families; Ancestors and Descendants of William Toy Shoemaker and Mabel Warren, Who Were Married in Philadelphia in 1895

Rolling with Patton: The Letters and Photographs of Field Director Gerald L. Hill, 303rd Infantry Regiment, 97th "Trident" Division, 1943–1945

The Home Front in World War II: From the Letters of Essie Mae Hill to Field Director Gerald L. Hill

War Letters, 1917–1918: From Dr. William T. Shoemaker, A.E.F., in France, and His Family in Philadelphia

Three Men in a Jeep Called "Ma Kabul," Script for a Movie: A True Story of High Adventure by Three Allied Intelligence Officers in World War II

Western Pilgrims: The Hill, Stockwell and Allied Families; Ancestors and Descendants of George J. Hill and Jessie Fidelia Stockwell, Who Were Married in Wright County, Iowa, in 1882

About the Author

George J. Hill is a fifth-generation Iowan. He graduated from high school in Sac City, Iowa, and then attended Yale University, where he majored in history. He graduated from Harvard Medical School, and after forty years as a practicing surgeon, he is now an Adjunct Professor of Surgery at the Uniformed Services University of the Health Sciences, and Professor Emeritus at Rutgers University. He served in the U.S. Marine Corps and the U.S. Public Health Service, and he was awarded the Meritorious Service Medal upon retirement as a Captain, Medical Corps, in the U.S. Navy. Dr. Hill also earned an M.A. in history at Rutgers University and a D.Litt. in history from Drew University. He has written or edited 17 books on medicine and surgery, family history and genealogy, environmental history and international relations.

Other Books by the Author

Medicine and Science

Leprosy in Five Young Men
Outpatient Surgery (3 editions; 2 translated into Spanish as *Cirugia Menor*)
Clinical Oncology, with John Horton

History

Edison's Environment (3 editions)
Intimate Relationships: The U.S. and Liberia, 1917-1947 (2 editions)
Proceed to Peshawar

www.ingramcontent.com/pod-product-compliance
Lightning Source LLC
LaVergne TN
LVHW061239100826
845148LV00008B/991

* 9 7 8 0 7 8 8 4 0 3 5 5 2 *